FRACTURE OF COMPOSITE MATERIALS

Proceedings of First USA-USSR symposium on

FRACTURE
OF
COMPOSITE MATERIALS

Held at the Hotel Jūrmala, Riga, USSR
September 4-7, 1978

Editors:

G. C. Sih

Institute of Fracture and Solid Mechanics
Lehigh University, Bethlehem, Pennsylvania, U.S.A.

V. P. Tamuzs

Institute of Polymer Mechanics
Academy of Sciences of the Latvian SSR
Riga, U.S.S.R.

SIJTHOFF & NOORDHOFF   1979
Alphen aan den Rijn – The Netherlands

© 1979 Sijthoff & Noordhoff International Publishers B.V.
Softcover reprint of the hardcover 1st edition 1979

Alphen aan den Rijn, The Netherlands

ISBN-13:978-94-009-9555-0   e-ISBN-13:978-94-009-9553-6
DOI: 10.1007/978-94-009-9553-6

CONTENTS

PREFACE

The current situation with regard to the use of high-performance composites is one of seeking a reliable means of characterizing the fundamental properties and/or behavior of composite materials. The employment of such materials in vital structural applications require criteria as part of the materials selection and design procedures. However, the principal limitation of the many currently available procedures, most of which were borrowed from the metal industry, is its intrinsic restriction to single phase materials. They cannot adequately account for the variations in the constituents or structure of the composite that can have a significant influence on its gross behavior. Although a considerable amount of research has been conducted in the past with the objective of understanding composite material behavior, for the most part, the research efforts tended to cover too wide a spectrum of problems with little penetration into the fundamental aspects. For this reason, the investigators in this field have long felt the need to bring together a small group of experts to review the fundamentals, discuss the problem areas, and display the current developments. The *USA-USSR Symposium on the Fracture of Composite Materials* was organized for this purpose. There is the additional need to share this information with the newcomers to the field as well as the oldtimers. This is accomplished by publishing the symposium proceedings contained in the present volume. A Russian translation of the papers in this volume will also be made available.

The Riga symposium was held on September 4-7, 1978 at the Hotel Jurmala in Majori, the central part of the popular health resort on the coast of Riga Bay, USSR. Approximately 60 scientists and engineers mostly from the United States and the Soviet Union attended the meeting. Participants from England, France and West Germany also presented papers. The original idea of organizing the symposium was due to Professor George C. Sih of the Institute of Fracture and Solid Mechanics at Lehigh University and Dr. Vitauts P. Tamuzs of the Institute of Polymer Mechanics of the Latvian Academy of Sciences. The planning was done jointly by the following members of the organizing committee:

N. S. Enikolopov
A. K. Malmeister (co-chairman)
S. T. Mileiko
Yu. N. Rabotnov
G. C. Sih (co-chairman)
V. P. Tamuzs (executive secretary)

The labor and time offered so generously by the staff members of the Institute of Polymer Mechanics are gratefully acknowledged. Special recognition is also

due to I. V. Knets, S. T. Mileiko, L. Nikitin, V. P. Tamuzs and A. Bogdanovich
for their patience in providing the simultaneous two-way translation of Russian
and English during the technical discussion. Despite the intensive three-day
long technical program, the participants also found time to visit the War Memo-
rial in Salaspils, the Open-Air National Ethnographic Museum, the city of Riga,
and the picturesque surroundings of Jurmala.

The organizing committee wishes to take this opportunity to thank the
authors, session chairmen, and discussers whose efforts have made this first
USA-USSR symposium on the fracture of composite materials a success. It is
hoped that this symposium series will continue at a biannual interval with its
second meeting to be held in the United States. Finally, credit must be given
to Mrs. Barbara DeLazaro who has retyped the majority of the manuscripts in
this proceedings.

November, 1978

G. C. Sih
V. P. Tamuzs
Editors

The organizing committee members at the opening session

Participants attending the technical session

A. Malmeister, President of the LSSR Academy of Sciences,
welcomed the participants at the early bird cocktail party

Social hour at the Hotel Jurmala

G. C. Sih from Lehigh University received a plaque from
V. P. Tamuzs, Scientific Vice-Director of the Institute
of Polymer Mechanics

G. C. Sih invites V. Latishenko, Director of the Institute of Polymer
Mechanics, to attend the second USA-USSR Symposium in the United States

Banquet dinner at the Hotel Jurmala

Lively discussion at the dinner; September 6, 1978

A visit to the monument of the Latvian Red Rifles

At the conclusion of conference in front of Hotel Jurmala

Section I

MICROFRACTURE

MICRO- AND MACROCRACKS IN COMPOSITES

S. T. Mileiko

Institute of Solid State Physics of the Academy of Sciences of the USSR
142432 Chernogolovka Moscow District, USSR

The present paper consists mostly of the author's published results and represents a brief summary of the work in [1-6].

Microcrack in a composite structure has a length ℓ which is of the order of magnitude of a characteristic size of the structure. For a fibrous composite $\ell \approx d$ where d is the fiber diameter. If $\ell \gg d$, then there is a macrocrack.

1. A microcrack in a composite with brittle fibers and a tough matrix arises as a result of fiber breaking at a point where the stress $\sigma' = \sigma_0$ reaches the value of the local strength $\sigma_f^*(x_0)$. Around the break δ-zone arises where the fiber stress σ' goes down to zero at $x = x_0$. At $x = x_0 \pm \delta$, the stress again is equal to the initial value σ_0, (Figure 1).

After the first break the following events are possible. First, the fracture of the entire composite corresponds to unstable microcracking. Second, the stress in the composite rises. In this case, the microcrack is stable. Third, the strain of the composite is rising while the stress is decreasing. The microcrack is again stable.

If the microcracks are stable the process of fiber breaking can go on until the whole length of fibers is covered by the δ-zones. Some of these zones can overlap. The calculation of the limit stress on the composite in such a state is a stochastic problem which has not been solved. A general form of the solution can be anticipated:

$$<\sigma_\infty^*> = \alpha<\sigma_f^*(\ell^*)>v_f + \sigma_m^* v_m$$

Here, apart from the quantities generally used, the value $<\sigma_f^*(\ell^*)>$ is the characteristic fiber strength on the length ℓ^* which is of the order of magnitude of the δ-zone in a limit state and α is the constant to be calculated in the problem mentioned above, which falls within the range $1/2 < \alpha < 1$.

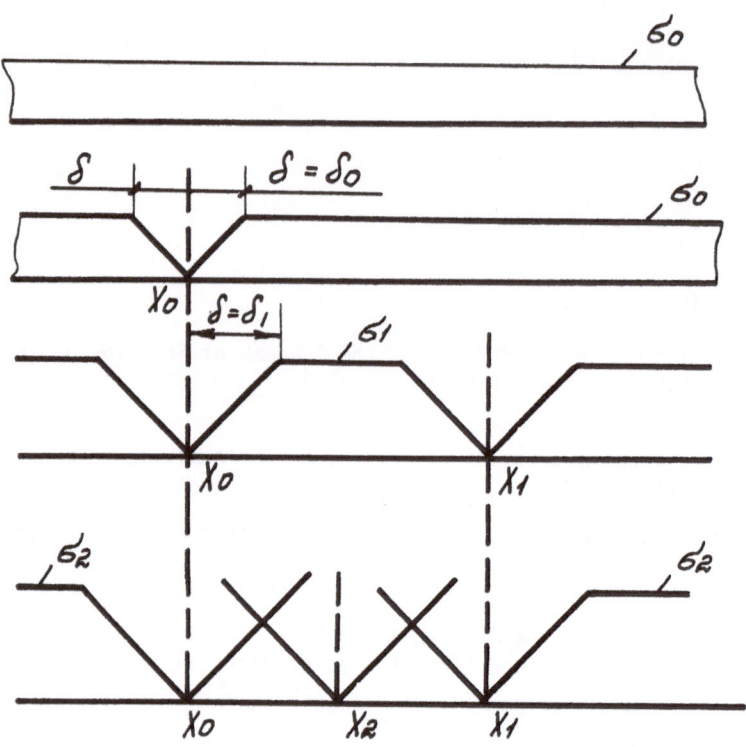

Fig. (1) - Stress distribution along the fiber perturbated
by fiber breaks

If after the first break, fracture in the composite takes place, then the expectation of the strength of the composite will be equal to the expectation of
the first break within the whole system of the fibers and then

$$\langle \sigma_1^* \rangle = \langle \sigma_f^*(L) \rangle v_f + \sigma_m^* v_m$$

where L is the whole length of the fibers stressed homogeneously in the composite.

Now, let us introduce the characteristic length nd of the microcrack in a
given structure. Here n is a number related to the quality of fiber distribution. At n=1, we have the case of the ideal distribution. For two fibers,
n=2 and so on. If, for example, in the composite with n=2, only a single break
occurs. This means that a microcrack of the length 2d will appear. In general,
microcracks of length nd will arise.

Refer to the $<\sigma*>$ - v_f plane in Figure 2. The curve labelled GO_n which corresponds to the ultimate stress by Griffith-Orowan theory and which together with the curves given by formulae for $<\sigma_\infty^*>$ and $<\sigma_1^*>$ written above gives the dependence of the mean strength of the composite on the fiber volume fraction with the resulting curve OABC.

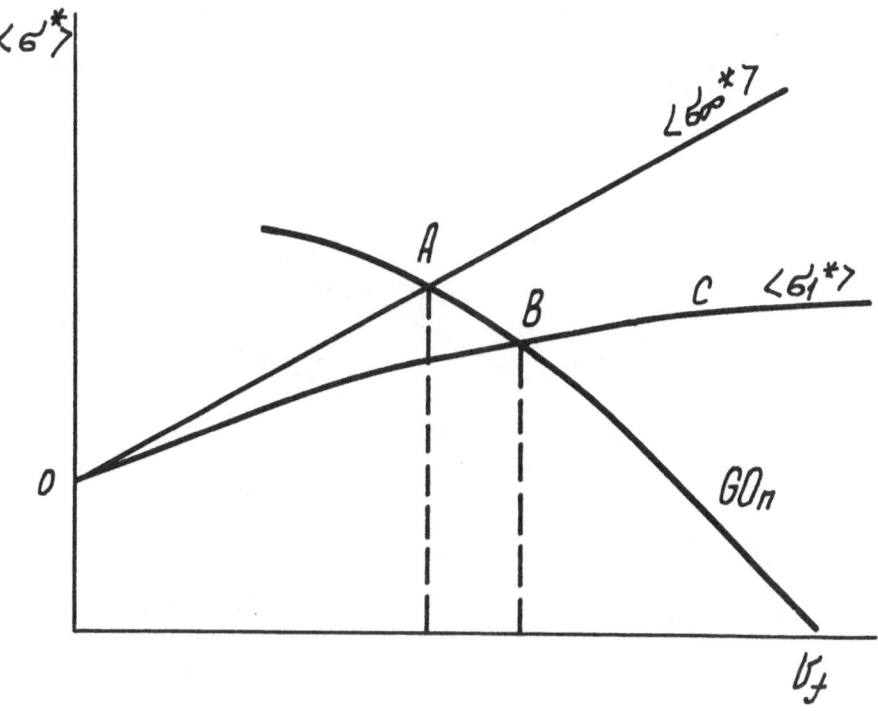

Fig. (2) - Curve OABC is the dependence of the mean strength of a composite on fiber volume fraction

The expression for the GO_n curve is

$$\sigma = \sqrt{\frac{GC}{\lambda nd}}$$

where the value G can be taken from Kelly and Cooper [7]:

$$G = \frac{v_m^2}{v_f} \sigma_m^* e_m^* d$$

This is valid at large enough v_f. At small values of v_f when the plastic constraint of matrix can be neglected,

$$G = G_m^\circ v_m$$

where G_m° is the work of fracture of the matrix. The constant C is equal to Young's modulus for an isotropic material and λ is a constant.

The relative position of the curves for $\langle\sigma_\infty^*\rangle$ and $\langle\sigma_i^*\rangle$ on the plane can be different. For example, refer to Figure 3.

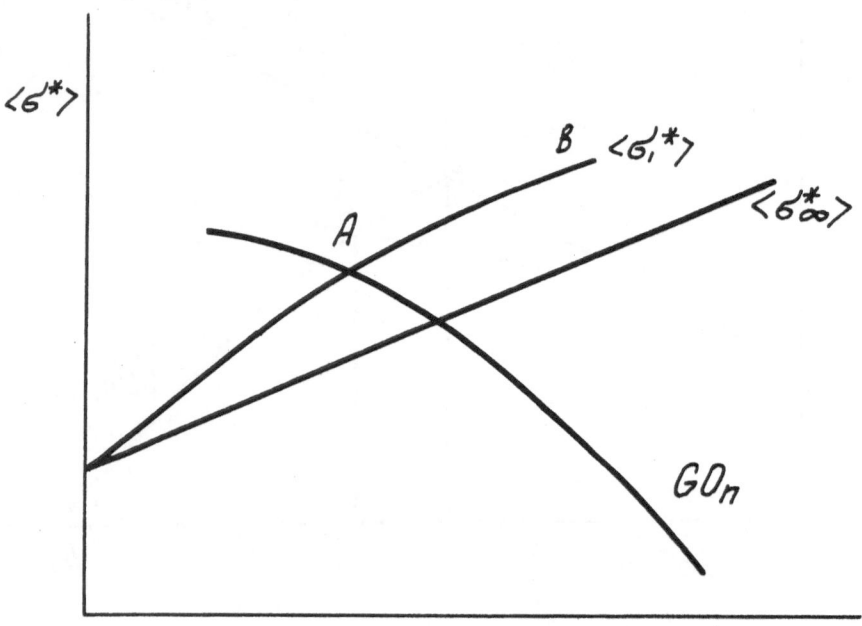

Fig. (3) - A possible dependence of the mean strength of a composite on fiber volume fraction (curve OAB)

It appears that the behavior of fiber reinforced plastics can be described in the same way. The only difference is the value of G which in this case should be determined mainly from a knowledge of the fiber-matrix interface and the relative position of the curves on the plane $\langle\sigma^*\rangle$ - v_f is such that the point B is shifted to the region of large σ_f. As a rule, $v_f > 1/2$.

2. Looking at the behavior of the macrocracks in the composites, it is possible to use well-known methods of fracture mechanics (see, for example [8]). It is important to determine the critical values K* of the stress intensity factor (or corresponding values of work of fracture). The measured values of K* for boron-aluminum composites are shown in Figure 4 (or the equivalent G-value).

The values of G for macrocracks and microcracks appear to be quite different.

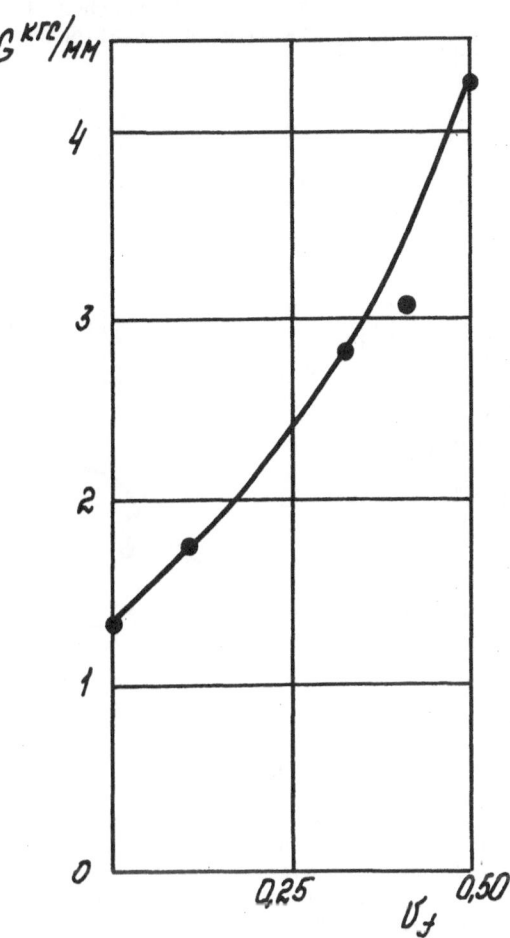

Fig. (4) - The work of fracture a boron-aluminum composite
versus the fiber volume fraction (experimental
data)

The possible reasons for such a large value of G for macrocrack have been
discussed [4,5]. When the load is applied to a specimen with a crack, the material in the crack tip zone is over-stressed, and within this zone the brittle
fibers break at the weak points, Figure 5. These breaks enlarge the over-
stressed zone in comparison with the size of this zone in a composite with fi-

bers of homogeneous strength. As the external load rises, the boundary of the process zone expands and around each microcrack there arises a plastically deformed matrix region. After some critical load is reached the whole process zone starts to move without changing its size in a stationary state. It signifies the beginning of crack movement.

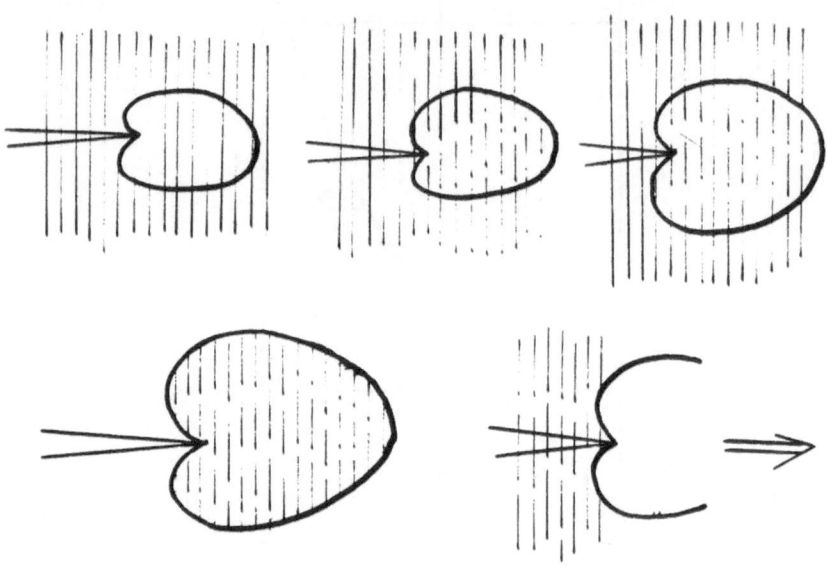

Fig. (5) - The initiation and propagation of the crack zone
in a composite

If this speculation is correct and the total plastic work on the whole system of microcracks is larger than the plastic dissipation at the crack tip in a unreinforced matrix then from the point of view of a moving crack, the composite structure would appear to act as an amplifier on the toughness.

So for a macrocrack, the characteristic value of K^* can be determined (see experimental data in [4,5] and also [9]. For a microcrack (at n=1), this value is determined in section 1. The dependence of K^* on the crack length ℓ obviously should be monotonic, and the general picture is shown in Figure 6. Note that the experimental data in [9] correspond to those in [3-4] in the region n≈1. Strictly speaking for characteristic microcrack with length nd, the value of K^* (or G) should be a bit higher than that of K^* at n=1 (Figure 6).

It is useful to note that for macrocracks in a ductile matrix with ductile fibers, the value of G can be calculated [2] as

$$G = G_f \frac{e^*}{e_f^*} v_f + G_m \frac{e^*}{e_m^*} v_m$$

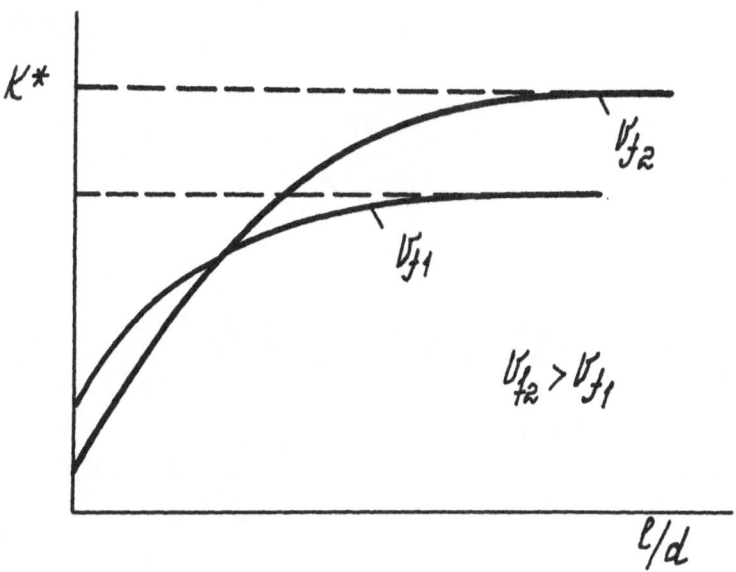

Fig. (6) - The schematical dependence of the fracture toughness
of a composite on the crack length

Here, G_f and G_m are the work of fracture of fiber and matrix materials, e_f^*, e_m^* and e^* are ultimate strains of the fiber, matrix and composite, respectively. The value of e^* is given in [1].

3. The fatigue behavior of composites with metal matrix is rather peculiar. A fatigue crack can be initiated and also arrested at the fiber-matrix interface at the microscopic level and also at the interface between individual foils if the matrix in a composite is formed out of foils. The latter has not received much attention by those authors who discuss the fatigue of composites. Figure 7 shows the result obtained by V. M. Anishinkov.

This fact led us to the analysis of the behavior of a fatigue crack in a simple laminate with variable bond between the laminae, which serves as a simple model of the composite [6]. The dependence of fatigue strength of such a composite on the interface bond strength appears to have a maximum, Figure 8. It is possible to interpret such a dependence if one assumes an island type of structure at the interface and to trace the growth of fatigue cracks from the defect randomly distributed on the free surfaces. This procedure has been used by P. A. Egin. Some of his results are presented in Figure 9.

Fig. (7) - The fatigue microcracks in the matrix of a steel-wire aluminum
composite arrested at the interfaces between the aluminum
foils

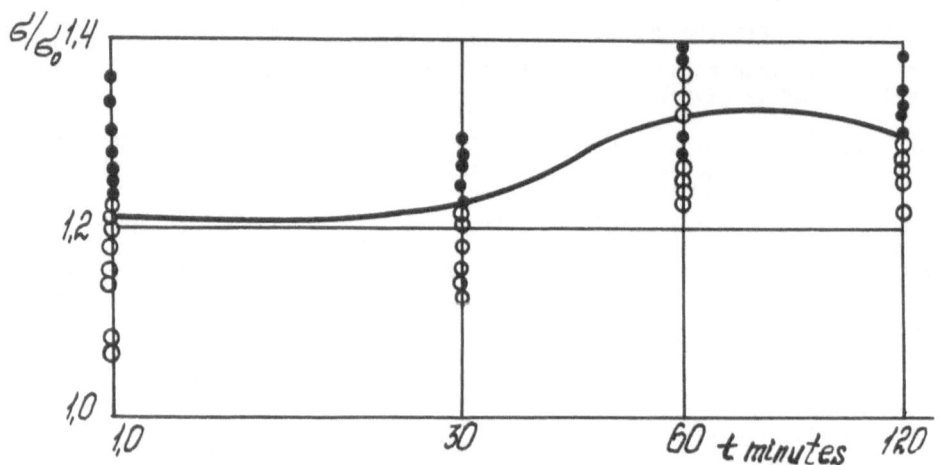

Fig. (8) - The fatigue strength of a laminated aluminum composite
versus the interface strength (the fatigue base is 10^6
cycles)

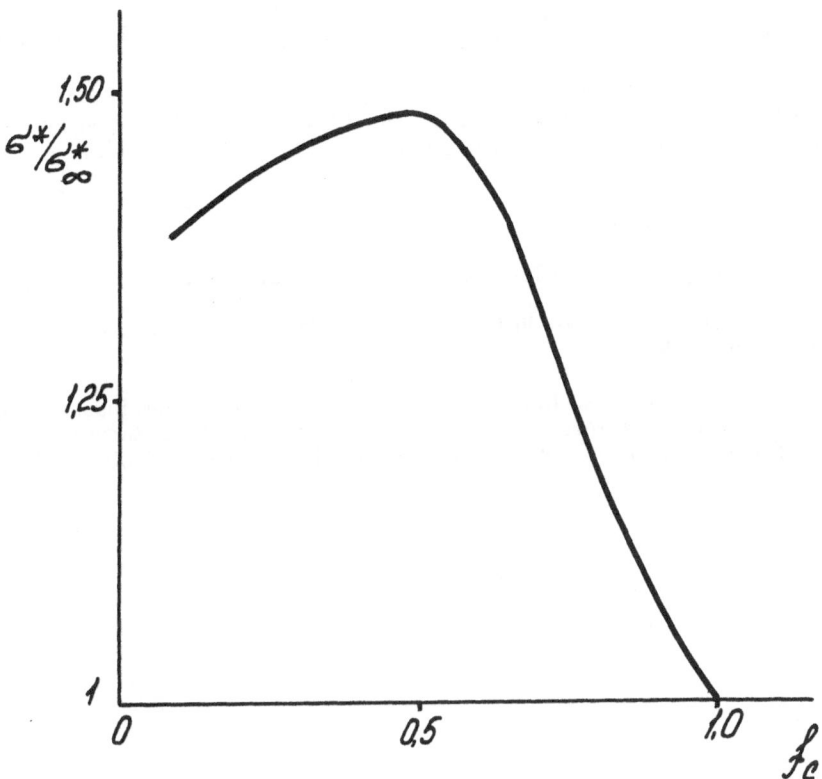

Fig. (9) - The calculated fatigue strength of a laminated composite σ^*
normalized by that of a monolytic specimen σ^*_o versus the
interface fraction of an ideal bond

REFERENCES

[1] Mileiko, S. T., "The tensile strength and ductility of continuous fiber composites", J. Mater. Sci., V. 4, N. 10, pp. 974-977, 1969.

[2] Archangelska, I. N. and Mileiko, S. T., "Fracture mechanics of metal matrix-metal fiber composites", J. Mater. Sci., V. 11, N. 2, pp. 356-362, 1976.

[3] Mileiko, S. T., Sorokin, N. M. and Zirlin, A. M., "The strength of boron-aluminium - a composite with brittle fibers", Polymer Mechanics, N. 5, pp. 840-842, 1973.

[4] Mileiko, S. T., Sorokin, N. M. and Zirlin, A. M., "The crack propagation in a boron-aluminium composite", Polymer Mechanics, N. 6, pp. 1010-1017, 1976.

[5] Mileiko, S. T., Sorokin, N. M. and Zirlin, A. M., "Fracture of boron-aluminium composites", In: Proceedings of the 1975 International Conference of Composite Materials, Vol. 1, AIME, New York, pp. 562-575, 1976.

[6] Anishenkov, V. M. and Mileiko, S. T., "The fatigue of a laminated composite", Docladi Akademii Nauk, Vol. 241, N. 5, pp. 1068-1069, 1978.

[7] Kelly, A., "Strong solids", Clarendon Press, Oxford, 1973.

[8] Corten, H. T., In: Fracture, Vol. 7, H. Liebowitz, ed., Academic Press, New York, London, 1972.

[9] Olster, E. F. and Jones, R. C., "Toughening mechanisms in continuous filament unidirectionally reinforced composites", In: Composite Materials: Testing and Design (Second Conference), ASTM STP 497, pp. 189-205, 1972.

DISPERSED FRACTURE OF UNIDIRECTIONAL COMPOSITES

V. P. Tamužs

Institute of Polymer Mechanics, Latvian SSR Academy of Sciences
Riga, USSR

DISPERSED FRACTURE OF HETEROGENEOUS MATERIALS

Since the advent of fracture inspection, two basic concepts have emerged. One is to treat fracture as the propagation of a macrocrack developed from the most critical defect in an undamaged material. This led to the development of the theory of fracture of linear mechanics [1,2]. The other is associated with "cumulative" damage and dates as far back as 1924 [3]. Refer to [4] for bibliographies on this subject.

For a long time, the concept of "damage" was not identified with the physical defects in a material. This is partly because of the lack of reliable methods for detecting the damage in the material. In general, such methods are still lacking. In some instances, however, reliable techniques for detecting microcracks in homogeneous materials under loads have been made available. First, we might mention the method of small-angle scattering of X-rays [4] which measures the density and orientation of submicrocracks in oriented polymers. Submicrocracks take the dimensions of 100 to 5,000 Å. Defects of large size, as a rule, cannot be detected by this method. The technique of acoustic emission has been used for detecting microcracks at the moment of their formation, but has failed to identify acoustic noises with the details of damage. Despite the drawbacks, acoustic emission still provides much information on the kinetics of cumulative damage in the material. A number of other methods for locating the damage zone are based on measuring the averaged physical and mechanical properties of a damaged material and their comparison with the basic characteristics of the undamaged material. Some recent works along this line have been discussed in [4,5]. The results of V. A. Latishenko and I. G. Matiss, P. O. Oldirev, V. M. Parfeyev and Tamužs will be presented in this symposium.

It follows from the studies of V. S. Kuksenko [4] and other investigators that all materials experience the stage of dispersed fracture depending on the degree of severity. Material heterogeneity tends to promote dispersed fracture. The volume fracture of heterogeneous materials such as rocks will be considered in greater length by Kuksenko in his paper.

In the study of composite materials, the presence of dispersed fracture and the statistical aspects of this process have attracted the attention of many investigators.

Before discussing the various fracture models proposed by the different investigators, it is important to recognize that composite materials possess well-oriented structures with distinct heterogeneity. The statistical scatter of the properties of the composite constituents may be determined on a basis by testing the individual constituents of the composite and recognizing the features of dispersed fracture that is a typical form of damage. As mentioned above, disperse fracture is a common mode of fracture in solids. For this reason, knowledge gained on the fracture of composite materials can also be applied to understand the fracture process in solids.

The first statistical model of composite material fracture is that of Rosen [6] and Zweben [7]. The composite material consisted of a chain whose links were formed by a bundle of fibers of critical length. Application of the probable relation between the strength of the chain and strength of the individual elements led to an estimate on the probability of the accumulation of defects in each layer and the strength of the composite. In the analysis, the stress concentration on the fibers owing to the broken fibers was considered which included a study of the fracture sequence consisting of the breakage of neighboring fibers. The results of Rosen and Zweben suggest that even a small number, from two to three, of the neighboring cracks can cause fracture of the entire reinforced composite.

The simplest kinetic model of fracture of a unidirectional composite has been proposed by V. V. Bolotin [8] who has also provided an approximate analysis on the cumulative damage [9] in a composite material by neglecting the effect of the fracture of one element on the strength of the other. As mentioned earlier, a more general approach to the fracture problem of composite materials is the kinetic statistical model. This approach permits the inclusion of non-stationary loading, temporary delay of fracture, accumulation of individual defects, the coalescence of small cracks into a macrocrack and the characteristics of crack propagation. All these events correspond to the actual process of fracture in a solid. The kinetic statistical model of fracture for oriented polymers under creep conditions [10,11] can describe all of the above mentioned features of fracture. The success achieved for this model as applied to oriented polymers has suggested the possibility of extending it to investigate the fracture of an oriented composite material.

The basic assumptions made in the aforementioned model [10,11] are

(1) A defect of any size is idealized to be a spheroid or a penny-shaped crack. The stress concentration on the elements in the neighborhood of a number of broken elements is calculated accordingly.

(2) The probability of fracture of an element under a specific level of overstress is given by a formula analogous to the radioactive decay which is supposed to be independent of the loading history.

KINETICS OF INITIATION AND ENLARGEMENT OF DEFECTS IN FIBROUS COMPOSITES

As in the study of Rosen and Zweben [7], the proposed model deals with a unidirectional fibrous composite material under tensile load along the fiber direction. Let us examine the kinetics of initiation and the enlargement of defects under constant load. To be assumed are that the reinforcing fibers are in tension and the matrix transmits the overload in the form of shearing force to the unbroken fibers in the vicinity of the defect. The stress analysis was calculated for the elastic fibers and an elasto-plastic matrix [12]. The time effects of fracture have been considered by introducing the long-term strength of the fibers as a time function. The other additional time effects caused by redistribution of the stresses due to creep of the matrix in the vicinity of the defect (similar as in [13]) are neglected. They can be treated separately.

Let the fracture of an element with a definite lifetime $\tau(\sigma_0)$ be a determined value. The probability of fracture of the element W is then given by

$$W[t,\tau(\sigma_0)] = H[t-\tau(\sigma_0)] \tag{1}$$

Owing to scatter of the composite properties, the lifetime of the element is a random value with the density $f(\tau,\sigma_0)$ and the distribution functions $F(\tau,\sigma_0)$. The determination of the statistical characteristics of the elements will be discussed in a later section. The probability of fracture of the element for the time moment t is

$$W(t) = \int_0^\infty f(\tau)H(t-\tau)d\tau = \int_0^t f(\tau)d\tau = F(t) \tag{2}$$

The expected number of single defects in the specimen is

$$NF(t) \tag{3}$$

where N is the number of elements in the specimen.

After the appearance of a single defect for the time moment t=x, an increased stress σ_1 originates on its neighboring elements. If the loading history of the neighboring elements is not taken into account, the probability of fracture of such an element will be

$$W(t,\sigma_1) = \int_0^t f[\tau(\sigma_1)]d\tau = F(t,\sigma_1) \tag{4}$$

where the time account has been restarted at the moment t=x. The probability of fracture of at least one neighboring element is

$$W_1^2 = 1 - [1-F(t,\sigma_1)]^{n_1} \qquad (5)$$

where n_1 is the number of the neighboring elements while the density of this probability equals to

$$p_1^2 = n_1[1-F(t,\sigma_1)]^{n_1-1} f(t,\sigma_1) \qquad (6)$$

The probability of the development of a double defect W_2 can be determined by the formula

$$W_2 = \int_0^t \frac{dW_1(x)}{dx} W_1^2(t-x)dx = \int_0^t W_1(x)p_1^2(t-x)dx \qquad (7)$$

The probability of nucleation of a triple defect and defects of size j in general can be determined in a similar manner:

$$W_j = \int_0^t W_{j-1}(x)p_{j-1}^j(t-x)dx \qquad (8)$$

In order to understand the loading history of the failing elements, let us examine a hexagonal packing of the fibers. At the moment of initial fracture, all the elements are subjected to the stress σ_0. After the fracture of a single element at the moment x_1, six neighboring elements (the overstresses are considered for the immediate neighboring elements) have the loading history $\sigma=\sigma_0$ at $0<t\le x_1$, $\sigma=\sigma_1$ at $t>x_1$ and so on. For the initiation of a double defect, the loading history of its neighbors in as follows: five elements are under the stresses $\sigma_0/0<t<x_1$; $\sigma_1/x_1<t<x_2$; $\sigma_2/t>x_2$ and three elements under $\sigma_0/0<t<x_2$; $\sigma_2/t>x_2$. Fracture of the third neighboring element entails a number of variants which are more conveniently shown in Table 1. It is easy to see that the loading history of any element falls within the loading of all stress levels

$$t \to x_1 \to x_2 \ldots \to x_{n-1} \to x_n$$
$$\sigma_0 \to \sigma_1 \to \sigma_2 \ldots \to \sigma_{n-1} \to \sigma_n \qquad (9)$$

and that of the lowest stress

$$t \to x_1 \to x_2 \ldots \to x_{n-1} \to x_n$$

$$\sigma_0 \to \sigma_0 \to \sigma_0 \ldots \to \sigma_0 \to \sigma_n$$

(10)

TABLE 1 - STRESS HISTORY OF NEIGHBORING ELEMENTS

Dimensions of Defects	Number of Neighboring Elements	Stress in Time Periods			
		$0 < t < x_1$	$x_1 \leq t < x_2$	$x_2 \leq t < x_3$	$t \geq x_3$
1	6	σ_0	σ_1		
2	5	σ_0	σ_1	σ_2	
	3	σ_0	σ_0	σ_2	
3 (one variant)	4	σ_0	σ_1	σ_2	σ_3
	3	σ_0	σ_0	σ_2	σ_3
	2	σ_0	σ_0	σ_0	σ_3
3 (another variant)	5	σ_0	σ_1	σ_2	σ_3
	2	σ_0	σ_0	σ_2	σ_3
	3	σ_0	σ_0	σ_0	σ_3
	and etc.				

Assigning the elements to one or the other extreme history of loading, we should obtain two limits between which lies the true kinetics of cumulative defects. Knowing the dependence of the distribution function $F(t,\sigma)$ on the parameter σ, the curves corresponding to the discrete stress levels σ_0, $\sigma_1 \ldots \sigma_n$ can be constructed, Figure 1.

We now introduce the assumption that the density of the fracture probability of an element is defined by the instantaneous stress and the fracture probability reached earlier. In other words, if at the moment $t = x_1$ the stress on the element changes from σ_0 to σ_1 and if by this time the fracture probability $F(\sigma_0, x_1)$ has been reached, the distribution function of the probability is then expressed in the following way:

$$F = F(\sigma_0, t) \text{ for } t \leq x_1 \tag{11}$$

$$F = F(\sigma_1, t - x_1 + x_1^1) \text{ for } t \geq x_1 \tag{12}$$

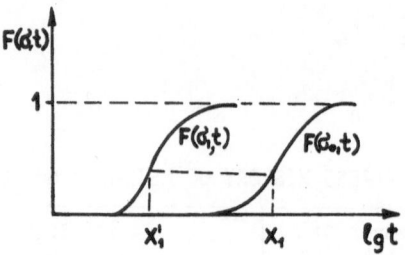

Fig. (1) - Curves of distribution of fracture probability
of the structural element in two-stress level
loading

where x_1^1 is determined from the continuity condition F when $t=x_1$, i.e., from
the equation

$$F(\sigma_0,x_1) = F(\sigma_1,x_1^1) \tag{13}$$

Equation (5) then transforms into

$$W_1^2 = 1 - [1-F(\sigma_1,t-x_1+x_1)]^{n_1} \tag{14}$$

while equation (7) changes to

$$W_2 = \int_0^t \frac{dW_1(x_1,\sigma_0)}{dx_1} W_1^2(\sigma_1,t-x_1+x_1^1)dx_1 = W_1(t,\sigma_0)W_1^2(\sigma_1,x_1^1)$$

$$+ \int_0^t W_1(x_1,\sigma_0)n_1[1-F(\sigma_1,t-x_1+x_1^1)]^{-n_1-1} f(\sigma_1,t-x_1+x_1^1)dx_1$$

$$= F(t,\sigma_0) \cdot \{1-[1-F(t,\sigma_0)]^n\} + \int_0^t F(x_1,\sigma_0)n_1[1-F(\sigma_1,t-x_1+x_1^1)]^{n_1-1}$$

$$\times f(\sigma_1,t-x+x_1^1)dx_1 \tag{15}$$

For the loading history in equation (10) and a defect of any size j, equation
(15) may be simplified:

$$W_j = \int_0^t W_{j-1}(x,\sigma_0)n_1[1-F(\sigma_j,t-x+x^i)]^{n_j-1} f(\sigma_j,t-x+x^j)dx$$

$$+ W_{j-1}(t,\sigma_0)W_{j-1}^j(\sigma_j,x^j) \tag{16}$$

where x_j is determined from the equation

$$F(\sigma_0, x) = F(\sigma_j, x^j) \tag{17}$$

If

$$\lim_{j \to \infty} W_j(t) = W^*(t)$$

is valid, then $W^*(t)$ defines the probability that a defect will nucleate in the element. The probability that at least one defect will nucleate in a specimen with N elements is expressed by the formula

$$W = 1 - [1-W^*(t)]^N \approx 1 - \exp[W^*(t)N] \tag{18}$$

Note that the loading history in equation (10) tends to underestimate the load on the structural elements in the vicinity of the crack edge. It follows that the present calculation will overestimate the lifetime of the composite.

STRESS CONCENTRATION AND STRUCTURAL ELEMENTS OF THE MODEL

As mentioned before, the stress averaged over the fiber cross section in the vicinity of the broken fiber has been given in [12]. The fibers were packed hexagonally and assumed to be elastic while the matrix is elasto-plastic. The analysis assumed that the matrix transmits shearing stresses and the fibers are kept in tension. The present calculations are based on the same assumptions as those in [12]. For the fracture of a number of neighboring fibers, the defect configuration is approximated by a penny-shaped crack as in [10]. The composite on both sides of the crack planes was substituted by a rod with mean characteristics of the former*. The number of the fibers around the defect were calculated on the basis of the diameter of defect and fibers.

It is natural to assume the size of the structural element in the fiber direction to be a doubled "ineffective" length δ. The obtained levels of overstress and the ratio of length of an element to the fiber diameter are given in Table 2. The shear modulus of epoxy resin G_m = 133 kg/mm^2; the tangential modulus in the plastic region G_T = 62 kg/mm^2; the elastic modulus of the fiber E = 7.5×10 kg/cm^2; the volume fraction of the fibers V = 0.6; and the stress applied to the fiber σ_0 = 80 kg/mm^2. We note in passing that the magnitude of δ is greater than that given by Rosen and increases with enlargement of the defect.

---

*The method is proposed by P. V. Tikhomirov while the numerical calculations were carried out by S. P. Yushanov.

TABLE 2 - OVERSTRESS IN TERMS OF FIBER GEOMETRY

j	σ_j/σ_o	δ/d
1	1.00	12.5
4	1.16	16.5
9	1.24	20.5
16	1.31	24.5
25	1.37	28.5

STATISTICAL CHARACTERISTICS OF STRENGTH AND LIFETIME OF STRUCTURAL ELEMENTS OF COMPOSITES

The cross sectional size of the structural element of a unidirectional fibrous composite is close to the fiber cross section while its longitudinal dimension is about ten times greater. In [6], it has been proposed that the characteristics of short-term strength of the short length fibers be obtained by extrapolation from the strength of the long fibers. The simplest extrapolation scheme is to assume that the strength distribution of the fiber of length L is determined by a Weibull distribution

$$F(\sigma,L) = 1 - \exp(-L\alpha\sigma^\beta) \tag{19}$$

where α and β are the distribution parameters. Dividing the fiber of length L into N parts, $L = N\Delta L$, and using the concept of "the weakest link", the strength distribution of the fiber of length ΔL can be obtained from

$$F(\sigma,\Delta L) = 1 - \exp(-\Delta L\alpha\sigma^\beta) \tag{20}$$

which is similar to equation (19). The experimental results on the long-term strength of fibers can be described by the doubled statistical Weibull distribution

$$F(\sigma,t,L) = 1 - \exp(-L\alpha t^\gamma \sigma^\beta) \tag{21}$$

where t, σ and L are the non-dimensional time, stress and fiber length, respectively. Referring to the fibers of length ΔL, the formula for the fracture probability distribution becomes

$$F(\sigma,t,\Delta L) = 1 - \exp[-\Delta L\alpha t^\gamma \sigma^\beta] \tag{22}$$

The relationship between the mathematical expectation of the lifetime $\bar{t}$ and the stress σ for the fibers of length L can be expressed with the aid of equation (21) as

$$\sigma \approx (\alpha L)^{-1/\beta} \bar{t}^{-\gamma/\beta} \tag{23}$$

The distribution parameters α, β and γ can be determined from testing the life-time of fibers of different length.

The determination of the statistical parameters of the elements as outlined above can yield unreliable results since there is no proof for the validity of extrapolation of the Weibull distribution for fibers of small length. In addition, we know well that the strength properties of fibers can be influenced by environmental conditions and the bonding. This means that the distribution parameters of the strength properties of structural elements should be determined from the composite long-term strength test and a knowledge of the accumulation of defects during loading.

ILLUSTRATIVE EXAMPLE

For illustration, consider the calculated results for the accumulation of defects in fiberglass reinforced plastics. The distribution function of the lifetime of the structural elements $F(\sigma$ and $\ell nt)$ for various stress levels is shown in Figure 2. The analytical expressions of the distribution curves in Fig-

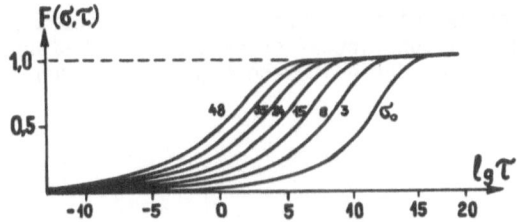

Fig. (2) - Rated curves of fracture probability of the elements with overstress on the fibers caused by defects of various size. The numbers on the curves designate the number of broken fibers with defect

ure 2 were obtained on the basis of extrapolation of the Weibull distribution for the fibers short-term strength on the effective length and the evaluation of the long-term strength by using the hypothesis of linear damage summation. Because of the complexity of the resulting expressions in comparison with equation (21), the details are not given here.

Each curve in Figure 2 corresponds to the rated maximum value of overstress on the fibers surrounding the defect. The number attached to the curves stands for the number of broken fibers with defect. The graphs for the mean stresses calculated by the method mentioned earlier are given in Figure 3. The kinetics of cumulative defects of different size in the composite not accounting for the loading history is displayed in Figure 4a-4d. This corresponds to the results obtained from equations (4) to (8). Note that the majority of single breaks occurs at the beginning of loading while the number of large defects increases with time up to final destruction. This result agrees qualitatively with the known experimental data on cumulative damage and is similar to the results in [10]. The formation probability of a defect of different size W_j is given in

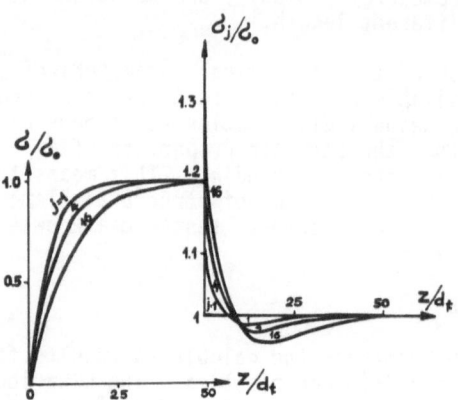

Fig. (3) - Rated curves of mean stress in broken fibers and neighboring
          fibers as a function of crack size and distance. The numbers
          on the curves designate the number of broken fibers

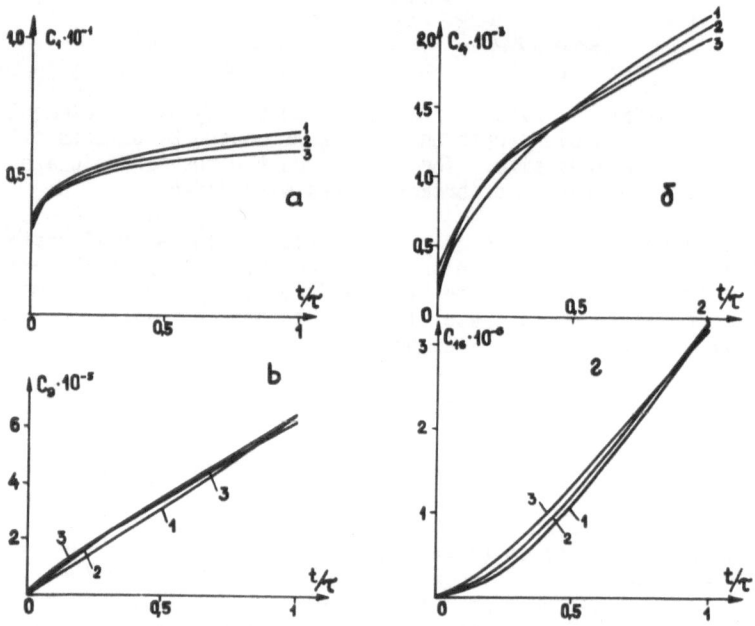

Fig. (4) - Kinetics of cumulative defects of various size: 1,4,9 and 10
          corresponding to (a), (b),...., (d), respectively. Curves 1, 2
          and 3 obtained under different stress levels $\sigma_1 < \sigma_2 < \sigma_3$

Figure 5. It is clear from the figure that the curves practically merge together when j≥16 giving the critical defect size. As to be expected, this is greater than that calculated for the two-dimensional model in [7].

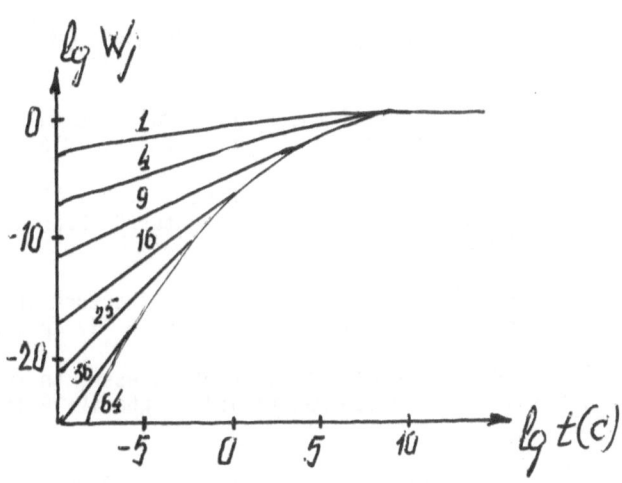

Fig. (5) - Curves of formation probability of a defect of various size constructed in logarithmic coordinates

It should be remarked that the given graphs display only the qualitative features of dispersed fracture in the heterogeneous composite material. Quantitative consideration of the parameters of the model necessitates a knowledge of the fracture history of the elements as considered by equations (13) to (16) and the kinetics of the cumulative damage to be determined experimentally.

REFERENCES

[1] Sih, G. C. and Liebowitz, H., "The mathematical theory of brittle fracture", In: Fracture, Vol. 2, Moscow, Mir, 1975.

[2] Cherepanov, G. P., "Mechanics of brittle fracture", Moscow, Naukha, 640 pp., 1974.

[3] Palmgren, A., "Die Lebensdauer von Kugellagern", "VDI-Z", Bd. 68, N. 14, pp. 339-341, 1924.

[4] Tamužs, V. P. and Kuksenko, V. S., "The fracture micromechanics of polymer materials", Riga, Zinatne, 294 pp., 1978.

[5] Latisenko, V. A., "Diagnictics of rigidity and strength of materials", Riga, Zinatne, 320 pp., 1968.

[6] Rosen, B. W., "Tensile failure of fibrous composites", AIAA Journal, 2, pp. 1985-1994, 1964.

[7] Zweben, C., "Tensile failure of composites", AIAA Journal, 12, pp. 2325-2331, 1968.

[8] Bolotin, V. V., "Some mathematical and experimental models to fracture processes", Problemi prochnosti, N. 2, pp. 13-20, 1971.

[9] Bolotin, V. V., "The statistical theory of cumulative damage of composite materials", Mekhanika Polimerov, N. 2, pp. 247-255, 1976.

[10] Tamužs, V. P., Tikhomirov, P. V. and Yushanov, S. P., "The fracture mechanism of materials having a heterogeneous structure", Fracture, Vol. 3, ICF4, Waterloo, Canada, 1977.

[11] Tikhomirov, P. V. and Yushanov, S. P., "Volume fracture of materials with nonhomogeneous structure", Mekhanika Polimerov, N. 3, pp. 462-469, 1978.

[12] Ovchinskiy, G. M., et al, "Redistribution of stresses in fracture of brittle fibers of metallic composite materials", Mekhanika Polimerov, N. 1, pp. 19-29, 1977.

[13] Lifshitz, G. M., "Decelerated fracture of fiber composites", In: Fracture and Fatigue, Ed. by L. Broutman, Moscow, pp. 267-332, 1978.

A CONCENTRATION CRITERION FOR MICROCRACK ENLARGEMENT IN HETEROGENEOUS MATERIALS

V. S. Kuksenko, D. I. Frolov and L. G. Orlov

A. F. Ioffe Physical Technical Institute of the Academy of Sciences of the USSR,
Leningrad

ABSTRACT

Nucleation and development of microcracks in polymeric and other materials with a heterogeneous structure are studied by small-angle X-ray diffraction, light scattering and acoustic emission. Stability of the incipient microcracks in stressed materials and transition from micro- to macro-fracture are investigated on the basis that the microcracks in the volume of a stressed specimen reach a critical concentration. The enlargement of microcracks in the specimen before rupture is examined by using statistical analysis. A concentration criterion for microcrack enlargement is also shown to be valid for nonpolymeric materials with a heterogeneous structure such as composite materials. Postulated are laws for governing the microfracture of heterogeneous materials that lead to the prediction of fracture at the macroscopic level.

INTRODUCTION

Fracture of solids is, as a rule, completed by the propagation of one or several main cracks. Nevertheless, both the formation of a main crack and its development involve a kinetic process when fracturing takes place in the stressed body. Fracture occurs through nucleation of a great number of the finest cracks under the influence of load. The geometry and conditions of loading on the solids can significantly affect the form, sizes and orientation of the initial incipient cracks. The accumulated damage leads to the eventual macroscopic failure of the specimen. This is a rather typical phenomenon [1-4].

Various methods, such as small-angle X-ray scattering, light scattering, acoustic emission, electron microscopy, dilatometry, etc., are employed for detecting the incipient cracks. An extensive investigation of the nucleation and accumulation of cracks in ionic crystals, metals, polymers and rock solids has been conducted by using these methods. In this work we shall report the experimental data on the cracks detected by small-angle X-ray scattering and sound emission methods.

CRACK NUCLEATION

The crack nucleation process has been studied extensively for polymeric materials. This is due to the fact that for these materials small-angle X-ray scattering has turned out to be a most reliable and efficient technique [3].

If there are cracks in the specimen interior, they scatter X-rays to give the central diffuse scattering. Measuring the angular distribution of the scattered radiation, we can estimate the crack sizes in different directions. Their concentration can be obtained from the absolute values of the scattered radiation intensity. This method can be used reliably for studying crack sizes ranging from 0.001 to 0.3-0.4μ.

The sound emission method is based on emission of elastic waves during crack formation. A connection between sound emission signal parameters and cracks is considered in [5].

An important peculiarity of crack formation is the fact that the different sizes are formed almost simultaneously. The time needed for the formation of an individual crack is negligibly small compared to time-to-rupture and cannot be measured by the diffraction methods. The radiation of elastic wave during crack formation is the evidence of the prevailing high velocity during crack formation. Without delving into the problems of "explosion-like" character of crack nucleation, we should like to note that the cracks become stable upon reaching definite sizes. Crack stabilization appears to result from the heterogeneous character of the polymeric materials which is inherent in pure polymers at the supermolecular level. This can also be artificially created in man-made composite materials. The main parameter in the crack formation process is, therefore, the crack concentration, which depends on the duration of loading, stress amplitude, etc.

Figure 1 shows the crack accumulation curves under constant tensile stress (curve 1) and under increasing tensile stress (curve 2) for an oriented nylon-6

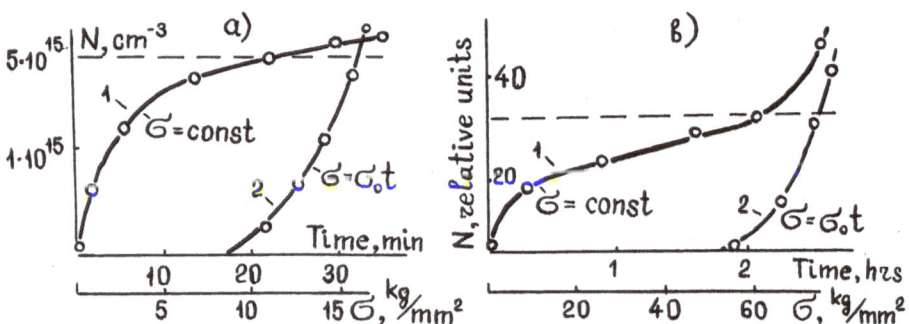

Fig. (1) - Crack accumulation curves under a uniaxial tensile stress (1) and under an increasing tensile stress (2). (a) a nylon-6 film; (b) fiberglass composite

film (a) and fiberglass composite (b). The data on the concentration of incipient cracks has been obtained by small-angle X-ray scattering for nylon-6 and sound emission for the fiberglass composite. The crack concentration in the fiberglass composite is given in relative units, since at present we cannot distinguish the acoustic signals resulting from breaking of glass threads, delamination between the matrix and threads, etc., in a unique manner. Thus, the crack accumulation curves for this case should be regarded as signals reflecting the total number of discontinuities. The accumulation curves differ sharply in their form, but the ultimate concentration for a given material turns out to be close to each other. When a definite crack concentration in a stressed specimen is reached, it loses stability and breaks down. The rupture occurs through the formation of a macroscopic crack in the vicinity where an increased concentration of the incipient submicrocracks were observed. To illustrate this observation, the following example can be used, Figure 2. Let the specimen be cut by

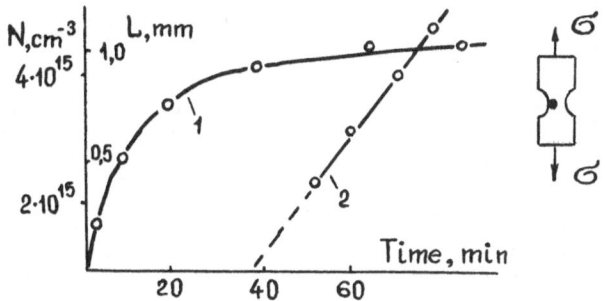

Fig. (2) - The kinetics of submicrocrack accumulation
(1) and macrocrack propagation (2) in a
nylon-6 film under a constant load

knives so that the fracture process can be localized. This simplifies the measuring procedure. The light or X-ray beam was directed to the line of the smallest cross-section near the edge of the specimen and the accumulation of cracks was traced. Refer to curve 1 in Figure 2. At the same time the filming of the specimen was carried out to detect the appearance of the macrocrack. The propagation rate is then measured and shown by curve 2 in Figure 2.

One can see from the very moment of the load application, an intensive accumulation of cracks take place in the site of the fracture localization. The macrocrack formation is seen to be preceded by an intense volume fracture associated with the microcrack accumulation. It is developed at a constant rate which justifies the extrapolation to a zero site.

CRACK DETECTION IN NYLON-6 FILM

It is interesting to inquire into the transition from a slow accumulation of stable cracks to form a macrocrack. Since the initial incipient cracks are stable and do not increase in size, it is quite natural to suppose that their enlargement results from a process of coalescence when the crack concentration reaches a critical value at which the distances between the cracks become commensurable with the crack sizes.

Such an enlargement can be examined by the same diffraction methods, say by the small-angle X-ray scattering technique. If we measure the radiation curves at the initial portion of the accumulation curves at t_1 in Figure 3a, both the quantitative and the qualitative difference can be revealed. The macrocrack is formed at time t_2. Figure 3b gives the radiation curves for the two cases. The

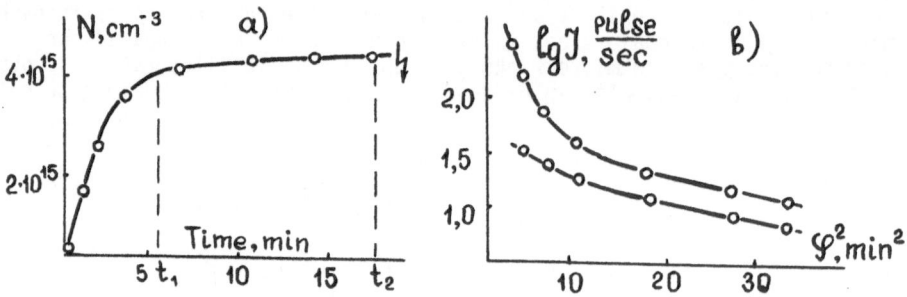

Fig. (3) - The kinetics of submicrocrack accumulation for a nylon-6 film under a uniaxial tensile stress (curve a). The X-ray diffuse scattering curves for a stressed nylon-6 film soon after the saturation in the submicrocrack accumulation is reached (1) and the moment of macrocrack formation (2) (curve b)

curves are seen to deviate sharply at small angles. The linear dependence in the first case indicates that the cracks are approximately of the same size. The next stage is characterized by the growth of the crack concentration region rather than by the enlargement of crack size. The steeper portion of the radiation curve represents the formation of larger cracks. The enlargement process leads to the eventual formation of a macrocrack. Unfortunately, this method cannot distinguish the transition stage during which time the fine cracks transform to a macrocrack, since the larger cracks cannot be detected by X-ray scattering. For this reason the third stage of fracture remains uninvestigated. In this respect, the acoustic emission method is more promising.

FRACTURE OF COMPOSITE MATERIALS

Figure 4 gives a crack accumulation curve for a composite material under uniaxial compression obtained by mixing quartz powder with an epoxy resin. The powder particle size is approximately 10μ in diameter. The crack accumulation curve is similar to that shown in Figure 1b (curve 1), which shows a common feature identifying the transition from the stable crack accumulation to a macroscopic fracture when a certain critical concentration of stable microcracks is reached.

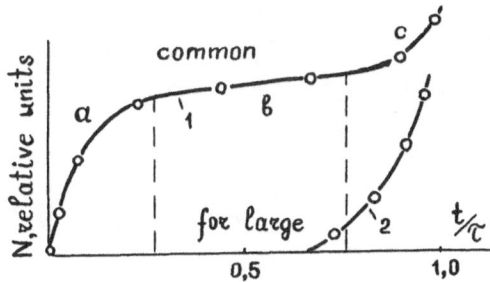

Fig. (4) - Crack accumulation curves for a uniaxially compressed
     specimen. Curve 1 - the total number of cracks, and
     curve 2 - large cracks

Let us now consider the transition from microfracture to macrofracture in
more detail. Information on crack enlargement can be obtained from an analysis
of the amplitude spectrum of sound emission signals at different stages of crack
accumulation. According to the model experiments [5], the sound emission signal
amplitude is proportional to the cross-section of the break or the crack size.
The analysis shows that the amplitude distribution is highly nonuniform at different portions of the accumulation curve. During the major portion of time,
say 80% of the total life, the pulses with small amplitudes not higher than certain values of A_m are generated in the specimen. Then, together with the small-
amplitude pulses, the pulses with the amplitude higher than A_m start to be generated. One can see that the period of the accumulation of the incipient cracks
constitutes the major period of time. The final stage of fracture then follows.
This is characterized by an increase in the amplitude of pulses generated by the
specimen before rupture. The final stage constitutes only a small part of the
specimen lifetime $\Delta t_3/\tau$ but is extremely important for understanding the transition from slow accumulation of the stable microcracks to the formation and propagation of large unstable macrocracks.

Crack enlargement can also be verified by a microscopic observation of the
larger cracks during the acceleration stage of fracturing. Figures 5a show the
photographs of the specimen accompanied by schematic diagrams interpreting the
various stages of material damage as illustrated in Figures 5b. Set I corresponds to the beginning of the stable stage; set II to the moment not long before the third stage takes place; and set III to the third stage before macrorupture. An appreciable number of large cracks can be seen to appear at the
third stage. These data point to the fact that the formation of macrofracture
is due to crack enlargement. However, the details of the transition from fine
incipient cracks to large microscopic and then macroscopic crack remain unknown.

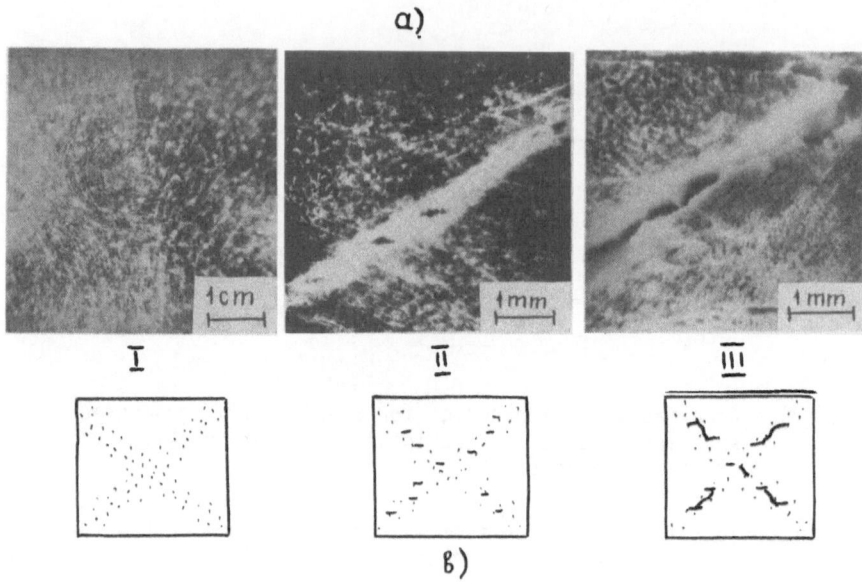

Fig. (5) - Photographs of the lateral surface of the specimen
for different duration of loading. (a) Schematic
diagrams identifying the different stages of crack
formation (b)

THEORETICAL AND EXPERIMENTAL CONSIDERATION

In [8], a theoretical model for crack enlargement during growth in discrete
steps is presented. It agrees with the experimental data qualitatively on the
kinetics of accumulation of cracks of different sizes. Another mechanism of
crack enlargement, considered in [1,2,7], considers the dynamic interaction and
coalescence of the finer incipient cracks in regions of high crack concentration. The critical condition is assumed to prevail when the average distances
between the cracks are commensurable with the sizes of the cracks. The two
mechanisms are not exclusive of one another. Moreover, we can assume that both
of these processes may occur simultaneously. It is, however, essential to know
which one of them is more dominant at a given stage of fracture. This is needed
for the construction of a quantitative theory of fracture. Information of this
type is also necessary for predicting macroscopic fracture on the basis of the
data of crack accumulation.

Nevertheless, it is worthwhile to report some experimental results confirm-
ing the concentration criterion of crack enlargement. Let us introduce a dimensionless parameter $K = N^{-1/3}/L$ characterizing the relative average distances be-

tween incipient cracks where L is the size of the initial incipient crack and $N^{-1/3}$ is the average distance between cracks.

In polymeric materials, the sizes of the initial cracks or discontinuities L vary in wide limits. The prebreaking concentrations vary even more appreciably. However, they are connected in a definite manner. Figure 6 gives a graphical comparison between the prebreaking crack concentration N_b and L_{cr}. All the

points fall nearly on a straight line described by the relation $N^{-1/3}L^{-1}$ from which the value of K can be estimated. It turns out to be 5/7.

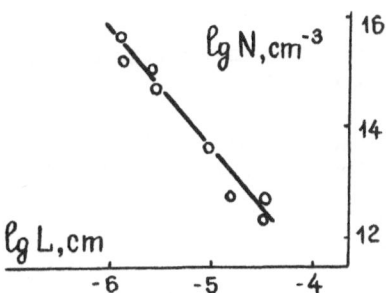

Fig. (6) - Relation between the number and sizes of cracks for various materials before rupture

First of all, the existence of a common dependence points to the fact that the phenomenon of volume fracture is typical for different types of solids. Further, it demonstrates that the critical crack concentration criterion appear to govern the transition from slow accumulation of stable microcracks to the formation of a main crack that separates the material.

A critical transition from volume crack accumulation to the formation of larger cracks can also be derived from the statistical laws [5]. As the cracks are distributed randomly, the regions with increased densities of incipient cracks will appear in a volume of the stressed specimen. An intense crack enlargement starts at K ≈ 3 [5]. A higher experimental value of K as compared to that predicted by the statistics of crack population is likely to indicate nonuniformity of the fracture process in a volume of the stressed specimen which tends to be localized in certain regions. Fracture localization confirms the concentration criterion of crack enlargement.

Figure 7 presents a dependence of the crack formation rate at a small part of the stable portion of crack accumulation. The kinetics of crack accumulation was followed by the sound emission method which permits the detection of the high speed of the process. The tests were conducted on biaxially compressed transparent specimen made of epoxy resin. The acoustic signals were measured by two gauges: one for the small-amplitude signals and the other for the large-amplitude signals. The specimen was filmed to identify the moment of a visual

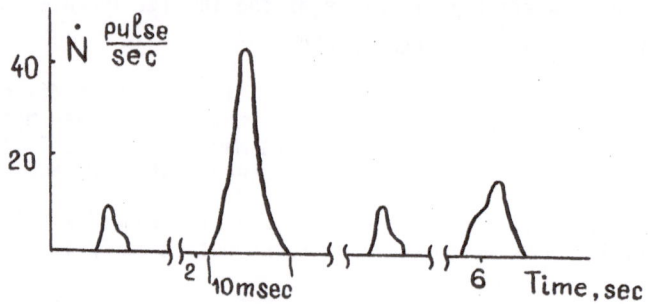

Fig. (7) - Crack accumulation rate at the stable stage of fracture

crack formation. The existence of sharp "bursts" of activity in the fracture process is the evidence of the fact that fracture takes place in a certain microregion of the specimen.

REMARKS ON CRACK FORMATION

Thus, the formation of large cracks does not result from a gradual growth of small cracks but is due to the fact that a microregion of the specimen loses stability when the concentration of stable cracks in it becomes critical. The rupture of this region leads to the formation of a large crack. The local character of the fracture process development makes it possible to consider the fracture pattern in more detail.

The external load introduces nonuniform stresses in a solid. The natural and artificially created heterogeneity of the solid further aggravates the concentration of stresses in various microregions, where the fracture process proceeds at higher rates. When the crack concentration in a certain microregion becomes critical, the microregion loses its mechanical stability and breaks down. This leads to the formation of a crack commensurable with the diameter of this microregion. The crack, thus formed, stops at the boundary of the ruptured region and remains quasi-stable. The formation of such enlarged cracks occur in large quantities and may cause the next stage of enlargement when several of the enlarged cracks may coalesce to form an even larger crack. Such a sequence of enlargement agrees with the discrete nature of the fracture process in time (see Figure 7). When a definite level enlargement is reached, the crack acceleration may appear. The development of one or possibly more of these enlargement processes leads to the eventual macrorupture. This development is responsible for a new increase in the crack accumulation rate, manifesting itself (see Figure 1) on the integral curve of crack accumulation.

It should be pointed out in conclusion that one and the same concentration-induced crack enlargement mechanism underlies the fracture transition process at all scale levels. It is such a fracture transition in steps from one scale level to another that is responsible for fracture occurring at the micro- and macro-levels.

REFERENCES

[1] Zhurkov, S. N. and Kuksenko, V. S., "Micromechanics of polymer fracture", Mech. Polym., No. 5, p. 795, 1974.

[2] Zhurkov, S. N. and Kuksenko, V. S., "The micromechanics of polymer fracture", Int. Journal of Fracture, 11, 62, 1975.

[3] Tamuzh, V. P. and Kuksenko, V. S., "Micromechanics of fracture of polymer materials", Zinatne, 1978.

[4] Kuksenko, V. S., Ryskin, V. S., Betechtin, V. I. and Slutsker, A. I., "Nucleation of submicroscopic cracks in stressed solids", Int. Journal of Fracture, 11, p. 829, 1975.

[5] Zhurkov, S. N., Kuksenko, V. S., Petrov, V. A., Saveljev, V. N. and Sultanov, U., "On the problem of predicting the fracture of rock solids", Izv. AN SSSR, Ser. Physics of Earth, No. 6, 8, 1977.

[6] Kuksenko, V. S., Slutsker, A. I. and Frolov, D. I., "Mechanism of nucleation and propagation of macrocracks in stressed polymers", Probl. Prochn., 11, 81, 1975.

[7] Zhurkov, S. N., Kuksenko, V. S., Slutsker, A. I. and Frolov, D. I., "The role of submicroscopic cracks in the main evolution of the main crack", Conf. Proc. 3rd Intern. Congress on Fracture, Munchen, p. 322, 1973.

[8] Tichomirov, P. V. and Yushanov, S. P., "Three dimensional fracture of materials of heterogeneous structure", Mech. Polym., 3, 1978.

STUDIES IN FRACTURE KINETICS OF COMPOSITE MATERIALS

V. R. Regel, A. M. Leksovskii and O. F. Pozdnyakov

A. F. Ioffe Physico-Technical Institute of the USSR Academy of Sciences, Leningrad, USSR

ABSTRACT

The effect of interface created by the various constituents on the mechanical properties of the composite is investigated. The accumulation of submicrocracks between the matrix and reinforcing fibers has been observed by small-angle X-ray technique for the loaded polymer under consideration. The role of delamination process associated with fracture is discussed. The detailed structure of the interface has also been studied by means of mass-spectroscopy. The interaction at the metal-polymer interface results in thermodestruction activation energy change for polymer layers with thicknesses in the range of 10 to 100 Å.

INTRODUCTION

Fracture of both homogeneous and heterogeneous bodies including composite materials is treated within the framework of the kinetic concept of strength [1]. The present work considers two aspects of composite fracture. First, the events of the fracture process leading to fracture will be studied systematically. Next, the influence of the surface condition of the composite constituents on the mechanical properties of the composite is considered. Experiments are carried out to investigate both of these effects for the artificially made heterogeneous composites.

The complete picture of composite fracture kinetics should include the description of the stress distribution in a matrix-fiber system. For instance, crack formation and growth beginning from the submicrosize (in matrix, fiber, and on the interface) should be followed up to the coalescence of all the small cracks until a macrocrack forms and traverses through the specimen. Unfortunately, there is no definite answer to the problem just mentioned. Referring to the recent publication of Broutman and Krock [2], it can be said that the problem is far from being settled.

Meanwhile, by knowing the sequence of events in the fracture process and the time scale of their occurrence, one can monitor the properties of the man-made composites. It is reasonable, therefore, to carry out further experiments. The bundles of oriented capron and polypropylene amorphous-crystal fibers were used as the model in our experiments with butvarphenolene resin and polyoxiethylene serving as the matrices. The volume fraction of the matrix was about 30%, whose strength is about 50 times less than that of the fibers.

Long time loading force reveals that the lifetime of fibers enclosed in the matrix is one to three orders of magnitude more than of those tested without the matrix, Figure 1. According to the rule of mixture, the strength increase

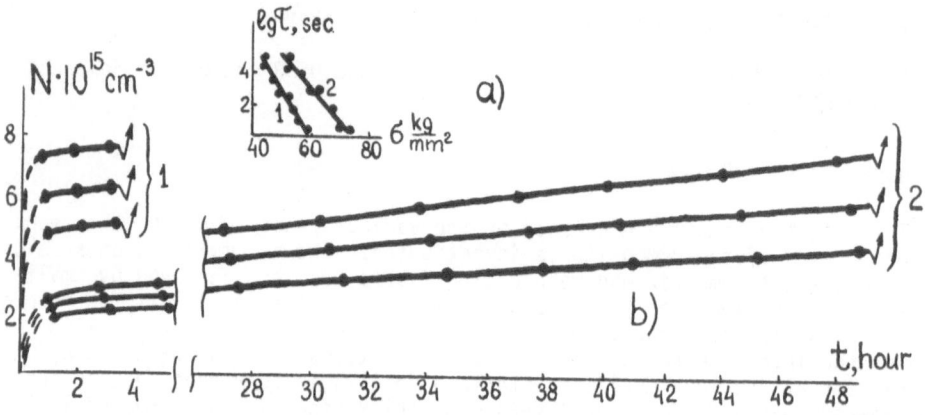

Fig. (1) - Comparison of the fracture kinetics for the capron fibers without matrix shown by curves 1 and with matrix shown by curves 2: (a) the dependence of lifetime upon force; and (b) the kinetics of submicrocracks accumulation

should be no more than by 1% and not 10 to 20%. The difference in the stresses with and without the influence of the matrix has been found earlier [3] by means of small-angle X-ray diffraction. This difference is found to be negligible. In the composite, the deviation of stresses in the individual fibers was not more than 5%. A 10 to 20% deviation was found for the isolated fibers.

DIFFUSE SCATTERING AND ACOUSTIC EMISSION

The change of concentration and of submicrocrack size were measured by long-time experiments under σ = const using a diffusion indicator in the range of small angles (2-30 angle minutes). The diffuse scattering results from the micro-defects in the fibers or the interface, but not in the matrix. The matrix is calibrated such that no diffusion is measured. This is checked experimentally each time.

The experiment showed that the submicrocracks up to certaim moments of time had the same size both in composites and in the fibers without the matrix, i.e., for the capron fibers the size was about 200-400 Å. As illustrated by Figures 1b and 2, the kinetics of accumulation of submicrocracks in the composite sam-

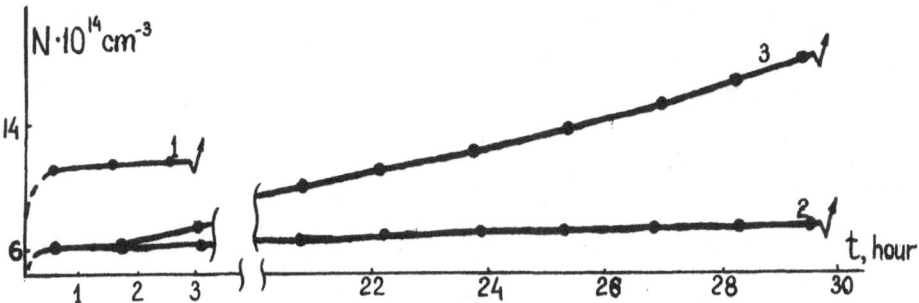

Fig. (2) - Kinetics of submicrocracks accumulation: Curve 1 -
propylene fibers without matrix and curves 2 and 3 -
polypropylene fibers inside butvarphenolene resin.
Curve 3 indicates the location of future rupture
and 2 the other locations.

ple is slowed down (curves 2). At the location of future rupture, however, the
larger cracks of size about 2000 Å and more were formed, curve 3 in Figure 2.
The special tests on the isolated fibers taken from that location reveal the
existence of commonly observed cracks of size about several hundreds Å. As is
shown by the scattering indicator, the sample is oriented with respect to the
X-ray microbeam, the large cracks are inclined not at 90° to the direction of
tensile stress but at much smaller angles. For these two reasons, one can suggest that the submicrocracks of the order of thousand Å prevail at the interface. This experimental evidence seems to be of interest and requires more detailed investigation of the so-called delamination cracks. It should be added
that by means of acoustic emission, the rupture of fibers may be detected. One
can reveal the onset of the rupture process and the formation of large submicrocracks. Refer to curve 2 in Figure 3. Delamination causes the fibers to act

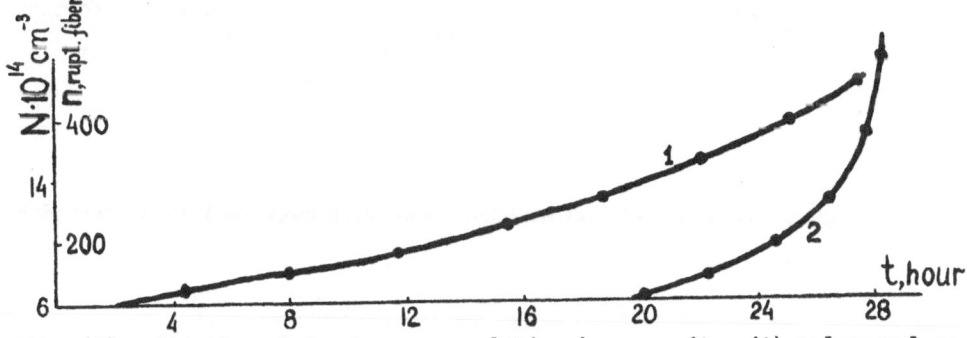

Fig. (3) - Kinetics of fracture accumulation in composite with polypropylene
fibers inside butvarpropylene resin. Curve 1 - delamination development detected by small-angle X-ray diffraction and curve 2 -
rupture accumulation of elementary fibers detected by acoustic
emission

individually and no longer functions as a composite. Finally, the accumulation of delamination and fiber breaking leads to the avalanche-type of rupture. The acceleration stage of crack growth has not been studied in detail. To solve this problem, it is also necessary to observe the development of the individual cracks.

In general, delamination plays an important role in the rupture process and the consideration of composite strength. The interaction of fibers with matrix can also influence the compressive behavior and the speed of the delamination process. For the case of polyvinyl alcohol fibers embedded in an epoxy matrix [4], delamination was slowed down approximately one hundred times by means of bifunctional diisocyanate molecules. This increased the long-time composite strength by several orders of magnitude.

The influence of matrix on fiber bonding as investigated in [5] shows that fuse scattering leads to the concentration of submicrocracks. The matrix not only protects the fibers from moisture but tends to redistribute and lower the bond stress. The details of the interface need further study in order to gain a better understanding of the interaction of the composite constituents and of the interface.

MASS SPECTROMETRY

The mass-spectrometry technique will now be discussed and applied to study thin film layers in the composite. Thin and ultra-thin polymer films precipitated on the surface of the composite reinforcement will be chosen to represent the interface. Here, adhesion and molecular absorption are considered as the primary reasons of adhesion bond formation.

The kinetics of thermal degradation of the absorbed films as a function of their thickness "δ" from several Å up to hundred Å were examined by mass-spectrometer. The activation energy of thermodestruction E was defined as a function of δ. The polymer films were precipitated from the dilute solution on the substrate and then put into the chamber of a time-of-flight mass-spectrometer. Thermodestruction of the two polymers, PS and PMMA, was investigated in [6]. Both were prepared by the anion polymerization technique. Their molecular mass is about 100,000. Tantal film was used as a substrate. The activation energy of thermal destruction for the bulk polymers PS and PMMA in vacuum equals 50 and 45 kcal/mol, respectively.

As an example, Figure 4 shows the temperature-time dependence for PMMA and the intensity of the yield of MMA monomer (peak with mass 100) in destruction. The treatment of such kinetic curves define the parameters of the process. The temperature dependences of the constants of monomer formation rate for various film thickness of PMMA and PS are represented in Figure 5. Figure 6 shows the dependence of activation energy E upon the layer thickness within the whole range of measurement. E was defined from the experimental data like those shown in Figure 5.

Experimental data concerning thermal destruction of thin polymer films (Figure 6) should be compared to the same data for bulk polymers. Judging from Fig-

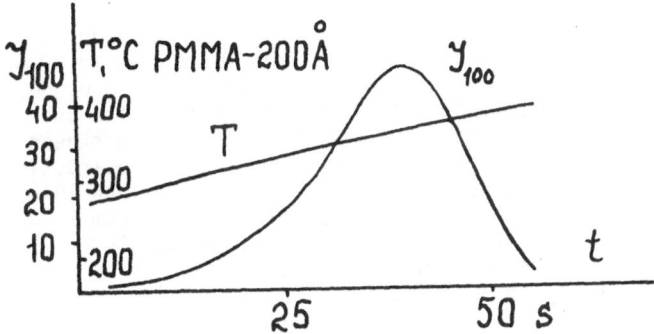

Fig. (4) - Time dependence of volatile products
yield intensity and temperature

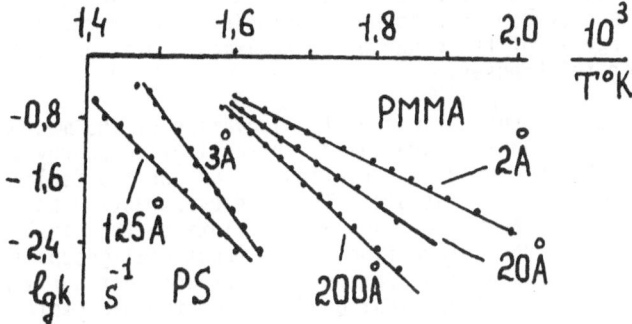

Fig. (5) - Reciprocal temperature dependence of thermodestruction
rate constants for film polymers

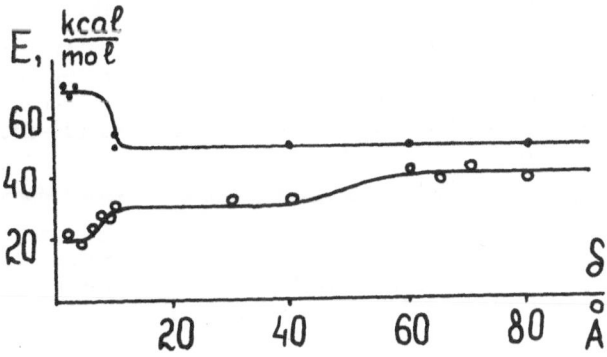

Fig. (6) - Dependence of thermodestruction activation energy
upon polymer film thickness

ure 6, there is a coincidence of E values for the bulk and thin layers within the range of thickness from the indicated maximum value down to 10 Å. This fact, together with the constancy of the degradation volatile products, reveal the constancy of the mechanism of thermal destruction for the PS films of any thickness down to 10 Å. Figure 6 shows that for film thickness less than 10 Å, the value of E increases to about 65 kcal/mol. The destructive mechanism of PS is changed by the increase of E points. For δ less than 10 Å, i.e., when the absorbent layer is less than the monomolecular one, macromolecules are then isolated [7]. In this case, the termination of macroradicals is eliminated due to their interaction and the breaking of C-C bonds followed by macroradical formation and their complete depolymerization. The activation energy of thermodestruction is determined by the C-C bond dissociation energy for PS, i.e., E = 65 kcal/mol. Thermodestruction reveals no catalysis by this substrate.

Another dependence of E on δ is observed for PMMA. Beginning from a layer thickness of 70 Å, the activation energy decreases with E values typical for destruction of this polymer in the bulk to 20 kcal/mol (see Figure 6). For the PMMA under study the decrease in E is not monotonic. Within the thickness range of 10-40 Å, the energy is of the order of 30 kcal/mol and is approximately 20 kcal/mol for smaller thicknesses.

If the above mentioned suggestions on the isolation of the macromolecules in PMMA for layers less than 10 Å are valid, the activation energy of 20 kcal/mol should be treated as the C-C bond dissociation energy for the isolated PMMA macromolecule adsorbed on the substrate. The thermodestruction activation energy decrease for PMMA in the ultra-thin layers points to the catalytic influence of substrate. Here, the interaction of the functional groups for PMMA (e.g., groups C=O) with some active centers of adsorbent surface is possible, which leads to the weakening of chemical bonds inside the macromolecule.

For thickness larger than the monomolecular, the thermodestruction must include the stage of macroradical termination by the "square law", E being the complex function of activation energy for the elementary stages (initiation, depolymerization, termination of radicals).

In the thickness range of 10-70 Å for PMMA, E is also less than that in the bulk. It could be suggested therefore that in this layer, the macromolecules must have only one direct contact with the substrate. This is enough to affect the total energy of the process E taking into account the radical-chain nature of the PMMA thermodestruction. The thickness of the transition zone for these two components can be approximated by 100 Å.

Thus, the interaction on the metal-polymer interface of PMMA tends to diminish the activation energy of thermodestruction for ultra-thin polymer films near the contact. In this case, the redistribution of the electron density of the initial molecules takes place owing to absorption on the active substrate centers.

Consider adhesion on the molecular level and eliminate the possible effect of macrodefects in the transient layers [8]. An increase in the adhesive strength caused by interaction with the substrate polymers such as PMMA can result from the reserve cohesive strength of the initial polymer micromolecules. The co-hesive strength of the macromolecules is proportional to the adhesive strength.

The two aforementioned aspects of composite strength involve both the kinet-ic approach to fracture and the modern experimental technique where the fracture process is studied from the atomic-molecular level up to the macromolecular level. The investigation revealed the primary influence of the interface me-chanical properties of the composite. Needed are additional investigations to further reveal the unusual fracture characteristics of these composites. The knowledge gained hopefully can be used to build new composites.

REFERENCES

[1] Regel, V. R., "Fracture", Vol. 4, ICF4, Waterloo, Canada, June 19-24, 1977.

[2] "Composite materials", Vol. 5, L. J. Broutman and R. H. Krock, eds., Aca-demic Press, New Jersey - London, 1974.

[3] Leksovskii, A. M., Orlov, L. Gh. and Regel, V. R., "Mechanics and technology of composite materials", Sofia, pp. 100-106, 1977.

[4] Regel, V. R., et al., Mechanika polymerov (in Russian), No. 5, pp. 815-818, 1976.

[5] Zurkov, S. N. and Kuksenko, V. S., Mekhanika polymerov (in Russian), No. 5, pp. 792-799, 1974.

[6] Madorsky, S. L., "Thermal degradation of organic polymers", Interscience publishers, New York - London - Sidney, 1964.

[7] Lipatov, Yu. S. and Sergeeva, L. N., "Adsorption of Polymers" (in Russian), Kiev, p. 199, 1972.

[8] Berlin, A. A. and Basin, V. E., "Fundamentals of polymer adhesion" (in Rus-sian), Moscow, p. 391, 1974.

Section II

STATISTICAL AND ANALYTICAL METHODS

VERIFICATION AND ESTIMATION OF STOCHASTIC MODELS OF FRACTURE

V. V. Bolotin

Academy of Sciences of the USSR
Moscow, USSR

ABSTRACT

Approaches to verifying the hypotheses on the stochastic models of fracture and estimating of the parameters involved are presented. Difficulties arising in the statistical nature of the fracture process such as the inability to fully reproduce the experimental results and the limited application of the law of large numbers are discussed. The ways to overcome these difficulties are discussed with emphasis placed on the composite and polymer materials.

STOCHASTIC PROCESS

The stochastic nature of fracture in solids arise at the molecular, microscopic as well as on the macroscopic scale levels. The strength of material and general character of fracture depend on the various random macroscopic factors such as cracks, inclusions and other flaws in the material structure. Random properties of the fibers and granules, random distribution of the fibers in the matrix, random flaws of the adhesive and cohesive type, etc., are particularly significant to the composite material behavior. Stochastic factors arising during the manufacturing process of a composite such as imperfections, instability of the process, limitation and tolerance of quality control, etc., should be taken into consideration. In studying the physical phenomenon of fracture, the scatter of experimental results may be reduced by improving the manufacturing process and by a careful selection of the test specimens. The aim of reliability and durability consideration is to design a safe and durable structure such that all stochastic factors including those in manufacturing are taken into account [1].

Experimental data scatter on fracture tests is analyzed statistically with special emphases placed on reliability and safe design. The present theory introduces a multidimensional random parameter, the strength vector r, for each sample (or elements) of the specimens (structural members, machine parts, etc.). The vector r is assigned to describe the intrinsic properties of an individual specimen. Theoretically, the vector r must include all the stochastic factors that will influence the strength and durability of the specimen. If the loading conditions are determined, the scatter of the experimental results is caused only by the stochastic nature of strength vector $\underset{\sim}{r}$.

Strictly speaking, fracture experiments are non-reproducible. This is because we cannot fracture the same specimen repeatedly. This fact imposes serious limitations on application of the law of large numbers to analyze fracture phenomena. For example, in fracture or yielding tests under combined stresses, the critical (ultimate) surface is random. Each specimen has its own individual critical surface depending parametrically on the random vector $\underset{\sim}{r}$. Loading a given specimen to fracture, we obtain only a single point on an individual surface. Figure la shows the critical surfaces which are numbered for each specimen. The individual critical surface, as a whole, remains undetermined.

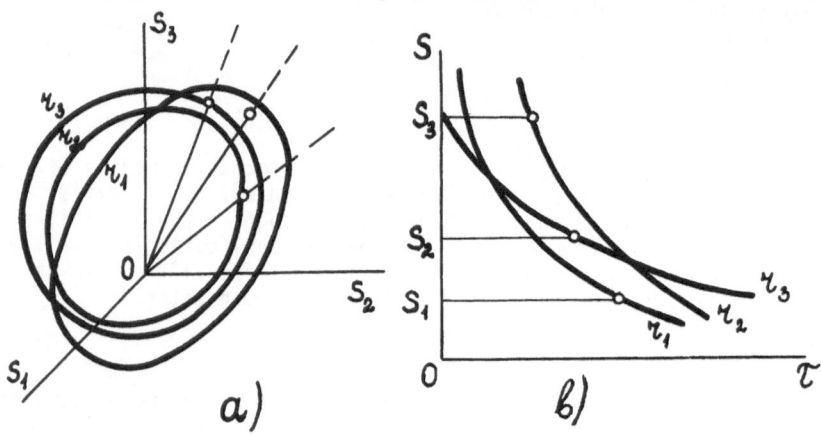

Fig. (1) - (a) - Critical surfaces under combined stresses and (b) - lifetime curves under constant loads

Another example is illustrated in Figure 1b. Let the lifetime τ of a set of specimens be studied experimentally as a function of stress S which is kept constant at different levels. The function $\tau(S)$ is random and varies from one specimen to another. Hence, the lifetime function may be written in the form $\tau = \tau(S|\underset{\sim}{r})$ where $\underset{\sim}{r}$ is the random strength vector. For a set of specimens with vector $\underset{\sim}{r}_1$, r_2,..., we obtain the set of functions $\tau = \tau(S|\underset{\sim}{r}_1)$, $\tau = \tau(S|\underset{\sim}{r}_2)$,..., which are shown schematically in Figure 1b. The test of a specimen at the constant stress level S leads to a single point on the curve $\tau = \tau(S|\underset{\sim}{r})$. The other points of the curve, in principle, cannot be found experimentally. As a result, we cannot estimate the two-point probability density $p(\tau_1,\tau_2; S_1,S_2|\underset{\sim}{r})$ and the two-point moments $<\tau(S_1|\underset{\sim}{r})\tau(S_2|\underset{\sim}{r})>$ relating the lifetime τ_1 and τ_2 corresponding to the stress levels S_1 and S_2. Hereafter, the operator $<\cdot>$ signifies the sample averaging as expected mathematically.

Consider an example on the linear cumulative damage hypothesis. Let the loading program be given by the vector function $\underset{\sim}{S}(t)$ where the vector $\underset{\sim}{S}$ accounts for the load, temperature and environment factors. The lifetime of an individual specimen under the constant loading condition S = const is denoted by $\tau_S(S|\underset{\sim}{r})$.

According to the linear cumulative damage rule, the lifetime τ under the continuous or piecewise continuous loading process $\underset{\sim}{S}(t)$ is determined from the functional equation [2]:

$$\int_0^\tau \frac{dt}{\tau_S[\underset{\sim}{S}(t)|\underset{\sim}{r}]} = 1 \tag{1}$$

It is essential that equation (1) contains the lifetime of an individual specimen whose complete probabilistic description cannot be found experimentally.

If the scatter is sufficiently small, these phenomena may not be taken into account. The scatter of the ultimate loads is often moderate but the moderate scatter of the ultimate loads is usually accompanied by significant scatter in lifetime. As an example, let the ultimate load S and lifetime τ satisfy the double Weibull distribution [3,4]:

$$F(S,\tau) = 1 - \exp[-(\frac{S}{S_c})^\alpha (\frac{\tau}{\tau_c})^\beta] \tag{2}$$

Here, $S_c>0$ is a load constant, $\tau_c>0$ is a time constant, and $\alpha, \beta \epsilon[1,\infty)$. When τ is considered as a parameter, equation (2) yields the probability distribution for the ultimate load at the given lifetime. When S is considered as a parameter, equation (2) represents the lifetime distribution. It follows from equation (2) that the lifetime equation for quantiles τ_γ of lifetime τ is

$$\tau_\gamma = \tau_c(\frac{r_\gamma}{S})^m \tag{3}$$

Here, r_γ is a characteristic (quantile) strength. Equation (3) is used often in the analytical processing of data from fatigue and creep fracture experiments. The exponent m is related to α and β as

$$m = \alpha/\beta \tag{4}$$

where m ranges from 4 to 12.

The variation coefficients with random values S and τ are used as a measure of the statistical scatter. Equation (2) yields

$$W_S = \sqrt{\frac{\Gamma(1+2/\alpha)}{\Gamma^2(1+1/\alpha)} - 1} \tag{5}$$

and

$$W_\tau = \sqrt{\frac{\Gamma(1+2/\beta)}{\Gamma^2(1+1/\beta)} - 1} \qquad (6)$$

These formulas yield a parametric functional relation between W_s and W_τ where the parameter is governed by the exponent m in the lifetime equation (3). The set of functions $W_\tau = f(W_s,m)$ is plotted in Figure 2. Increase in the exponent m results in an increase of the scatter of lifetime for a given scatter of the ultimate load S. Hence, a high rate of scatter is involved when considering lifetime. Under these circumstances, special requirements are needed for the

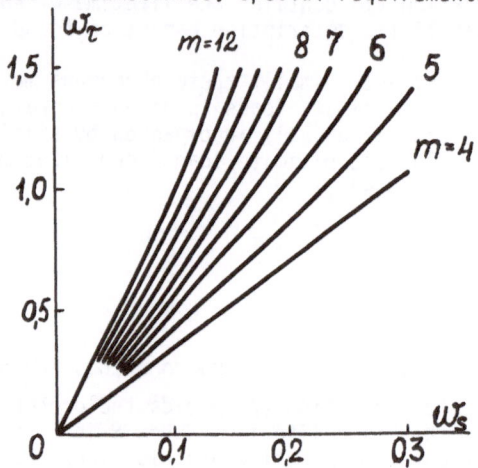

Fig. (2) - Variation coefficient of the lifetime versus variation
coefficient of ultimate load

design of experiments and the treatment of experimental data. The choice of sample size, the techniques of data processing, and the hypotheses used in the testing procedure are decided by appealing to the appropriate probabilistic analysis and the methods of mathematical statistics.

TWO-STEP TESTING PROCEDURE

We consider in detail the two-step testing program. Each specimen is being subjected to the load S_1 during the lifetime segment τ_1. It is then subjected to the load S_2 until fracture occurs. The lifetime of the specimen is determined by $\tau = \tau_1 + \tau_2$ where τ_2 is random varying from one specimen to another. Let the linear cumulative damage rule be valid. The values of τ_1 and τ_2 satisfy the equation

$$\frac{\tau_1}{\tau_S(S_1|r)} + \frac{\tau_2}{\tau_S(S_2|r)} = 1 \qquad (7)$$

which follows from equation (1). We can find $\tau_2(\underset{\sim}{r})$ from the experiments. However, equation (7) contains two additional unknowns $\tau_S(S_1|\underset{\sim}{r})$ and $\tau_S(S_2|\underset{\sim}{r})$ of the same specimen. In order to confirm or discard equation (7), it is necessary to fracture each specimen three times.

Instead of equation (7), experimentalists have endorsed similar expressions containing mathematical expectations of the random functions $\tau_2(\underset{\sim}{r})$, $\tau_S(S_1|\underset{\sim}{r})$ and $\tau_S(S_2|\underset{\sim}{r})$. Introducing the nondimensional lifetime parameters

$$\theta_1 = \frac{\tau_1}{<\tau_S(S_1)>}, \quad \theta_2 = \frac{\tau_2}{<\tau_S(S_2)>} \tag{8}$$

Equation (7) gives

$$f(\theta_1,\theta_2,\underset{\sim}{S}_1,\underset{\sim}{S}_2) = 1 \tag{9}$$

It is of special significance that equation (9) does not, in general, coincide with the linear equation

$$\theta_1 + \theta_2 = 1 \tag{10}$$

which follows directly from the simplification of equation (7).

An additional complication arises from the truncation of samples during experiments. For example, all specimens fractured before entering the second step will be rejected as "defective" ones. Specimens whose lifetime exceeds a certain base time τ_b will not be counted in the tests. Hence, the truncated sample of the specimens is defined as

$$A = \{\underset{\sim}{r}:\tau_S(S_1|\underset{\sim}{r}) \geq \tau_1, \ \tau_1 + \tau_2(\underset{\sim}{r}) \leq \tau_b\} \tag{11}$$

Some numerical results showing the influence of the above mentioned factors are presented in Figures 3 and 4. The linear cumulative damage rule was applied to the stochastic model. The relation between the lifetime τ and stress level S is taken in the form

$$\tau = \tau_c(\tfrac{r}{S})^m \tag{12}$$

where τ_c is a time constant, $m\varepsilon[1,\infty)$ and r is a random scalar with the distribution function

$$F(r) = 1 - \exp[-(\frac{r-r_0}{r_c})^\alpha] \tag{13}$$

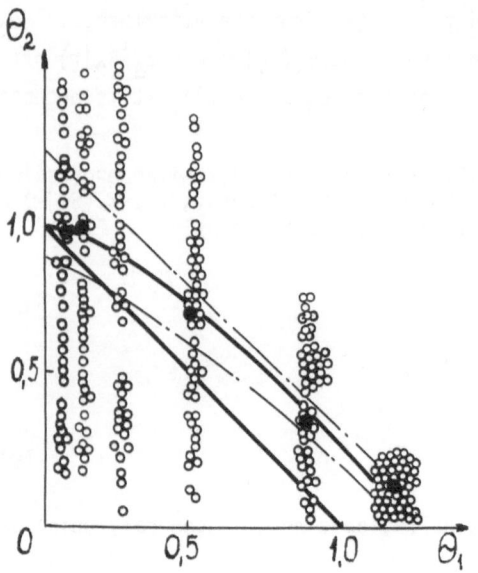 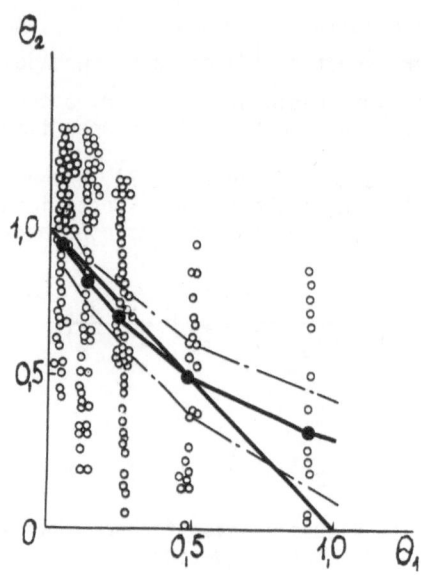

Fig. (3) - Probabilistic analysis and the Monte-Carlo simulation results for the two-step ascending program

Fig. (4) - The same for the two-step descending program

Here, r_0 and r_c are positive constants such that $\alpha\varepsilon[1,\infty]$. This model includes one random scalar parameter, namely the characteristic strength of the specimen r. Figure 3 corresponds to the two-step program when $S_1 < S_2$ and Figure 4 shows a plot for the inverse loading program. The relation between the mathematical expectation $\langle \tau_2 \rangle$ and the duration of the first step τ_1 reveals the effect of apparent strengthening when the load program is ascending and the effect of apparent weakening is associated with the inverse loading program. These effects are caused by the truncation of samples and are not connected with a special damage accumulation rule [3]. Sample data which are plotted in Figures 3 and 4 were obtained by the Monte-Carlo method. The average of 50 tests is represented by the black circles. Also, the 90% confidence intervals for these average values are drawn. The intervals are rather wide including partly the straight line in equation (10) as well as the regression curves in equation (9). Usually, not more than 10 to 12 specimens for each loading are selected.

ARBITRARY LOADING

Another approach to the linear cumulative damage rule for arbitrary loading history $S(t)$ is considered. Assuming that the linear cumulative damage rule is true, equation (1) is satisfied for the observed lifetime τ. Here, the function $\tau_S(S|r)$ is random and practically unknown. In fact, we calculate the value

$$\theta = \int_0^\tau \frac{dt}{<\tau_S(S)>} \tag{14}$$

where the random function $\tau_S(S|r)$ in equation (1) is substituted by an appropriate estimation of its mathematical expectation $<\tau_S(S)>$. Hence, the value of θ is random. It is not equal to unity even if the linear cumulative damage rule is valid. There is no reason to compare θ with unity or to assume other values of θ.

A probabilistic and statistic analysis will now be considered. The problem consists of finding the probability distribution of θ. In certain cases, it is possible to find an explicit solution. Let, for example, $\tau_S(S|r) = f(S)g(r)$ where $f(S)$ and $g(r)$ are the scalar functions. Equations (1) and (14) result in

$$\theta = \frac{g(r)}{<g(r)>} \tag{15}$$

If the probability density $p(r)$ is given, it is easy to obtain the distribution function $F_\theta(\theta)$ and the value $\bar\theta$.

Let the stochastic model presented by equations (12) and (13) be valid. In this case $g(r) = r^m$. If $r_0=0$, we obtain from equations (13) and (15) that

$$F_\theta(\theta) = 1 - \exp[-(\frac{\theta}{\theta_c})^\beta] \tag{16}$$

where $\theta_c = [\Gamma(1+1/\beta)]^{-1}$ and $\beta=\alpha/m$. The mathematical expectation of θ is unity. The variation coefficient W_θ coincides with the variation coefficient W_τ given by equation (6). Under these assumptions, the distribution function in equation (16) does not depend on the loading history. It may be used as the base for verifying the linear cumulative damage rule. It should be emphasized that the scatter of θ has the same order of magnitude as the scatter of lifetime τ under constant loading. Only a very significant deviation from the theoretical distribution $F_\theta(\theta)$ can justify the rejection of the linear cumulative damage rule.

ESTIMATION OF PARAMETERS

Having selected the stochastic model, there remains the problem of estimation of the parameters used in the model. A distinction between the parameters referring to the mechanical state equation, such as equation (12) and parameters of the strength distribution, such as equation (13) needs to be made. In general, the state equation is given by

$$u = f(S,a,r) \tag{17}$$

where u is the vector of the observed values. The load vector S is given by the loading history. The vector parameter a is by definition determined and characterizes the general properties of the~specimens. The strength vector r is random. Its sample values correspond to the intrinsic properties of an ~ individual specimen. The probability density $p(r,b)$ depends on the deterministic vector b which represents the second group of parameters to be determined.

The conventional treatment of the fracture experiments includes separate estimation of the model parameters and the parameters relating to the stress distribution. In this paper, a unified approach is proposed. The model parameters and the parameters of the strength distribution are estimated by using the same procedure.

Consider the following assumptions. First, the dimensions of the vectors u and r are the same. The intrinsic properties of the specimens may be determined by measuring the phenomenological quantities such as plastic work, residual strain etc. Next, existence of the reciprocal function

$$r = f^{-1}(S,a,u) \tag{18}$$

is assumed. Sample values of the vector u obtained from various loading histories are denoted by u_j. Using these data, we calculate from equation (18) the set of corresponding sample values r_j depending parametrically on the unknown vector a. The set of empirical distributions that also depend on vector a is denoted by $\tilde{p}(r;a)$. On the other hand, the set of corresponding theoretical distributions $p(r;b)$ depends parametrically on the vector b. We require that the distributions $p(r,a)$ and $\tilde{p}(r,b)$ are in agreement in the sense of the method of least squares. This requirement results in the equation

$$\sum_K W_K[\tilde{p}(r_K,a) - p(r_K,b)]^2 \to \min \tag{19}$$

where r_K are the sample values of r and W_K are the weight coefficients.

This approach has certain advantages over the conventional methods. First, it provides a suitable compromise when one needs to adjust both the model parameters and the strength distribution with a limited number of models and distributions available. Second, this approach is based substantially on the unification of the experimental data of various load histories. The unified sample of vector r is quite representative even when the samples of the vector u corresponding to each program are poor. As the vector $\underline{r}$ is by definition representing all properties of the specimen related to strength and fracture, it is possible to use experimental data from various different experiments such as standard short-time tests, fatigue tests, etc.

The proposed approach will be illustrated by an example, Figure 5. The observed data are generated by the Monte-Carlo technique with parameters taken

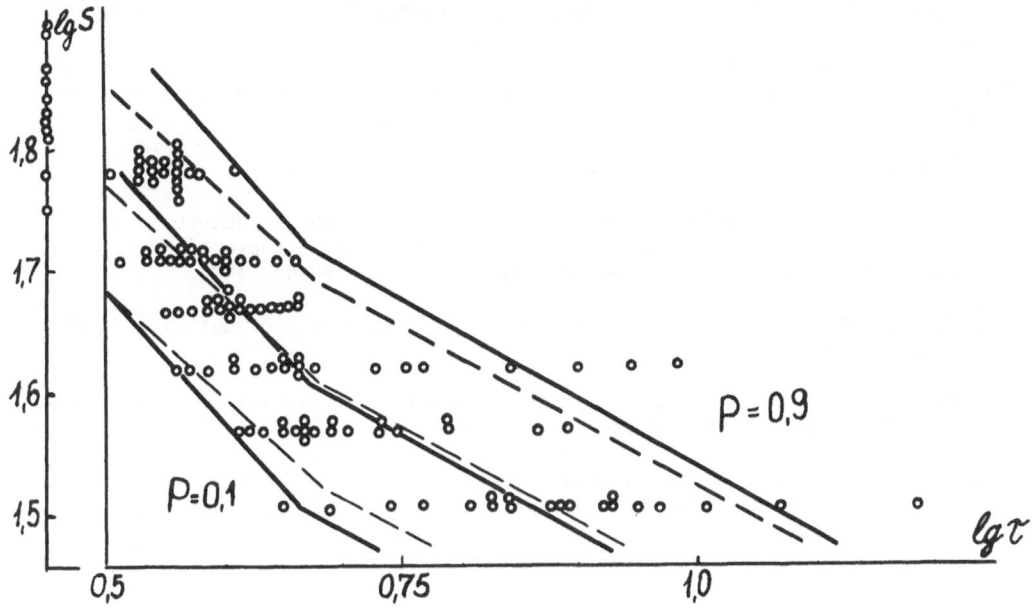

Fig. (5) - Observed data and predicted quantiles for the unified
lifetime and short-time loading programs

from the tests of a composite material. The assumed model is one-dimensional, i.e., u and r are scalars. The lifetime equation is

$$
\tau = \begin{cases} \tau_c \left(\frac{r}{S}\right)^{m_1} & (S \geq r) \\ \\ \tau_c \left(\frac{r}{S}\right)^{m_2} & (S < r) \end{cases} \tag{20}
$$

where the time τ_c corresponds to the angle point on the lifetime curves and the exponents m_1 and m_2 satisfy the inequality $m_2 \geq m_1$. The value of r is random according to the distribution function in equation (13). Adding the standard short-time tests to the long-time tests, the additional relation is obtained:

$$S = \gamma r \tag{21}$$

for the ultimate stress S where γ is an additional parameter to be determined. Hence, the vectors to be estimated are $\underset{\sim}{a} = (\tau_c, m_1, m_2, \gamma)$ and $\underset{\sim}{b} = (r_0, r_c, \alpha)$. In this particular case, it is more convenient to deal with the distribution functions $\tilde{F}(r,a)$ and $F(r,b)$ rather than the probability densities. In Figure 5, the corresponding quantiles are given by the dotted lines calculated with the help of the estimated parameters. The "true" quantiles are depicted by the solid lines. Here, the good agreement is mostly a consequence of the fact that the "true" model equations are known. Nevertheless, good prediction of rare events is of significance.

REFERENCES

[1] Bolotin, V. V., "Mechanics of composite materials and structures", in V. V. Bolotin, I. I. Goldenblatt and A. F. Smirnov, Structural Mechanics, its Presence and Perspektives, M., Stroizdat, pp. 65-98, 1972.

[2] Malmeister, A. K., et al., "The strength of rigid polymer materials", FDT-MT-24-475-74, NTIS.

[3] Bolotin, V. V., "Statistical methods in structural mechanics", Holden-Day, Inc., San Francisco-Cambridge-London-Amsterdam, p. 240, 1969.

[4] Freudenthal, A. M., "Statistical approach to brittle fracture", Fracture, (ed. H. Liebowitz), Vol. 2, Academic Press, New York, pp. 592-621, 1968.

COMPUTER SIMULATION OF COMPOSITE FRACTURE PROCESS WITH BONDING STRENGTH DEFECTS BETWEEN COMPONENTS

I. M. Kopyov, A. S. Ovchinsky an⋅ I. K. Bilsagayev

A. A. Baikov Institute of Metallu y
USSR Academy of Sciences

ABSTRACT

Computer simulation of the fracture rocess in composite materials enables a realistic modeling of the physical problem. Simulation of fiber fracture and delamination of fiber from matrix enables one to trace the failure by formation of cracks in the composite. This elucidates the transition from damage accumulation to the point to total destruction. The strength properties of a composite material can thus be predicted from properties of its constituents and the way they interact within the composite.

Since the strength properties of the fibers are scattered considerably, fiber breaking can occur at the early stage of loading. Depending on the properties of the constituents, the volume fraction and bonding strength of fiber to matrix, localized fiber fracture can act as zone of damage that leads to the final fracture of the composite specimen. As a rule, final fracture is preceded by partial delamination of the fiber-matrix boundary and the progressive failure of fibers owing to stress redistribution. The experimental data and theoretical analysis confirm this.

In order to predict the strength of composites reinforced by brittle fibers, it is necessary to know the critical stress that causes the transition of individual fiber failure and fiber-matrix delamination to the "avalanche-like" process of final fracture. Computer simulation of these processes enables the evaluation of this critical stress.

COMPUTER SIMULATION OF THE FRACTURE PROCESS FOR COMPOSITES WITH IDEAL BONDING

The model proposed is based on following assumptions [1,2]:

1. The fibers are divided into segments with length ℓ_c. Strength of segments and fiber-matrix bonding are "ideal". The fiber junction zones contain defects.

2. The fiber junctions contain defects and are distributed in a plane of the composite, Figure 1. The strength of the defected zones (DZ) based on

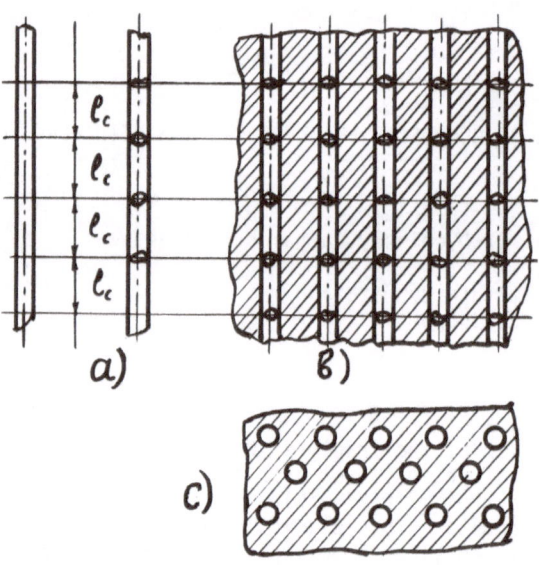

Fig. (1) - The discrete model of a composite: a) segments and defected zones of a fiber; b) defected zones distributed in the material; and c) plane section view of defects

statistical distribution can be described as

$$F(\sigma_f) = 1 - [1-G(\sigma_{fb})]^{1/n} \qquad (1)$$

where $n = L/\ell_c$. In equation (1), σ_f is the fiber stress and $G(\sigma_{fb})$ the integral statistical function with strength distribution σ_{fb} for fibers with length L. $F(\sigma_f)$ refers to short fibers with length ℓ_c. A special program is used for the randomization of statistical distribution.

The material described is "tested" in a digital computer. The following operation sequence is carried out:

1. A stress level is programmed in the DZs of fibers.

2. The strength level of every DZ is compared with this stress. When the strength of the DZs is less than this stress level, these DZs are considered as fractured by loading. Then the stress in each of the DZ zone is redistributed among the 6 neighboring DZs with a hexagonal structure. The additional stress is equal to $k_p\sigma_o$ where σ_o is the fracturing stress and k_p the overload coefficient correcting for the matrix loaded in the vicinity of fractured zone. If there are prefractured DZs among those 6 DZs, the secondary stress redistribution then takes place. Additional stress tends to decrease the secondary redistribution process. This process ceases when additional stresses can be disregarded.

3. When the strength of some other DZs is exceeded, new fractures are observed by overloading. New stress redistribution takes place until all DZs are loaded below their strength. These new stresses will be equal to or above the initial stress level.

4. When additional "load" is applied to the DZs of the specimen, the computation can be repeated.

In this way, the kinetics of fiber fracture in the composite can be traced by repeatedly increasing the stress level. The volume fractions of the constituents are introduced as the overloading coefficient k_p. The development of the fracture processes is characterized by the special damaging functions. The number of fractured DZ is a function of the rising stress. Refer to Figure 2.

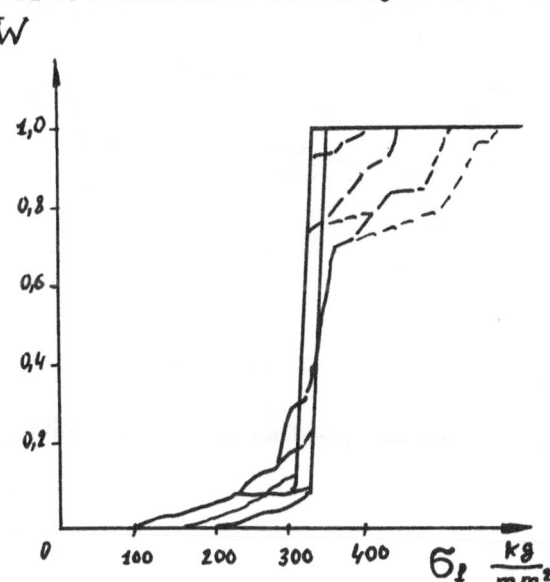

Fig. (2) - Damaging functions of different cross-sections in a composite specimen with σ_{fb} = 250 kG/mm², β=3, n=10 and k_p = 0.1

It is possible to distinguish two characteristic stages of a fracture process. Below some stress level, there is a gradual accumulation of damage. Above this level, the damaging function can jump up to 1 which signifies the avalanche-like process of total fracture. As an example, consider the simulation of the fracture process of a composite with matrix made of aluminum alloy D16-M reinforced by boron fibers having strength about 250 kG/mm². The Weibull's distribution parameter β=3 and β=6 are used. The fiber's number is 480, the volume fractions V_f are 0.03, 0.10 and 0.45, and the number of the defected sections or joints is 10. A good agreement between the computer simulation and experimental data [3] has been found. This is shown in Figure 3.

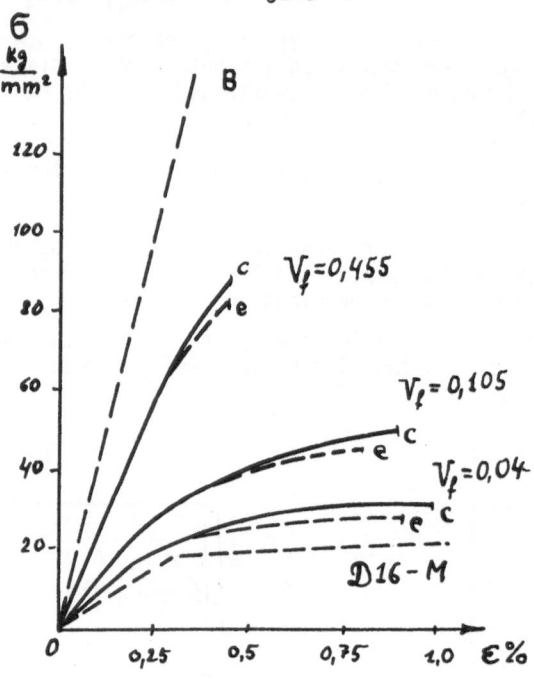

Fig. (3) - Theoretical (c) and experimental (e) tensile diagrams

In the case of low fiber volume fractions, the cumulative type of failure takes place. High fiber volume fractions lead to the avalanche-like process of sudden fracture. The critical volume fraction is around 0.1 where the number of DZs fractured by loading become less than that fractured by overloading. These features are especially more pronounced in materials with more compact fiber strength distribution (β=6). The damage functions are used for the construction of deformation diagrams proposed in [4].

SIMULATION OF COMPOSITE FRACTURE WITH BONDING STRENGTH OF A STATISTICAL NATURE

In this composite model, the DZs are not distributed in a plane but can occur in a random fashion. Each fiber with segment of length ℓ_c can contain one defect, Figure 4. The statistical nature of the bonding strength is taken into

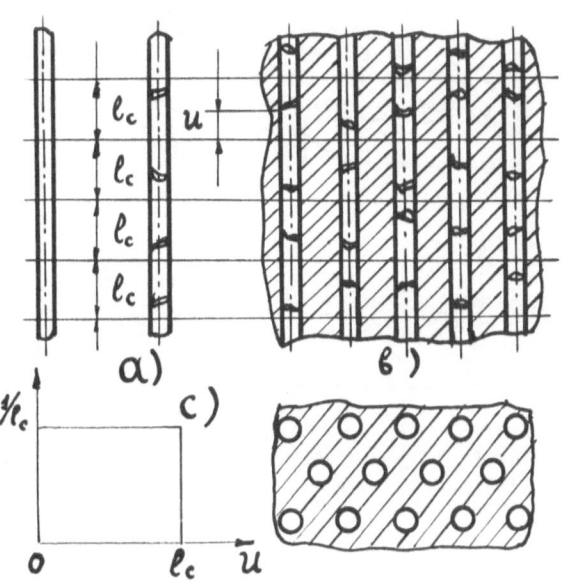

Fig. (4) - The discrete model of a composite: a) segments and DZs of a fiber; b) DZ distributed in the material; and c) statistical function of DZ transferring lengths within ℓ_c segments

account. Experiments on the extrusion of fiber segments from composite plates make it possible to construct an appropriate probability function of the bonding shear strength distribution as shown in Figure 5.

Three-dimensional consideration describing the strength of short fibers, distribution of DZs and bonding strength are stored in the computer memory. The calculation sequence differs from that for the "ideal bonding" model in the following aspects. Shear stress stimulated by every DZ fracture is compared with the bonding shear strength of that DZ and the condition for the onset of delamination is estimated. The working algorithm is based on the existence of a relationship between tensile stress in the fibers and the delamination shear stress. The relative length of the delaminated zone is

$$z_{del}/d_f = (\sigma_o - \sigma_{del})/\tau_f \tag{2}$$

where σ_o is the fiber tensile stress at the moment of delamination, σ_{del} the

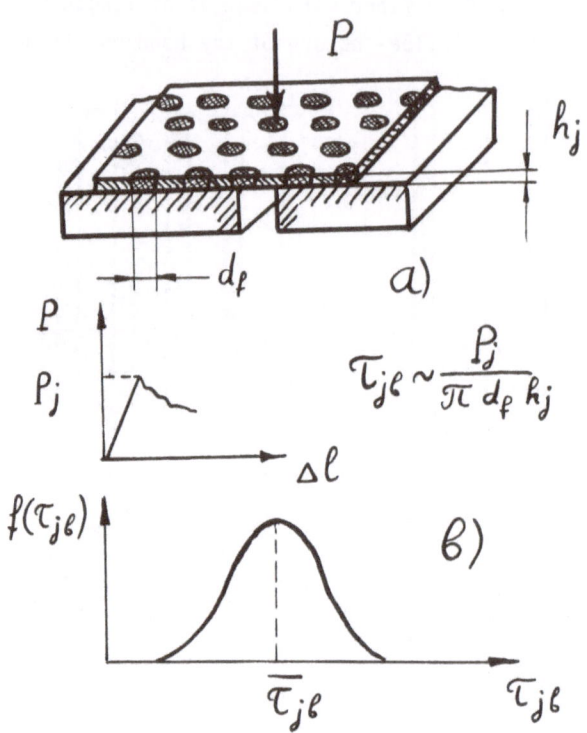

Fig. (5) - a) Estimation of shear strength at fiber-matrix bonding and
b) the bonding strength distribution

tensile stress at the onset of delamination, τ_f the frictional stress between
fiber and matrix in the delaminated zone and d_f the fiber diameter. The fractured DZ with the adjacent zone being delaminated bear no load and their stresses
are redistributed to the 6 neighboring fibers. In other respects, the simulation process is the same as the previous one.

In contradiction to the previous model, the delamination of a fiber from the
matrix over a certain length leads to overloading of the neighboring DZs not only
in the fracture plane but also along a portion of the fiber length. A special
situation may arise when several DZs are overloaded and destroyed on the same fiber. It is also possible to obtain a model of any fracture process with different division surface between normal and parallel to direction of loading.

It is important to make a check on the qualitative character of the present
model. The capacity of the computer limits the calculation to three planes with

defects. As a consequence, the contribution of the delamination length to the composite fracture is rather weak, Figure 6. Improvement can be made by using a computer with greater capacity.

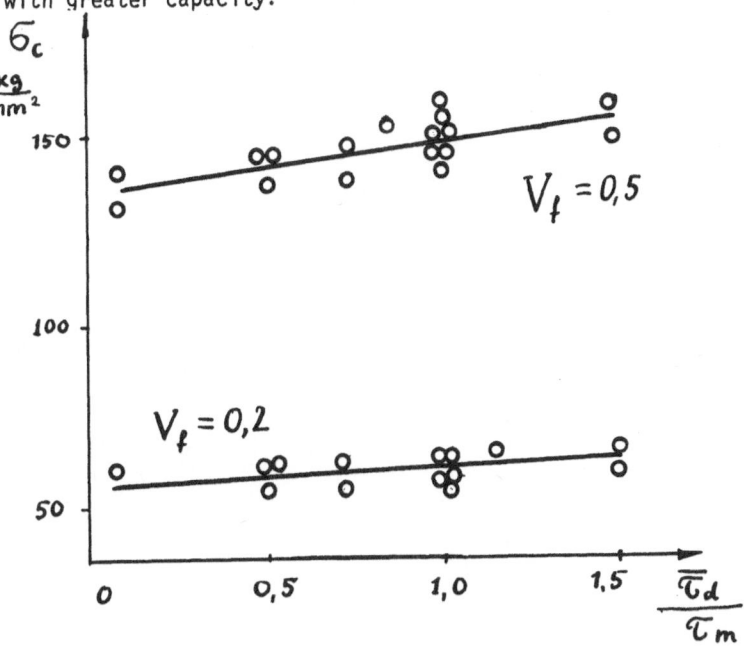

Fig. (6) - The composite strength as a function of fiber matrix bonding strength when delamination length is less than $2l_c$

REFERENCES

[1] Ovchinsky, A. S., Nemcova, S. A. and Kopjev, I. M., "Mathematical modelling of fracture processes of the composites reinforced by brittle fibers", Mekhanika Polimerov (Polymer Mechanics), No.5, pp. 800-808, 1976.

[2] Ovchinsky, A. S., Staselko, T. V. and Kopjev, I. M., "The influence of scattering of the strength characteristics of fibers and the nonuniformity of fiber's distribution on process of fracture of composite materials", Tzvestija VLIZou. Mashinostrojenije, No. 10, pp. 5-8, 1975.

[3] Mileiko, S. T., Sorokin, N. M. and Zirlin, A. M., "The strength of boron-aluminum - a composite with brittle fibers", Polymer Mechanics, No. 5, pp. 840-842, 1973.

[4] Ovchinsky, A. S., Kopjev, I. M. and Bilsagajev, N. K., "Method for calculation of stress-strain curves of composite materials taking into account the statistical distribution of reinforcing fibers strength", Mekhanika Polimerov (Polymer Mechanics), No. 6, pp. 1021-1031, 1975.

FAILURE ANALYSIS OF COMPOSITES WITH STRESS GRADIENTS[*]

Edward M. Wu

Lawrence Livermore Laboratory, University of California
Livermore, California 94550

ABSTRACT

In technical applications of composites, there exists a need to analyze composite strength in terms of stress gradients such as cracks, cutouts, and concentrated loads. Traditional continuum stress analysis cannot fully reconcile the interaction of the stress gradient and the local heterogeneity of fiber matrix and lamination. We present an adaptation of the statistical strength theory to analyze the effect of stress gradients by explicit relation to the intrinsic strength variability of the composite. This method is suitable for failure analysis of composites under a general state of stress without employing an arbitrary averaging parameter.

INTRODUCTION

The development of current composite materials is the result of engineering combination (as distinguished from physical and chemical combination) of several materials in macroscopically multiphase form to achieve certain physical properties not realizable by the constituent materials individually. The resulting composite properties of such multiphase combinations often are anisotropic. These anisotropic physical properties may be partitioned into two categories: (1) those related to averaged global responses (such as stiffness, thermo, conductive, and transport properties), and (2) those related to local phenomena (such as strength, fracture, and interfacial properties). For the first category, continuum analysis and numerical modeling have provided quantitatively accurate predictions. For the second group, a satisfactory reconciliation between micromechanism and continuum analysis has not been made. A most conspicuous departure from the traditional continuum analysis is the current treatment of strength and fracture as isolated phenomena.

The work presented here is an attempt to develop a physical association (which can be statistically quantitative) between strength theories and fracture mechanics. Such a quantitative understanding of the parameters that govern composite failure is imperative to the implementation of fail-safe design and the inspection of critical load-bearing composite structures. Our results also may be useful for predicting the size effect of scaling up laboratory samples to larger size structures in the presence of stress concentrations and stress singularities.

---

[*]This work was performed under the auspices of the U. S. Department of Energy, at Lawrence Livermore Laboratory, under contract No. W-7405-Eng-48.

In current research practices, characterization of the strength of anisotropic multiphase composites is usually separated into two broad categories: (1) composite strength in the absence of macroscopic flaws, and (2) composite strength in the presence of macroscopic flaws (and stress risers). These two categories are referred to, respectively, as anisotropic failure criterion characterization and fracture mechanics; usually, they are treated as separate physical phenomena. Clearly, such arbitrary categorizing is a consequence of attempts to identify the critical paths of composite strength characterization through association with those experiences gained from isotropic solids. The one-parameter nature of isotropic fracture follows directly from the physical observations that isotropic crack extension is always perpendicular to the direction of maximum tension and that energy dissipation always occurs via a crack-opening mode. Thus, the similarity between mathematical model and physical observation is easily maintained.

In contrast, composites, particularly in the laminated form, exhibit a large range of instability conditions involving various amounts of slow crack growth. In composites, the modes of energy dissipation are not limited to the crack-opening mode; they also include forward sliding and out-of-plane shear. Thus, the crack trajectories seldom follow the maximum tensile stress direction and often lead to nonself-similar crack extension with complex branching. The effects of external loads (symmetric and skew-symmetric to the crack) and combined loading on crack instability need to be documented for composites. Also, the size effect of flaws is far more dominant in composites than in homogeneous isotropic materials.

Whereas the one-dimensional nature of isotropic fracture lends itself to experimental quantification in the form of a single critical stress intensity factor or fracture toughness parameter, the multiple-parameter nature of crack extention in composites precludes empirical permutation of the parameters. For anisotropic composite laminates, there are at least seven primary parameters controlling the fracture characteristics:

(1) Deformational and strength responses of the constituent lamina.
(2) Lamination geometry.
(3) Crack orientation with respect to the material axis of anisotropy.
(4) Crack length.
(5) Nature of the applied stresses.
(6) Energy dissipation associated with the three kinematically admissible modes of crack extension.
(7) Crack trajectory.

Because of these many parameters, experimental quantification by systematic permutation of the parameters must realistically be viewed as intractable. In this paper, we present an analytical model that reduces the above parameter list from seven to merely the constituent lamina failure criterion and the inherent statistical variability parameter m.

THEORETICAL MODEL

The theoretical model is based on the postulate that:

In the case of qusai-static rupture, the failure of a volume element can be characterized by a weakest link analysis of the local stress.

In particular, the local stress can range from a homogeneous state (as in uniform tension, Fig. 1a), to a stress concentration (Fig. 1b), and finally to a stress singularity in the presence of a crack (Fig. 1c). This postulate provides the bridge between strength theories and fracture mechanics.

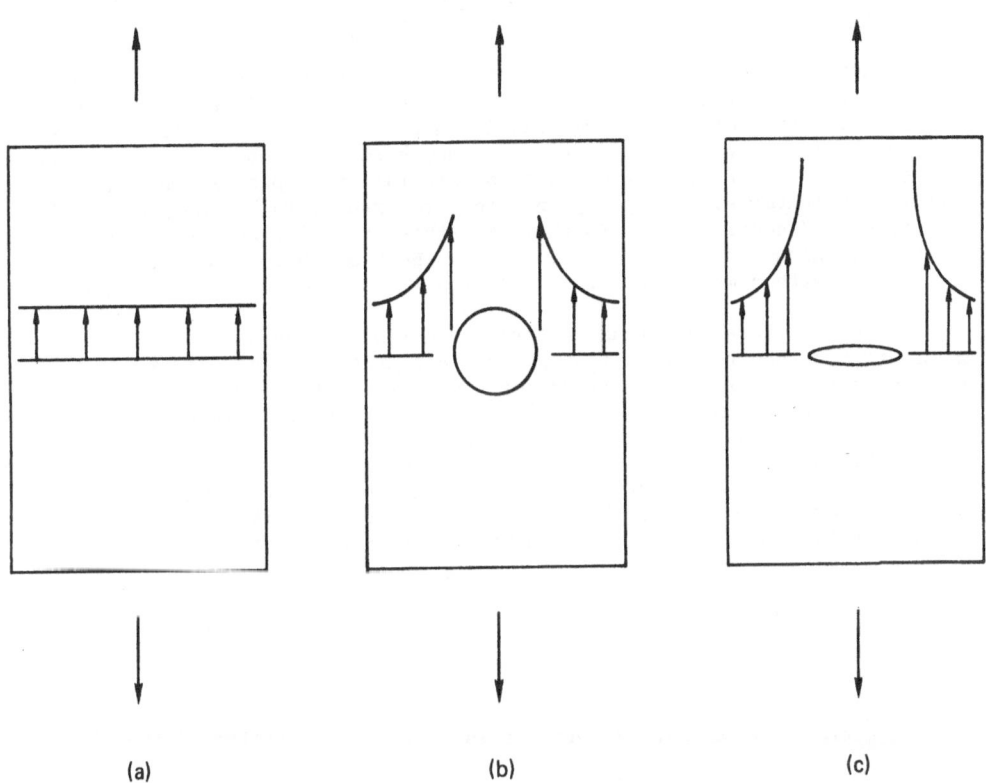

(a)   (b)   (c)

Fig. 1. Range of stress gradients: (a) homogeneous state in uniform tension, (b) stress concentration, and (c) stress singularity in the presence of a crack.

One of the most familiar forms of weakest link characterization of the strength of materials is the Weibull statistic strength theory. The probability of survival P_s, for a material of volume V, and subjected to a spatially dependent stress $\sigma(x_i)$, is represented as:

$$P_s = \exp\left\{-\int_V \left(\frac{\sigma(x_i) - \sigma_u}{\sigma_o}\right)^m dv\right\} \qquad (1)$$

where σ_u is the stress threshold below which the probability of failure is zero, σ_o is a normalization parameter, and m is the Weibull parameter that characterizes the variability of observed strength scatter. This Weibull statistical strength theory has been successfully employed in the characterization of brittle ceramics and carbon structures.

Although some *ad hoc* adaptation of this theory to composites has been reported, none has focused on the basic limitations of this Weibull form. From Eq. (1), the implicit assumption is that failure is a one-dimensional process. This implies that identical strength would be observed regardless of whether the material is subjected to uniaxial or complex states of stress. Furthermore, strength properties are assumed to be independent of directions. Generalizations to eliminate these restrictions are needed for a rational characterization of composites. Both of these restrictions can be resolved with a mathematically operational anisotropic failure criterion.

In recent years, numerous failure criterion have been proposed. Examination of their formulations [1] reveals that they are mathematically awkward; some even lack consistency of conversion between stress and strain. Tsai and Wu [2] found that the tensor polynomial failure criterion encompasses maximum flexibility without redundancy and, furthermore, that this criterion lends itself to the design of critical experiments [3]. The tensor polynomial failure criterion is used here, although we emphasize that other experimentally verified criteria may be substituted. The tensor polynomial failure criterion, when expressed in terms of stress, takes the form in contracted notation:

$$f(\sigma_i) = F_i\sigma_i + F_{ij}\sigma_i\sigma_j + F_{ijk}\sigma_i\sigma_j\sigma_k + \ldots = 1, \qquad i = 1,2, \ldots 6 . \qquad (2)$$

For a typical engineering composite (e.g., graphite/epoxy), the linear and quadratic terms in Eq. (2) provide sufficient correlation of the experimental data as shown in Fig. 1. These experimental data were obtained from tubular samples tested under combined stress conditions along radial loading paths on an axial-rotary-internal pressure mechanical testing machine that is controlled by an on-line digital computer. The experimental details are reported in [4]. The data actually populate a three-dimensional space in $\sigma_1\sigma_2\sigma_6$, but they have been convoluted (or projected) onto the $\sigma_1\sigma_2$ plane for easy comparison. In Fig. 2, the same set of experimental data is convoluted onto the $\sigma_1\sigma_2$ plane by three different failure criteria. Better correlation by the tensor polynomial criterion is exhibited visually and by the lowest RMS (root-mean-square) deviation of experiment from theory.

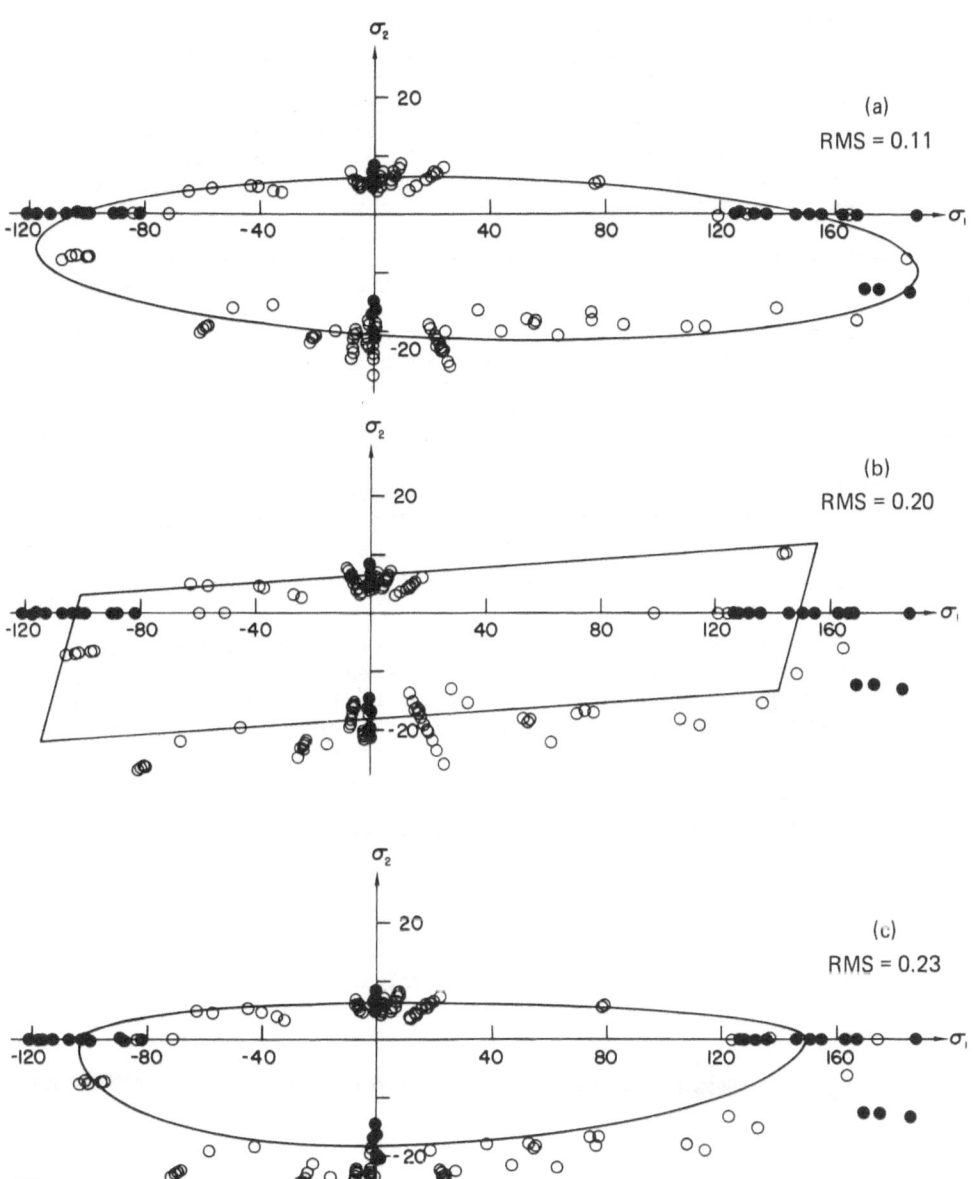

Fig. 2. Failure data of graphite/epoxy lamina convoluted on the $\sigma_1\sigma_2$ plane
(root mean square stresses are in ksi): data convoluted by (a) the tensor
polynomial failure criterion, (b) the maximum strain failure criterion, and
(c) the modified Mises-Hill failure criterion.

The physical interpretation of the failure envelope requires some attention. The composite is assumed to be homogeneous and anisotropic, and to contain a population of randomly distributed microscopic flaws C_1, C_2, ... C_j. Although the flaws are small compared to the characteristic dimension D of the body as depicted in Fig. 3a, continuum analysis reveals that, under arbitrary loads P_i, the state of stress is unbounded at the location of the geometric singularities C_1, C_2, ... C_j, and thus would lead to immediate failure even for extremely small P_i. This is contrary to physical observations. The stresses appearing in Eq. (1) should therefore be interpreted as the average stress acting on a small but finite characteristic volume (specified by a dimension r_c,[*] Fig. 3a) that fully encapsulates one microscopic flaw. Thus, although the stress is singular inside this characteristic volume r_c, the average stresses external to r_c are bounded and may be used to characterize the failure of this volume through a failure criterion of the form

$$\mathscr{S} \leq \mathscr{F} \quad .$$
(3)

Here, $\mathscr{S}$ is the average stress vector acting external to the characteristic volume and is defined in terms of the unit vector $\vec{e}_i$ in the stress of Fig. 3b,

$$\mathscr{S} = \sigma_i \vec{e}_i, \quad i = 1, 2, \ldots 6 \quad .$$
(4)

Also, $\mathscr{F}$ is the strength vector to the failure surface $f(\sigma_i)$ as determined by Eq. (1) and as illustrated in Fig. 3b. Under an arbitrary loading P_i, the stress vector $\mathscr{S}$ at any location of the body can be determined through continuum analysis or numerical techniques. It follows that, when criterion $f(\sigma_i)$ is known, the location of a prevalent failure condition can be anticipated by considering the probability of survival of each volume element within the body. For a given volume element V_i (where $V_i > r_c^3$) having a flow density per unit volume ρ subjected to the action of a stress vector $\mathscr{S}$, the probability of survival is

$$P_s = g\left(\frac{\mathscr{F}}{\mathscr{S}}\right)^{\rho V_i} \quad .$$
(5)

For the total volume V consisting of V_i volume elements, the cummulative probability of survival is

$$P_s = \prod_{i=1}^{n} g\left(\frac{\mathscr{F}}{\mathscr{S}}\right)^{\rho V_i} \quad .$$
(6)

Equation (6) can be rewritten in the integral form,

$$P_s = \exp \int_{V_c}^{V} \ln g\left(\frac{\mathscr{F}}{\mathscr{S}}\right) dV \quad , \quad i \frac{\mathscr{F}}{\mathscr{S}} \geq 1 \quad ,$$
(7)

where the lower limit of integration is the characteristic volume $V_c = 0(r^3)$.

---

[*]The explicit determination of this characteristic volume will be discussed after we develop the general form of the statistical failure model.

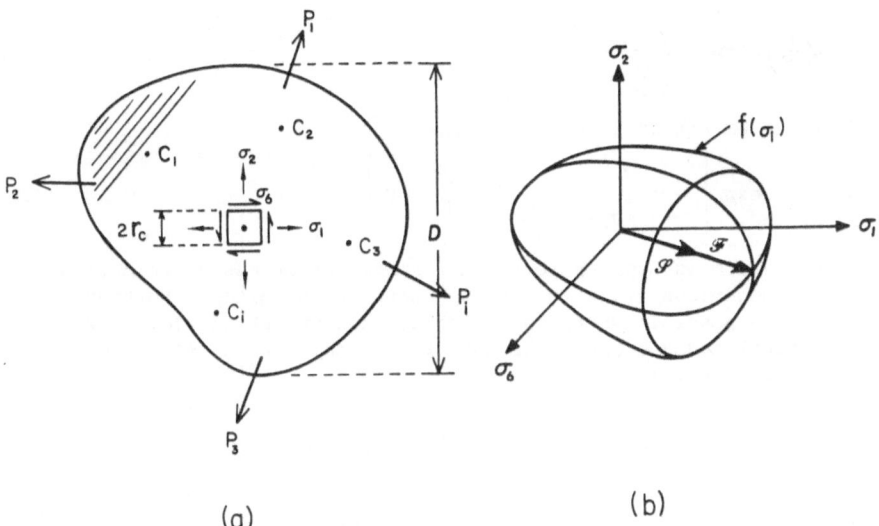

(a)

(b)

Fig. 3. Homogeneous anisotropic body with randomly distributed microscopic
flaws in (a) and diagram of the criticality of stress vector $\mathscr{S}$ acting on
the characteristic dimension r_c, the failure surface $f(\sigma_i)$, and the
strength vector $\mathscr{F}$ in (b).

In this generalized representation of the probability of survival of an anisotropic body, the only restrictions are in the limitation conditions of $g(\mathscr{F}/\mathscr{P})$. To ensure no failure under zero stress,

$$\lim g\left(\frac{\mathscr{F}}{\mathscr{P}}\right) = 1 , \qquad \frac{\mathscr{F}}{\mathscr{P}} \to \infty . \tag{8a}$$

To ensure no survival under limiting stress,

$$\lim g\left(\frac{\mathscr{F}}{\mathscr{P}}\right) \to 0 , \qquad \frac{\mathscr{F}}{\mathscr{P}} \to 1 . \tag{8b}$$

Any function that satifies Eqs. (8a) and (8b) may be considered as a candidate for characterizing a given composite. In particular, we may choose an exponential form, as did Weibull; i.e.,

$$g\left(\frac{\mathscr{F}}{\mathscr{P}}\right) = \exp - \left(\frac{1}{\frac{\mathscr{F}}{\mathscr{P}} - 1}\right)^m . \tag{9}$$

This leads to an eminently tractable form,

$$P_s = \exp\left\{ - \rho \int_{V_c}^{V} \left(\frac{1}{\frac{\mathscr{F}}{\mathscr{P}} - 1}\right)^m dV \right\} . \tag{10}$$

This particular form is applicable for all ranges of stress distributions ranging from homogeneous state to stress concentration sites. Further simplification is possible where severe stress concentrations (e.g., sharp notches or cracks) cause drastic strength reduction; i.e., $\mathscr{P} \ll \mathscr{F}$. Under such circumstances, Eq. (10) becomes

$$P_s = \exp\left\{ - \rho \int_{V_c}^{V} \left(\frac{\mathscr{P}}{\mathscr{F}}\right)^m dV \right\} \qquad \text{for } \mathscr{P} \ll \mathscr{F} \tag{11a}$$

We note that in the one-dimensional case under an applied stress $\mathscr{P} = \sigma$, $\mathscr{F} = X$ where X is the tensile strength, Eq. (11a) reduces to an equation of the Weibull form (Eq. (1)) where the stress threshold σ_u is set to zero:

$$P_s = \exp\left\{ - \rho \int_{V_c}^{V} \left(\frac{\sigma}{X}\right)^m dV \right\} = \exp\left\{ - \int_{V_c}^{V} \left(\frac{\sigma}{\sigma_0}\right)^m dV \right\} . \tag{11b}$$

Hence we see that the Weibull form implies severe stress risers which account for its success in characterizing "brittle" materials.

In our generalization (Eq. 10), we not only introduce the generality of anisotropic strength, we provide for the extended range of application to local stress site from mild stress concentrations to seven stress singularities. In this generalization, we also specify a lower limit of integration in the computation of the cummulative probability of survival (the characteristic volume V_c or the characteristic dimension r_c). This characteristic dimension r_c is the limit of the continuum and may be explicitly determined by exploring the effect of a stress gradient on the probability of survival.

Traditionally, in the deterministic correlation of local point stress to strength, only the stress magnitude is taken into account. As a consequence, the correlation is not able to treat cases where stress becomes singular, e.g., around crack tips and dislocation sites. The stress gradient effects are implicitly taken into account in the Weibull form (Eq. 1). However, in the actual computation involving stress singularity, the integral becomes ill behaved.

We therefore desire to examine explicitly the effect of stress magnitude and stress gradient by a second postulate:

> For a given material, there exists, at a characteristic stress magnitude $\mathscr{S}_c$, a limiting stress gradient $\mathscr{S}'_c$, above which no stress gradient effect on strength can be measurable.

We can carry out this exploration using the scalar stress component with no loss of generality; i.e., let $\mathscr{S}_c \to \sigma_c$ and $\mathscr{S}'_c \to \sigma'$, where $\sigma' = (d\sigma/dX)$ (see Fig. 4).

For a small (by definition) characteristic dimension r_c, we take the stress gradient to be constant $0 \leq X < r_c$. Hence the stress distribution within r_c is

$$\sigma = \sigma_c + \sigma' X \quad . \tag{12}$$

Because we seek the effect of a severe stress concentration, we can utilize Eqs. (11a) or (11b) to compute the probability of survival of this element within r_c (accounting for stress gradient effect):

$$P_s \bigg|_{\mathscr{S}'} = \exp\left\{ -\int_0^{r_c} \left(\frac{\sigma_c + \sigma'X}{\sigma}\right)^m dX \right\} ,$$

$$= \exp\left\{ - \frac{1}{\sigma_0^m} \frac{(\sigma_1 + \sigma'r_c)^{m+1} - \sigma_c^{m+1}}{\sigma'(m+1)} \right\} . \tag{13}$$

The corresponding probability of survival of a homogeneous stress σ within an identical volume is

$$P_s \bigg|_{\sigma \text{ homogeneous}} = \exp\left\{ - \int_0^{r_c} \left(\frac{\sigma}{\sigma_0}\right)^m dX \right\} ,$$

$$= \exp\left\{ - \left(\frac{1}{\sigma_0^m}\right) \sigma^m r_c \right\} \quad . \tag{14}$$

For equal probability of survival, the ratio of an identical volume, we can equate Eq. (13) to Eq. (14) and obtain

72

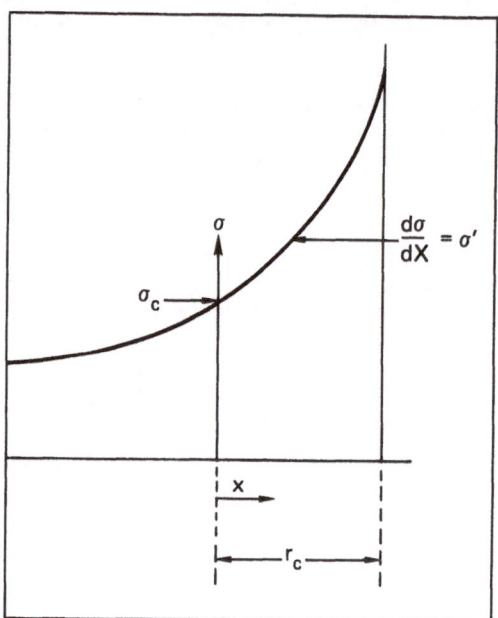

Fig. 4. Plot of the scaler stress component where the characteristic stress magnitude $S_c \rightarrow \sigma_c$, and the limiting stress gradient $S'_c \rightarrow \sigma'$, where $\sigma' = (d\sigma/dX)$, r_c is the characteristic dimension, and X is tensile strength.

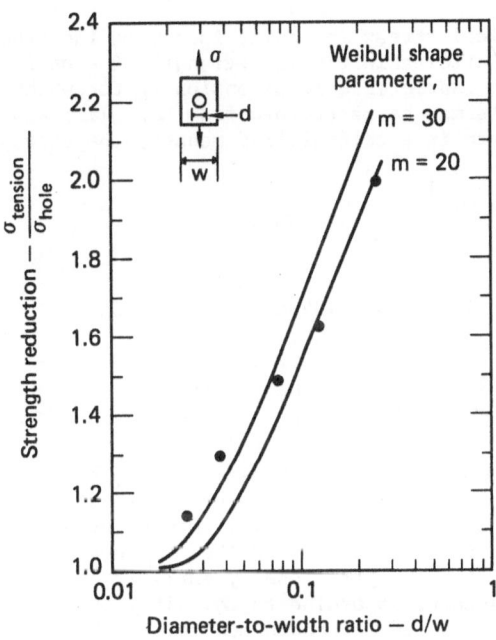

Fig. 5.   Plot of strength reduction ($\sigma_{tension}/\sigma_{hole}$) of a quasi-isotropic
fiberglass composite with a small hole.

$$\frac{\sigma_c}{\sigma} = \left[\frac{(m + 1) \dfrac{\sigma' r_c}{\sigma_c}}{\dfrac{(1 + \sigma' r_c)m + 1}{\sigma_c} + 1} \right]^{\frac{1}{m}} . \tag{15}$$

We note that the highest stress gradient occurs in the continuum analysis of cracks. Thus, the strength reduction in Eq. (15) will have a lower limit equal to the strength reduction associated with the presence of a crack. For the isotopic case, this strength reduction limit is

$$\frac{\sigma_c}{\sigma} \geq \left(\frac{\dfrac{k_c}{\sqrt{2r}}}{X} \right) , \tag{16}$$

where k_c is the critical stress intensity factor of the fracture of a crack and X is the highest attainable tensile strength of a uniformly loaded sample. The limiting characteristic dimension r_c for which the limiting stress gradient exists may be determined from Eqs. (15) and (16). For an isotopic crack fracture in a self-similar manner, the stress gradient is

$$\sigma' = \frac{d}{dX} \left. \frac{k_c}{\sqrt{2X}} \right|_{X=r_c} = \frac{-k_c}{2\sqrt{2}} r^{-3/2} . \tag{17}$$

Substituting Eq. (17) into Eqs. (15) and (16) yields

$$r_c = \frac{1}{2} \left[\frac{2(1 - \frac{1}{2})^{m + 1}}{(m + 1)} \right]^{2/m} \left(\frac{k_c}{X} \right)^2 . \tag{18}$$

For anisotropic composites, the tensile strength X needs to be replaced by the strength vector $\mathscr{F}$ in the direction of crack extension. With this limiting dimension known, Eqs. (10) or (11) may be used to compute the statistical strength of the composite in the presence of stress gradients. This formulation is not only operationally explicit, it also is physically meaningfull. We note that, according to Eq. (15), the characteristic dimension r_c is related to the strength scatter of the material as characterized by the Weibull parameter m, and that

$$m \to 0 , \qquad r_c \to \infty ;$$

$$m \to \infty , \qquad r_c \to \frac{1}{2} \left(\frac{k_c}{X} \right)^2 . \tag{19}$$

Because the strength scatter is inversely proportional to the Weibull parameter m, the first limiting condition coincides with the intuitive notion that large scatter and inhomogeneity require a large characteristic volume. The second limiting condition is the case of deterministic strength in which we recover the deterministic formulation of r_c as proposed by Wu [5].

We tested this limiting strength gradient posulate with experimental data on the size effect of circular holes in quasi-isotropic glass/epoxy composites. It has been reported [6] that the strength reduction due to circular holes is dependent on hole size and that, for small hole sizes, the strength reduction is considerably smaller than the theoretically predicted elastic stress concentration of 3. Using the elasticity solution of stress distribution around a circular hole and Eq. (10), we plotted the strength reduction as a function of hole size, together with experimental measurements from Ref. [6] (Fig. 5). For these limited experimental data, we observe an encouraging correlation with the predictions for Weibull parameters between 20 and 30. This agrees well with the literature value of m = 25 for quasi-isotropic fiberglass composites in tension.

CONCLUSION

We have generalized Weibull's statistical strength theory to account for complex states of stress and anisotropic strength so that the failure analysis of composites can include statistical strength and size effects. In addition, we have postulated the existence of a limiting strength dependence on the stress gradient. From this postulate, we have derived a relation that enables the explicit evaluation of a critical dimension which defines the limit of the continuum as a function of material variability. This approach offers a rational link between continuum analysis and local failure sites that is quantitative in terms of established statistical parameters. Tentative but encouraging correlations were observed between Weibull predictions and limited experimental results on strength reduction due to circular holes. Further confirmation of this correlation will require additional experimental results on the stress concentration site with different stress gradients.

REFERENCES

[1] Wu, E. M. "Phenomenological Anisotropic Failure Criterion" in Composite Materials, (Academic Press, 1974), Vol. 2., Vol. Ed., G. P. Sendeckyj, Series Eds., L. J. Broutman and R. H. Krock.

[2] Tsai, S. W. and Wu, E. M., "A General Theory of Strength for Anisotropic Materials", Journal of Composite Materials, Vol. 5 (1971).

[3] Wu, E. M. "Optimal Experimental Measurement of Anisotropic Failure Tensors" Journal of Composite Materials, Vol. 6 (1971).

[4] Wu, E. M. and Jerina, K. L. "Computer-Aided Mechanical Testing of Composites", Materials Research and Standards, Vol. 12 (1978).

[5] Wu, E. M. "Strength and Fracture of Composites" in Composite Materials (Academic Press, 1974), Vol. Ed., L. J. Broutman, Series Eds., L. J. Broutman and R. H. Krock.

[6] Whitney, J. M. and Nuismer, R. J. "Stress Fracture Criterion for Laminated Composites Containing Stress Concentrations," Journal of Composite Materials, Vol. 8 (1974).

METHODS FOR PREDICTING FAILURE BEHAVIOR OF COMPOSITE MATERIALS

SHUN-CHIN CHOU

Army Materials and Mechanics Research Center
Watertown, MA 02172 USA

ABSTRACT

Various methods are described in two groups based on their basic assumptions; the "Material Modeling" and "Mathematical Modeling". Typical studies of these two approaches are highlighted with particular attention to a recently developed finite element technique which has a potential to improve our understanding of failure behavior of composite materials.

INTRODUCTION

Composite materials have been widely used in structures in recent years to meet the demand of weight saving. However, their potential has not been fully explored due to the lack of methodology for predicting the failure behavior. The reason for this handicap is that there are major deficiencies in our ability to determine the stress field within a laminate, particularly when the laminate contains a stress riser. In principle, the use of finite element method in conjunction with the three dimensional theory of elasticity provides a powerful tool to analyze a laminate. Unfortunately, to analyze a multilayered structure containing a steep stress gradient with this method requires a large computational storage capacity and long computing time which have usually restricted our investigation to a small region; and the full nature of structural response cannot be studied. In other words, the current technology does not provide us a practical means to analyze the stress field in a multilayered composite material containing some form of stress concentrations. As a result of this deficiency, the investigation of failure modes in a laminate and formulation of failure criteria have been impeded even though many approximate theories and numerical techniques have been developed [1-12]. This paper is intended to highlight some of these approaches and also to give a brief description of a finite element technique which is current being developed with the objective of improving our capability in analyzing the stress field of a laminate.

ANALYTICAL METHODS

In general there are two basic categories of methods of studying the failure behavior of multilayered composite materials; namely, macroscopic and microscopical. This paper will only discuss methods of failure analysis in the macroscopic level. Among those methods developed for this purpose, we can further classify them into two groups based on their differences in the basic assumptions. These two groups for the convenience of discussion in this paper are called "Material Modeling" and "Mathematical Modeling", and they will be discussed in the next two sections.

A. Material Modeling

It is known that the elastic response of a laminate can be predicted very well by laminated plate theory with laminar properties as inputs. However, for nonlinear stress-strain responses, many models have been suggested to explain experimental results. Most of these models have been established based upon the "Material Modeling" approach which assumes that certain material properties of a ply vanish or vary in some manner when the stresses (or strains) in a ply satisfy one of the failure criteria used in that particular study such as maximum strain, maximum stress on tensorial polynomial failure criterion. In the following, two typical studies will be used as examples to illustrate the procedures of this "Material Modeling" approach.

Chou, Orringer and Rainey [11,13,14] used a so-called "Tangent Modulus Analysis" in conjunction with a finite-element model of a laminate to determine the elastic constants of each ply at every stress and strain level along the experimentally measured stress-strain curve obtained by testing a finite width laminate under tension. Once the elastic constants of each ply at each stress level were determined, they were used to predict the responses of specimens containing stress concentrations. Certainly the specimens were made of the same laminate as of that which was used to determine the elastic constants of each ply. Comparisons of the theoretically predicted responses with the experimental results indicated that the material modeling approach worked very well in their studies. The procedures of this "Tangent Modulus Analysis" can be described briefly as follows. First, a linear elastic analysis of the intact laminate is performed and stresses are calculated for each ply in each grid of the finite element model by using the laminated plate theory. The ply stresses are then combined with the failure surface to scale the results and to find the first-ply failure (FPF) stress resultant N_{FPF} for the laminate. This determines point A (see Fig. 1). Next, the finite element model is

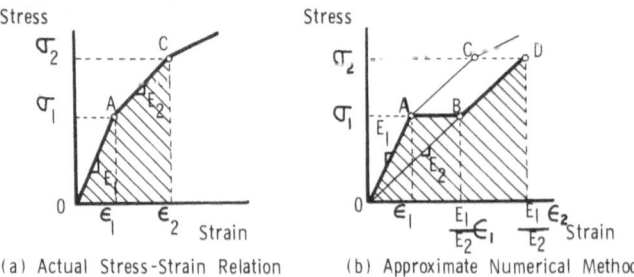

(a) Actual Stress-Strain Relation   (b) Approximate Numerical Method

Fig. 1 Schematic Illustration of Tangent Modulus Method

reassembled with stiffness changed appropriately in those plies of those elements in which FPF occurs, and N_{FPF} is applied to this model to determine point B. Again, stresses are calculated in every ply of every element to define the initial state for the next increment. Using the same value of N_{FPF} but different stiffnesses implies that experiments are performed under constant load-rate control. The increment in strain from point A to point B is a result of reduction in stiffnesses in certain plies when failure occurs in those plies. This analysis also provides stress increments which can be added to the initial stresses σ_1 and compared with the failure surface to determine point D, i.e., the next ply-failure event and the corresponding laminate stress resultant. The procedure is repeated in this manner, allow-ing the stress calculations and the failure surface to determine the locations of the points A, C, ---, step-by-step until the total stress resultant applied to the laminate or the strength of the laminate is reached. Figure 2 shows the finite element meshes used for the analysis of laminate with elliptical hole (a/b=4) and circular hole. One-quarter of the laminate was analyzed with appropriate symmetry conditions along the centerline boundaries of the model. Figs. 3 and 4 show the comparison of the predicted load-strain curves with experimental results of E-glass/epoxy [0/±45/90]s laminate containing elliptical hole (a/b=4) and circular hole respectively. It is noticed that they are in excellent agreement with each other.

Another study which may be cited in the "Material Modeling" approach is completed recently by Chamis and Sullivan [15]. They used the laminated plate theory in conjunction with a tensorial polynomial strength criterion [16] and the method of least square to determine the in-situ ply strengths from a large number of experimentally determined laminate strengths obtained by testing coupon specimens in either simple tension, compression or sheer. In other words, Chamis et al assumed that the ply strengths were initially unknown and were to be determined from laminate strengths. This assumption implies that all plies in the laminate fail simultaneously. Their approach [15] can be briefly quoted as follows: The quadratic two-dimensional ten-sorial polynomial strength criterion is given as

$$A_1\sigma^2_{\ell 11} + A_2\sigma^2_{\ell 22} + A_3\sigma^2_{\ell 12} + A_4\sigma_{\ell 11}\sigma_{\ell 22} + A_5\sigma_{\ell 11}\sigma_{\ell 12} + A_6\sigma_{\ell 22}\sigma_{\ell 12}$$
$$+ A_7\sigma_{\ell 11} + A_8\sigma_{\ell 22} + A_9\sigma_{\ell 12} = 1.$$

where the σ_ℓ's denote ply stresses at a load condition, the A's denote uni-axial strength coefficients which are the unknowns. The subscripts refer to material axes with 1 taken along the fiber direction. The A's are determined using the method of least squares matrix equation which is given as

$$[\sigma_\ell]^T \ [\sigma_\ell] \ \{A\} = [\sigma_\ell]^T \ \{1\}$$
$$\text{9xm} \quad \text{mx9} \quad \text{9x1} \quad \text{9xm} \quad \text{mx1}$$

where the σ_ℓ's are the ply stresses at laminate fracture and are computed using the laminated plate theory. Once these coefficients are known, the ply

80

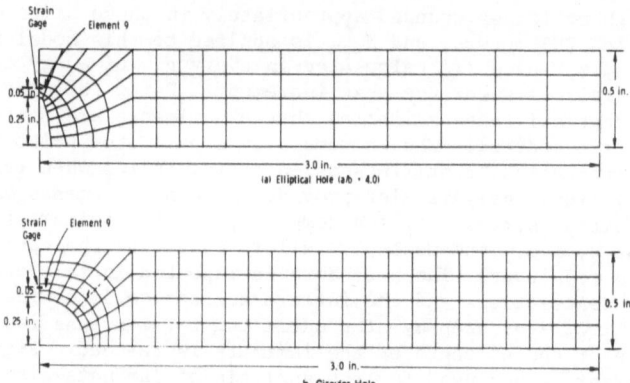

Figure 2. Finite element mesh and location of strain gage

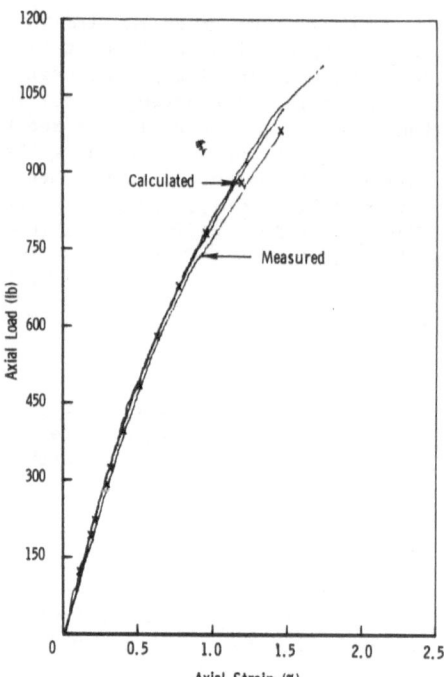

Figure 3. Measured Axial Load-Strain Curves of
Laminate (0/±45/90)s with Elliptical Hole of a/b = 4

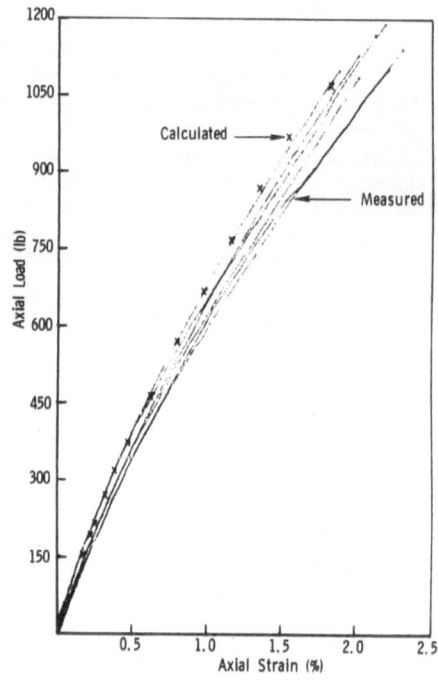

Figure 4. Comparison of Calculated and Measured Axial
Load-Strain Curves at Element No. 9

strengths are determined by solving the tensorial polynomial strength criterion for the various special cases (determine its roots from the quadratic equation) that is by assuming that the ply is subjected to one stress only. Their findings indicate that in-situ ply strengths could be drastically different from the uniaxial test data.

Both studies mentioned above are for laminates subjected to simple loading conditions such as tension, compression, or shear. Nuismer [17] presented an analytical result of failure behavior in biaxially loaded laminated composites containing a stress concentration. He assumed that certain stiffnesses vanish when the stresses in a ply satisfy a failure criterion. The stresses were calculated by using the laminated plate theory. No attempt was made to compare the analytic results with experimental results.

It should be stated that in the "Material Modeling" approach, failure modes and failure criteria cannot be examined since the classical laminated plate theory is used which is formulated without the consideration of the effect of transverse stress components which have strong influences on the behavior of a laminate as experimental results indicated. However, the simplicity and feasibility of the "Material Modeling" approach to study a complicated structure make it a very attractive means to engineers and designers.

B. Mathematical Modeling

Most structures in common practices contain some forms of stress risers such as cutouts, joints, edges and discontinuities in layers or thicknesses. Experimental results indicate that failure usually initiates from the region where a stress riser exists. Therefore, in order to predict the strength of such a structure, it is imperative to have the ability to analyze the stress field of a structure particularly in regions of steep stress gradients. It is recognized that the problem could be analyzed using three dimensional finite elements; however, in a multilayered composite material the finite element would include subdivisions in all three directions including the direction through the thickness of a laminate. This requires a large computer storage and long computing time which limit the application of the method to a detailed study of small regions. If a large mesh size was used, the gridwork would be so coarse that the nature of the stress field in regions of steep gradient could not be fully evaluated. This limitation on the application of three dimensional finite elements had led many researchers to develop approximate theories for solving some manageable problems. One of the problems which has been studied by many investigators is to determine the stress field in a finite width laminate with straight free-edges subjected to one-dimensional uniform stretch. Puppo and Evensen [2] studied the free-edge effect in terms of interlaminar shear stresses by modeling the laminate as a set of anisotropic layers separated by isotropic shear layers. Isakson and Levy [9] used a finite element approach to analyze the same problem as that of [2]. Pipes and Pagano [3] using the finite difference technique to solve the plane strain elasticity equations. Wang and Crossman [8] used a finite element method to examine the same problem as that of [3]. Hsu [6] used a perturbation method to obtain an

analytical solution for the plane strain elasticity equations derived by
Pipes and Pagano. A number of other approximate techniques have also been
developed to solve the same problem in an attempt to obtain a better accuracy.
Since these approximate methods mentioned above have been developed to solve
a particular class of problems, it may be difficult to extend these methods
to solve other problems such as the effect of free-edges of a cutout on the
strength prediction of a laminate. Obviously, a more versatile method is
needed to improve our ability in determining the stress field in a laminate
under such conditions so that the failure modes can be determined and failure
criteria may be proposed.

Lo, Christensen and Wu [18] developed a high order theory of plate
deformation which includes the effects of transverse shear deformation, transverse normal strain and warpage of the cross section. They began with an
assumed displacement form which allowed for cubic variation through the thickness for inplane displacements, u and v, and quadratic variation through the
thickness for the transverse displacement, w, i.e.,

$$u = u^{\circ}(x,y) + z\psi_x(x,y) + z^2\zeta_x(x,y) + z^3\phi_x(x,y)$$
$$v = v^{\circ}(x,y) + z\psi_y(x,y) + z^2\zeta_y(x,y) + z^3\phi_y(x,y)$$
$$w = w^{\circ}(x,y) + z\psi_z(x,y) + z^2\zeta_z(x,y)$$

The reason for this truncation of the power series in terms of z-coordinate
was to achieve the consistency of z variations in terms of shear strains, i.e.,
the transverse shear strains due to in-plane displacements u and v were of
the same order in z as that determined by the transverse displacement w.
This representation of displacements was used with the principle of stationary
potential energy to develop the high order plate theory. The problem of an
infinite plate of thickness, h, subjected to a sinusoidal pressure on the top
surface z=h/2 was solved with this high order plate theory. The results were
compared with the exact solution available from the theory of elasticity and
other lower order plate theories. It was found that the high order plate
theory was superior than other approximate theories when h/L≤2 where L was
the half wave length of the sinusoidal loading. For h/L>2, all approximate
theories diverged from the exact solution.

This theory was extended to a high order laminated plate theory by
Lo, Christensen and Wu [19] to solve problems of bidirectional and angle-ply
laminates subjected to sinusoidal loadings. The displacements and flexural
stress obtained were compared with the results given in [20,21]. The agreement was very good. However, Lo et al neither presented results for interlaminar stresses nor discussed the problem of free traction boundary conditions.

Spilker [22] chose the high order plate theory developed by Lo et al
as a basis for the development of a hybrid-stress based three-dimensional
finite element for thick plate analysis. The assumed stress hybrid model is
based on a modified statement of complimentary energy in which equilibrating
stresses are assumed in the interior of the element, and independent compatible
displacements are assumed along the boundary of the element. The element
developed is a four-noded plate elements of quadrilateral planform and total

thickness 2h, which is shown in Fig. 5. The displacement behavior along the

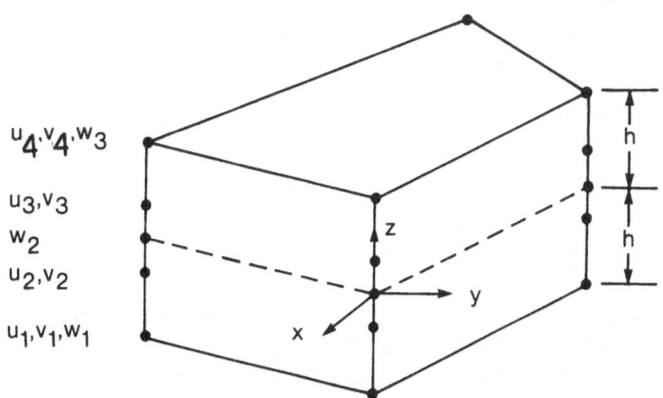

u_4, v_4, w_3

u_3, v_3
w_2
u_2, v_2

u_1, v_1, w_1

Figure 5. Geometry, nomenclature, and degrees of freedom
for a thick plate element (from Ref. [22]).

boundaries was assumed to be linear, and the stress parameters were chosen so
that the equilibrium equations were satisfied. In the hybrid-stress
model there are many ways to make the stress assumption. Spilker tried
several stress assumptions and found one which not only satisfied the equilibrium equations, but also satisfied the traction free conditions exactly at
the top and bottom surfaces of a plate. This stress assumption will be
mentioned here. In the high order plate theory of Lo et al [18], the through
thickness displacement variation were chosen so as to achieve consistency in
the z variation of all strains. When the in-plane strain components are cal-
culated from the displacements, it is seen that εx, εy and γxy are of order
z^3. Spilker assumed that σ_x, σ_y and σ_{xy} were also of order z^3 and contain
arbitrary sufficiently high x-y variations, then from the equilibrium equations

$$\partial\sigma_x/\partial_x + \partial\sigma_{xy}/\partial_y + \partial\sigma_{xz}/\partial_z = 0$$
and
$$\partial\sigma_{xy}/\partial_x + \partial\sigma_y/\partial_y + \partial\sigma_{yz}/\partial_z = 0$$

σ_{xz} and σ_{yz} should be of order z^4, and from

$$\partial\sigma_{xz}/\partial_x + \partial\sigma_{yz}/\partial_y + \partial\sigma_z/\partial_z = 0$$

σ_z should be of order z^5. Furthermore, Spilker imposed the requirement that
σ_{xz} and σ_{yz} be zero at upper and lower surfaces; and σ_z be zero at the lower
surface only. A stress assumption containing the above discussed orders of
z variation, satisfying the equilibrium equations, and also satisfying the
traction free conditions exactly was given as follows

$$\sigma_x = \beta_1 + \beta_2 x + \beta_3 y + h^2\beta_4 xy + z(\beta_5 + \beta_6 x + \beta_7 y + \beta_8 xy)$$
$$+ z^2(\beta_9 + \beta_{10}x + \beta_{11}y - 3\beta_4 xy)$$
$$+ z^3(\beta_{12} + \beta_{13}x + \beta_{14}y + \beta_{15}xy)$$

$$\sigma_y = \beta_{16} + \beta_{17}x + \beta_{18}y + h^2\beta_{19}xy + z(\beta_{20} + \beta_{21}x + \beta_{22}y + \beta_{23}xy)$$
$$+ z^2(\beta_{24} + \beta_{25}x + \beta_{26}y - 3\beta_{19}xy)$$
$$+ z^3(\beta_{27} + \beta_{28}x + \beta_{29}y + \beta_{30}xy)$$

$$\sigma_z = (1/3)(z^3 - 3h^2 z - 2h^3)\,\beta_{31} + (1/10)(z^5 - 5h^4 z - 4h^5)\beta_{32}$$
$$+ (1/2)(2h^2 z^2 - z^4 - h^4)\beta_{33}$$

$$\sigma_{xy} = \beta_{34} + (-\beta_{18} + h^2\beta_{35})x + (-\beta_2 + h^2\beta_{36})y + h^2\beta_{33}xy$$
$$+ z(\beta_{37} + \beta_{38}x + \beta_{39}y + \beta_{31}xy)$$
$$+ z^2[\beta_{40} - (\beta_{26} + 3\beta_{35})x - (\beta_{10} + 3\beta_{36})y - 3\beta_{33}xy]$$
$$+ z^3(\beta_{41} + \beta_{42}x + \beta_{43}y + \beta_{32}xy)$$

$$\sigma_{xz} = (1/2)(h^2 - z^2)(\beta_6 + \beta_{39} + \beta_{31}x + \beta_8 y) + (1/4)(h^4 - z^4)(\beta_{13} + \beta_{43}$$
$$+ \beta_{32}x + \beta_{15}y) + (z^3 - h^2 z)(\beta_{36} + \beta_{33}x + \beta_4 y)$$

$$\sigma_{yz} = (1/2)(h^2 - z^2)(\beta_{22} + \beta_{38} + \beta_{23}x + \beta_{31}y)$$
$$+ (1/4)(h^4 - z^4)(\beta_{29} + \beta_{42} + \beta_{30}x + \beta_{32}y)$$
$$+ (z^3 - h^2 z)(\beta_{35} + \beta_{19}x + \beta_{33}y)$$

where h is the half thickness of the plate. With this assumed stress components and displacements along the boundaries,,the stiffness matrix was evaluated and a problem similar to that studied by Lo et al was solved by using this element (see Fig. 6). Comparisons of the finite element solution with those from the high order plate theory and exact elasticity solution are shown in Figs. 7 to 13. The finite element solution seems to be in excellent agreement with the exact elasticity solution. This element at the present form is used to analyze a single layered homogeneous plate; however, based on the conclusion from [22] and the successful experience of Lo et al [19] in extending their high order plate theory to high order laminated plate theory, it is expected that this hybrid-stress element can be extended to a multilayered plate element which includes the effect of transverse stress components. This task is currently being carried out by Spilker and Chou and results will be reported in the future. We hope that this multilayered plate element will improve our ability to analyze the stress field of a laminate containing steep stress gradients, and in turn give us a better understanding of the failure behavior of composite materials.

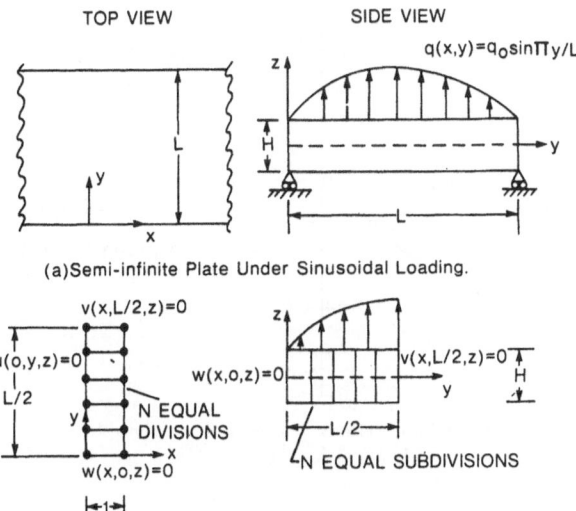

(a)Semi-infinite Plate Under Sinusoidal Loading.

(b)Finite-Element Model of Half-Strip.

Figure 6. Problem definition and finite-element modeling (from Ref. [22]).

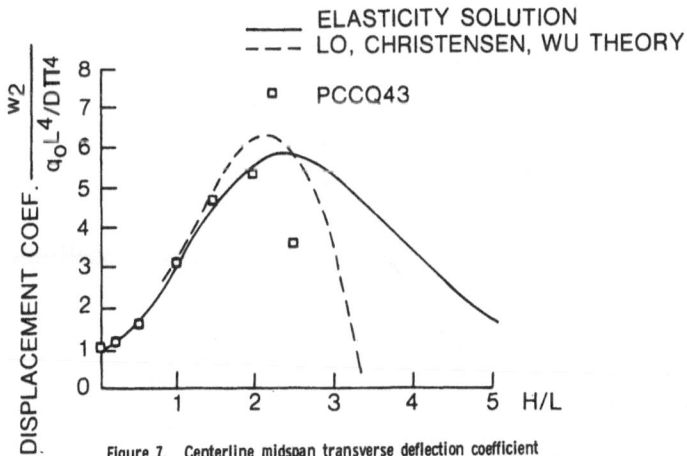

Figure 7. Centerline midspan transverse deflection coefficient
for various ratio of H/L (ten element mesh) (from Ref. [22]).

86

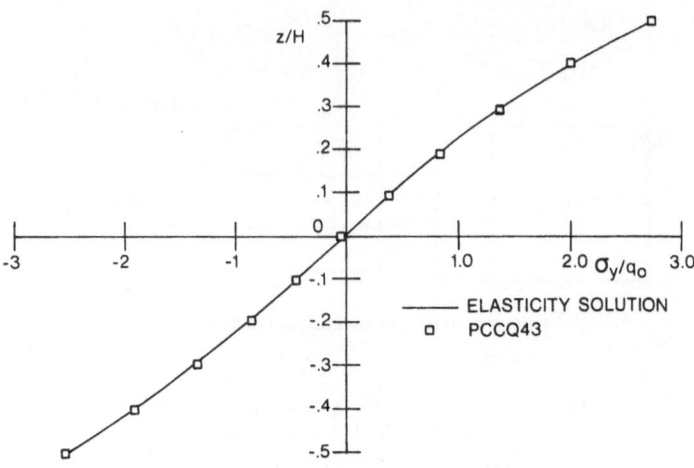

Figure 8. Through thickness variations of flexural stress at
y = L/2 for H/L = 0.5 (from Ref. [22]).

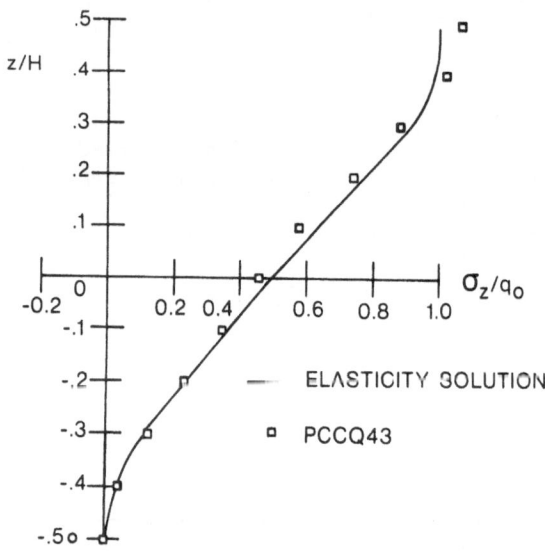

Figure 9. Through thickness variation of transverse normal
stress at y = L/2 for H/L = 0.5 (from Ref. [22]).

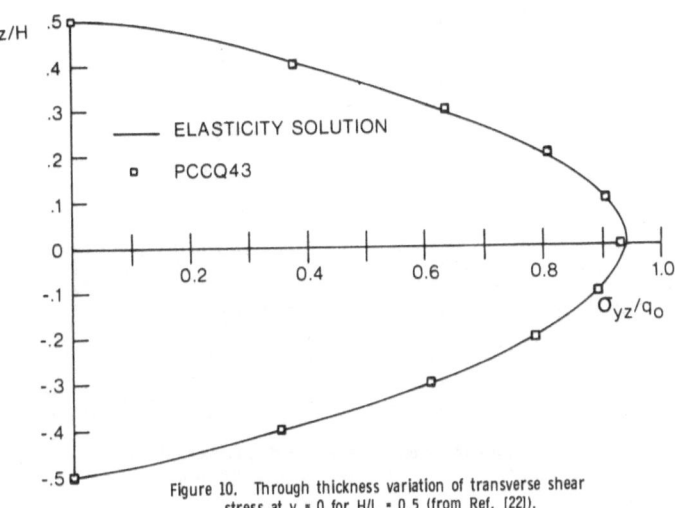

Figure 10. Through thickness variation of transverse shear stress at y = 0 for H/L = 0.5 (from Ref. [22]).

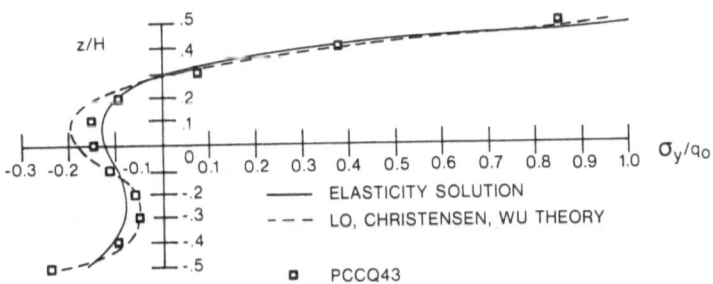

Figure 11. Predicted distribution of σ_y through thickness at y = L/2 for H/L = 1.5 (from Ref. [22]).

88

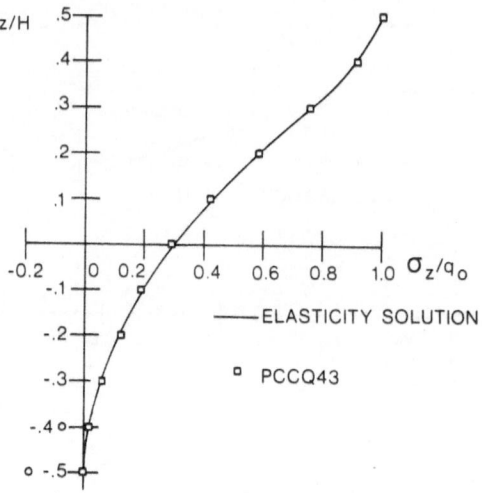

Figure 12. Predicted distribution of σ_z through thickness at y = L/2 for H/L = 1.5 (from Ref. [22]).

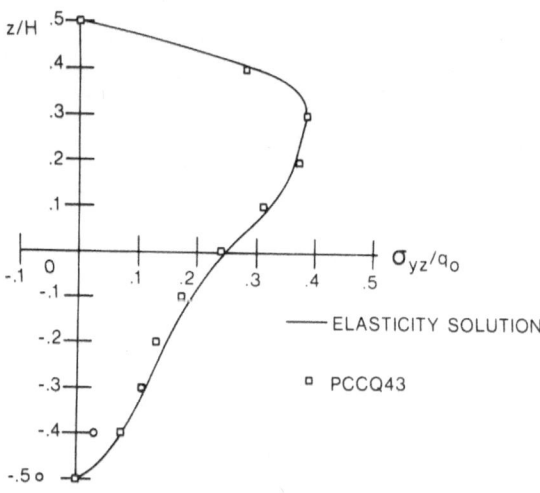

Figure 13. Predicted distribution of σ_{yz} through thickness at Y = 0 for H/L = 1.5 (from Ref. [22]).

SUMMARY

Various methods developed for studying failure behavior of composite laminates have been briefly discussed. Based on their basic assumptions, these methods may be divided into two categories; namely, the "Material Modeling" and "Mathematical Modeling". The "Material Modeling" approach assumes that the ply stiffness or strengths are unknowns and are determined by using the classic laminated plate theory to match the experimentally measured stress-strain relations of laminates. Although the failure modes cannot be investigated with this approach, the simplicity of its application to analyze a structure of composite materials provides the designers a practical means for predicting the strength of a structure of composite materials. Two typical studies were used to illustrate the "Material Modeling" approach; Chou et al [11,13] used the tangent modulus analysis to determine the ply stiffnesses from the experimentally measured stress-strain response of a laminate and Chamis et al [15] used the laminated plate theory to determine the in-situ ply strength from the laminate strength.

The "Mathematical Modeling" approach simplifies the problem of analyzing the stress field of a laminate containing a steep stress gradient to either a plane stress problem or a plane strain problem, and keeps the properties of a ply intact to determine the failure criteria from experimental results. Some recent developments in the stress analysis were mentioned; Lo et al [18,19] developed a high order plate theory and extended to a high order laminated plate theory and Spilker [22] used the high order plate theory [18] as a base to develop a high order three-dimensional hybrid-stress element for thick plate analysis. This finite element is currently being extended to formulate a multilayered plate element which includes the effect of transverse stress components. It is anticipated that this finite element will provide us a means of obtaining a better understanding of the failure behavior of a laminate.

REFERENCES

[1] Pagano, N. J., "Stress fields in composite laminates", AFML-TR-77-114, Air Force Materials Laboratory, Wright-Patterson Air Force Base, Ohio, 1977.

[2] Puppo, A. H. and Evensen, H. A., "Interlaminar shear in laminated composites under plane stress," J. Composite Materials, Vol. 4, p. 204, 1970.

[3] Pipes, R. B., and Pagano, N. J., "Interlaminar stresses in composite laminate under uniform axial tension," J. Composite Materials, Vol. 4, p. 538, 1970.

[4] Pipes, R. B., and Pagano, N. J., "Interlaminar stresses in composite laminates - an approximate elasticity solution," J. Applied Mechanics, Vol. 41, p. 668, 1974.

[5] Tsai, S. W., "Strength characteristics of composite materials," NASA CR-224, National Aeronautics and Space Administration, 1965.

[6] Hsu, P. W., "Interlaminar stresses in composite laminates - a perturbation analysis," Ph.D. Thesis, VPI and State University, Blacksberg, Virginia, 1976.

[7] Rybicki, E. F., "Approximate three-dimensional solutions for symmetric laminates under in-plane loading," J. Composite Materials, Vol. 5, p. 354, 1971.

[8] Wang, A. S. D., and Crossman, F. W., "Some new results on edge effect in symmetric composite laminates," J. Composite Materials, Vol. 11, p. 92, 1977.

[9] Isakson, G., and Levy, A., "Finite-element analysis of interlaminar shear in fibrous composites," J. Composite Materials, Vol. 5, p. 273, 1971.

[10] Tang, S. and Levy, A., "A boundary layer theory - Part II: Extension of laminated finite strip," J. Composite Materials, Vol. 9, p. 42, 1975.

[11] Chou, S. C., Orringer, O. and Rainey, J. H., "Post-failure behavior of laminate I - No stress concentration," J. Composite Materials, Vol. 10, p. 371, 1976.

[12] Spilker, R. L., Chou, S. C., and Orringer, O., "Alternate hybrid-stress elements for analysis of multilayer composite plates," J. Composite Materials, Vol. 11, p. 51, 1977.

[13] Chou, S. C., Orringer, O. and Rainey, J. H., "Post-failure behavior of laminates II - stress concentration," J. Composite Materials, Vol. 11, p. 71, 1977.

[14] Chou, S. C. and Rainey, J. H., "Modeling of Failure behavior of a laminate with elliptical hole," Fracture Mechanics and Technology, Vol. 1, G. S. Sih and C. L. Chow, ed., Noordhoff International Publishing, Leyden, pp. 271-284, 1977.

[15] Chamis, C. C. and Sullivan, T. L., "In-situ ply strength: an initial assessment," NASA TM-73771, Lewis Research Center, Cleveland, Ohio, 1977.

[16] Wu, E. M., "Phenomenological anisotropic failure criterion," Composite Materials, Vol. 2, G. P. Sendeckyj, ed., Academic Press, New York, pp. 353-431, 1974.

[17] Nuismer, R. J., "Prediction of failure in biaxially loaded composites containing stress concentrations," presented at 5th ASTM Committee D-30 Symposium on Composite Materials Testing and Design, New Orleans, Louisiana, 1978.

[18] Lo, K. H., Christensen, R. M. and Wu, E. M., "A high order theory of plate deformation - Part 1: Homogeneous plates," J. Applied Mechanics, Vol. 44, p. 663, 1977.

[19] Lo, K. H., Christensen, R. M. and Wu, E. M., "A high order theory of plate deformation - Part 2: Laminated plates," J. Applied Mechanics, Vol. 44, p. 669, 1977.

[20] Pagano, N. J., "Exact solutions for composite laminates in cylindrical bending," J. Composite Materials, Vol. 3, p. 398, 1969.

[21] Pagano, N. J., "Influence of shear coupling in cylindrical bending of anisotropic laminates," J. Composite Materials, Vol. 4, p. 330, 1970.

[22] Spilker, R. L., "High order three-dimensional hybrid-stress elements for thick plate analysis," submitted to Int. J. Numerical Methods in Engineering, 1978.

THE INTERFACE CRACK IN RETROSPECT AND PROSPECT

J. Dundurs

Department of Civil Engineering
Northwestern University, Evanston, Illinois, USA

Maria Comninou

Department of Applied Mechanics and Engineering Science
The University of Michigan, Ann Arbor, Michigan, USA

ABSTRACT

The conventional formulation used in the past for problems involving
interface cracks leads to oscillatory singularities and a physical contradiction, because the solutions predict interpenetration of material. These
unsatisfactory features can be eliminated by incorporating in the analysis
the appropriate inequality constraints. The results show that no direct
transition from adhesion to separation is generally possible, and that there
must be intervening contact zones at the tips of an interface crack. The
paper first discusses the nature of the singularity at the transition from
adhesion to contact, and reviews the two solutions worked out recently for
frictionless contacts at the tips. Next, the problem of an interface crack
in a monotonically increased shear field is formulated taking into account
friction in the contact zones, and some results are given. Finally, the
paper projects the work that will be needed in the future to evaluate correctly the large number of previous solutions containing oscillatory singularities.

INTRODUCTION

It has been known since the study of the first problems in the middle
sixties [1-4] that the elastic analysis of the interface crack leads to a
dilemma: The two sides of the crack are assumed to be free of tractions in
the formulation, but the crack faces are seen to overlap after the solution
is constructed [2,5]. The overlapping of the crack faces is caused by the
oscillatory singularities that appear in the elastic fields. The oscillatory
singularities are in turn a consequence of the assumption that the transition
from adhesion to separation is direct. The fact that solutions involving oscillatory singularities predict interpenetration of material means simply
that this assumption is not justified, that no direct transition from

adhesion to separation is generally possible, and that there must be intervening contact zones. In other words, the interface crack cannot be completely open, and its faces must be in contact near the tips. It may be noted
that a similar situation is also encountered in contact problems in which,
unless the combination of materials is special, a slip zone always exists
between stick and separation zones [6].

The oscillatory singularities were recently eliminated from the formulation of the interface crack by allowing intervening contact zones to be
present and incorporating in the analysis the appropriate inequalities (the
gap between the two faces may not be negative, the normal tractions in the
contact zones must be compressive) [7-9]. The present article is a continu-
ation of this line of study. We first review and investigate in more detail
the asymptotic nature of the stresses at a closed tip of an interface crack.
Next we recapitulate the results for interface cracks loaded in tension and
shear, but analyzed by assuming a frictionless contact. Subsequently we
formulate the problem and give some results for an interface crack in a shear
field when there is friction in the contact zones. Finally we project some
work that is needed in the future.

THE ASYMPTOTIC NATURE OF THE STRESS FIELD NEAR A CLOSED TIP OF AN INTERFACE
CRACK

Referring to Fig. 1 for the placement of the coordinate axes and labeling
of the two materials, the boundary conditions at the closed tip of the inter-

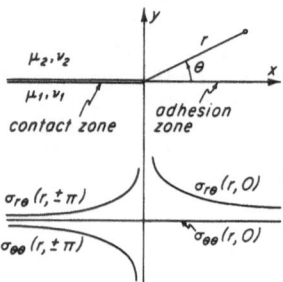

Fig. (1) - Configuration used in the
asymptotic analysis

face crack are

$$u_r^{(1)}(r,0) = u_r^{(2)}(r,0); \qquad\qquad u_\theta^{(1)}(r,0) = u_\theta^{(2)}(r,0) \qquad (1,2)$$

$$\sigma_{r\theta}^{(1)}(r,0) = \sigma_{r\theta}^{(2)}(r,0); \qquad\qquad \sigma_{\theta\theta}^{(1)}(r,0) = \sigma_{\theta\theta}^{(2)}(r,0) \qquad (3,4)$$

$$u_{\theta}^{(1)}(r,-\pi) = u_{\theta}^{(2)}(r,\pi); \qquad\qquad \sigma_{\theta\theta}^{(1)}(r,-\pi) = \sigma_{\theta\theta}^{(2)}(r,\pi) \qquad (5,6)$$

$$\sigma_{r\theta}^{(1)}(r,-\pi) = \sigma_{r\theta}^{(2)}(r,\pi) = -f\,\sigma_{\theta\theta}(r,\pm\pi); \quad \sigma_{\theta\theta}(r,\pm\pi) \le 0 \qquad (7,8)$$

In (7), f denotes the coefficient of friction, and both directions of slip are treated by letting f assume also negative values. Thus, positive f corresponds to the upper material slipping with respect to the lower material in the positive x-direction, while negative values of f indicate slip in the opposite direction. If the body is subjected to loads that start from zero and increase monotonically, it can be reasoned that, regardless of the level of friction, slip in the contact zone must extend to the very tip of the interface crack [6]. What happens when the loads are reduced is an entirely different matter, and it is more than likely that back slip does not immediately penetrate to the tip. Our considerations in this article are restricted to loads that start from zero and increase monotonically.

The expressions for the singular parts of the stresses at the closed tip of the interface crack are [8]

$$\sigma_{rr}^{(1)} = -\frac{1}{2}\,Cr^{-\lambda}\{(2+\lambda)(1-\beta)\sin\lambda\theta - [2-\lambda(1-\beta)]\sin(2-\lambda)\theta\} \qquad (9)$$

$$\sigma_{rr}^{(2)} = -\frac{1}{2}\,Cr^{-\lambda}\{(2+\lambda)(1+\beta)\sin\lambda\theta - [2-\lambda(1+\beta)]\sin(2-\lambda)\theta\} \qquad (10)$$

$$\sigma_{r\theta}^{(1)} = \frac{1}{2}\,Cr^{-\lambda}\{\lambda(1-\beta)\cos\lambda\theta + [2-\lambda(1-\beta)]\cos(2-\lambda)\theta\} \qquad (11)$$

$$\sigma_{r\theta}^{(2)} = \frac{1}{2}\,Cr^{-\lambda}\{\lambda(1+\beta)\cos\lambda\theta + [2-\lambda(1+\beta)]\cos(2-\lambda)\theta\} \qquad (12)$$

$$\sigma_{\theta\theta}^{(1)} = -\frac{1}{2}\,Cr^{-\lambda}\{(2-\lambda)(1-\beta)\sin\lambda\theta + [2-\lambda(1-\beta)]\sin(2-\lambda)\theta\} \qquad (13)$$

$$\sigma_{\theta\theta}^{(2)} = -\frac{1}{2}\,Cr^{-\lambda}\{(2-\lambda)(1+\beta)\sin\lambda\theta + [2-\lambda(1+\beta)]\sin(2-\lambda)\theta\} \qquad (14)$$

where C is free constant which may be viewed as a stress intensity factor, and

$$\beta = \frac{\mu_2(\kappa_1-1)-\mu_1(\kappa_2-1)}{\mu_2(\kappa_1+1)+\mu_1(\kappa_2+1)}; \qquad -\frac{1}{2} \le \beta \le \frac{1}{2} \qquad (15)$$

with $\kappa = 3 - 4\nu$ for conditions of plane strain (μ = shear modulus, ν = Poisson's ratio). The constant λ determining the strength of the singularity is a root of the equation

$$\cot\lambda\pi = f\beta; \qquad 0 < \lambda < 1 \tag{16}$$

It follows from (9-14) and (16) that the singularity is weaker than inverse square root ($\lambda < 1/2$) for $f\beta > 0$ and stronger ($\lambda > 1/2$) for $f\beta < 0$. The inverse-square-root singularity is recovered for frictionless contact, or $f = 0$.

Equations (9-14) show that the singular stress field at the closed tip is in a sense antisymmetric with respect to the interface and contains only one free constant (stress intensity factor). The latter aspect is not surprising: The faces of an open crack can have a relative displacement in two directions. This is why two stress intensity factors are obtained for the crack in a homogeneous material. If the tip of the crack is closed, however, the two faces can undergo only a relative tangential shift, and the number of free constants in the asymptotic analysis drops from two to one.

The nature of the singularity can be assessed further by considering the tractions on $y = 0$. From (11-14) it follows that

$$\sigma_{r\theta}(r,0) = Cr^{-\lambda}; \qquad\qquad \sigma_{\theta\theta}(r,0) = 0 \tag{17,18}$$

$$\sigma_{r\theta}(r,\pm\pi) = Cr^{-\lambda}\cos\lambda\pi; \qquad \sigma_{\theta\theta}(r,\pm\pi) = -C\beta r^{-\lambda}\sin\lambda\pi \tag{19,20}$$

The shearing tractions are singular on both sides of the crack tip. The normal tractions vanish ahead of the crack tip, but are singular in the contact zone of the tip. The distribution of the tractions is indicated schematically in Fig. 1, and it is clear that the spreading of an interface crack is more intimately related to interface failure in shear than tension.

From (17), (19) and (20) it follows also that the tractions in the contact zone are connected with the shearing tractions ahead of the crack tip through the relations

$$\sigma_{r\theta}(r,\pm\pi) = \cos\lambda\pi\ \sigma_{r\theta}(r,0) \tag{21}$$

$$\sigma_{\theta\theta}(r,\pm\pi) = -\beta\sin\lambda\pi\ \sigma_{r\theta}(r,0) \tag{22}$$

Since $\cos\lambda\pi$ can be either positive or negative depending on the direction of slip, nothing can be said immediately about the sign of $\sigma_{r\theta}(r,\pm\pi)$, but an important result follows from (22) because $\sin\lambda\pi$ is positive: The materials can always be labeled so that $\beta > 0$. As we must have $\sigma_{\theta\theta}(r,\pm\pi) \leq 0$, it is seen that $\sigma_{r\theta}(r,0) \geq 0$ for the placement of coordinate axes used. Hence $C \geq 0$ regardless of the nature of the applied loads. This result was used to reason for the case of tensile loading that the singularity must be weaker than inverse square root [8]. A more striking conclusion evolves, however, when the interface is loaded in shear, as C will be of the same sign for both possible directions of the applied shear load. This might at first appear as paradoxical, but the conclusion has been confirmed by solving the full boundary value problem for a crack of finite length and no friction [9]. It might be noted that, if not the sign, the magnitude of C depends on the direction of the applied shear load when the crack has a finite length.

An important question is whether, upon loading, the interface crack will spread along the interface, or whether it would have the tendency to veer into one of the solids. It will not be easy to answer this question. First, the problem of the veering crack is one of considerable difficulty even for the homogeneous solid [10]. Second, the answer obviously will depend on the properties and relative strengths of the solids and the bond between them. Some assessment of the situation can be made, however, by simply plotting the angular variation of the singular stresses $\sigma_{\theta\theta}^{(1)}(r,\theta)$ and $\sigma_{\theta\theta}^{(2)}(r,\theta)$ from (13) and (14) divided by $\sigma_{r\theta}(r,0) = Cr^{-\lambda}$. There is not much variation in the curves in the range $0 < \beta \le \frac{1}{2}$ and $0 \le f \le 1$, and we show in Fig. 2 the

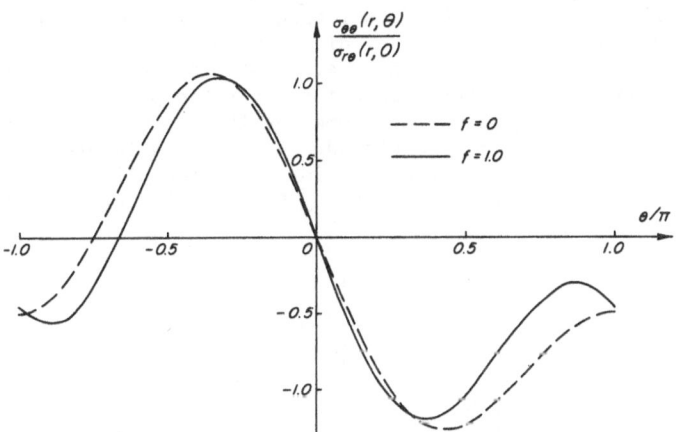

Fig. (2) - Variation of $\sigma_{\theta\theta}(r,\theta)/\sigma_{r\theta}(r,0)$ from the asymptotic analysis for $\beta = 0.5$

results only for $\beta = 0.5$, and $f = 0$ and 1 (negative values of $f\beta$ and the corresponding strong singularities with $\lambda > \frac{1}{2}$ have been encountered for neither tensile nor shear loading). In case of frictionless contact at the tips, the stresses $\sigma_{\theta\theta}^{(1)}$ and $\sigma_{\theta\theta}^{(2)}$ reach extremum values for the angles θ_1^* and θ_2^* (the subscripts 1 and 2 refer to the two materials) determined by

$$\cos\theta_1^* = \frac{1 + \beta}{3 + \beta} \; ; \qquad - \pi/2 < \theta_1^* < 0 \qquad\qquad (23)$$

$$\cos\theta_2^* = \frac{1 - \beta}{3 - \beta} \; ; \qquad 0 < \theta_2^* < \pi/2 \qquad\qquad (24)$$

The angle θ_1^* ranges from $-70.5°$ for $\beta = 0$ to $-64.6°$ for $\beta = 0.5$, whereas θ_2^* changes from $70.5°$ for $\beta = 0$ to $78.5°$ for $\beta = 0.5$. An interesting aspect of the stress distribution evolving from Fig. 2 is that $\sigma_{\theta\theta}$ is tensile in only one of the materials. More specifically, a tensile cleavage stress exists only in the material with the lower value of $\mu/(\varkappa - 1)$, or $\mu/(1 - 2\nu)$ in plane strain. The asymptotic analysis indicates, therefore, that the interface crack is unlikely to veer into the material with the higher value of $\mu/(\varkappa - 1)$.

INTERFACE CRACK WITH FRICTIONLESS CONTACTS AT THE TIPS

Considerable advantage in the formulation and solution is gained by viewing the interface crack as an array of distributed edge dislocations. This allows one to write the governing integral equations practically at sight, and provides guidance toward their solution under the proper inequality constraints. We recall for this purpose some results for discrete edge dislocations [11]: Suppose that a discrete edge dislocation is placed at the point $(\xi,0)$ of the interface (see Fig. 3 for the orientation of the coordinate axes with respect to the two solids). An edge dislocation with a

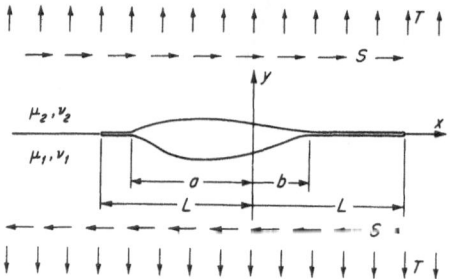

Fig. (3) - Interface crack with contact zones

Burgers vector in the x-direction (interface glide dislocation) induces the following traction components at the interface:

$$\sigma_{xy}(x,0;\xi) = -\frac{Cb_x}{\pi}\frac{1}{\xi - x} \; ; \qquad \sigma_{yy}(x,0;\xi) = -\beta Cb_x \delta(\xi - x) \qquad (25,26)$$

If the Burgers vector is in the y-direction (interface climb dislocation), the traction components are

$$\sigma_{xy}(x,0;\xi) = \beta Cb_y \delta(\xi - x); \qquad \sigma_{yy}(x,0;\xi) = -\frac{Cb_y}{\pi}\frac{1}{\xi - x} \qquad (27,28)$$

where δ denotes the Dirac delta function, and b_x and b_y are the components of the Burgers vector. Furthermore,

$$C = \frac{2\mu_1(1 + \alpha)}{(\varkappa_1 + 1)(1 - \beta^2)} = \frac{2\mu_2(1 - \alpha)}{(\varkappa_2 + 1)(1 - \beta^2)} \qquad (29)$$

with

$$\alpha = \frac{\mu_2(\varkappa_1 + 1) - \mu_1(\varkappa_2 + 1)}{\mu_2(\varkappa_1 + 1) + \mu_1(\varkappa_2 + 1)} \; ; \qquad -1 \le \alpha \le 1 \qquad (30)$$

and β defined by (15).

Consider a crack of length 2L lying in the interface of two elastic solids with the shear moduli μ_1, μ_2 and Poisson's ratios ν_1, ν_2 as indicated in Fig. 3. Under the action of the applied loads, the crack opens in the interval (- a, b), where a and b are unknowns which are to be determined in the course of solution. As a direct change from adhesion to separation is not possible, the crack is closed, and its two faces are in frictionless contact in the intervals (-L, - a) and (b,L). The boundary conditions on the tractions require that the shearing tractions vanish on (-L,L) and that the normal tractions vanish on (- a, b). Furthermore, the gap between the two crack faces must be nonnegative in the separation zone (- a, b), and the normal tractions must be compressive in the contact zones (-L, - a) and (b,L).

Suppose that, in the absence of the crack, the interface transmits the uniform shearing tractions S and tensile normal tractions T. The dislocation distributions then must modify the stress fields so as to satisfy the boundary and auxiliary conditions outlined above. Using (25-28), the boundary conditions lead immediately to the two integral equations

$$S + C\left\{\beta B_y(x)[H(x + a) - H(x - b)] - \frac{1}{\pi}\int_{-L}^{L}\frac{B_x(\xi)}{\xi - x}\,d\xi\right\} = 0; \qquad -L < x < L \qquad (31)$$

$$T - C\left\{\beta B_x(x) + \frac{1}{\pi}\int_{-a}^{b}\frac{B_y(\xi)}{\xi - x}\,d\xi\right\} = 0; \qquad -a < x < b \qquad (32)$$

where $B_x(x)$ is the density of glide dislocations in the interval $(-L,L)$ and $B_y(x)$ the density of climb dislocations in the interval $(-a, b)$. Moreover, $H(x)$ is the Heaviside unit step function. Denoting by $g(x)$ the gap between the solids and by $h(x)$ their relative tangential shift, or

$$g(x) = u_y^{(2)}(x,0) - u_y^{(1)}(x,0); \qquad -a < x < b \qquad (33)$$

$$h(x) = u_x^{(2)}(x,0) - u_x^{(1)}(x,0); \qquad -L < x < L \qquad (34)$$

we note that

$$B_x(x) = -\frac{dh(x)}{dx}; \qquad B_y(x) = -\frac{dg(x)}{dx} \qquad (35,36)$$

To ensure single-valued displacements, or no net dislocation, we must require that

$$\int_{-L}^{L} B_x(\xi)d\xi = 0; \qquad \int_{-a}^{b} B_y(\xi)d\xi = 0 \qquad (37,38)$$

It may be noted that $B_y(x)$ must vanish in the contact zones.

The two cases of $S = 0$ and $T = 0$ were considered separately [7,9]. For a tensile loading of the interface ($S = 0$) we have $a = b$ on account of symmetry. It is seen from (36) that $B_y(x)$ must be continuous in $(-a, a)$, and since the transition from contact to separation at $-a$ and a has to be smooth, $B_y(x)$ must vanish at these points. Then (31) can be considered as a Cauchy singular integral equation for the unknown function $B_x(x)$ and solved by formally treating $B_y(x)$ as known. The solution is found in Muskhelishvili [12, §118]. Normalizing the interval $(-L,L)$ by the change of variables

$$x = Ls; \qquad \xi = Lr \qquad (39)$$

and retaining the same symbols for the functions in the new variables,

$$B_x(s) = -\frac{\beta}{\pi} w(s) \int_{-\gamma}^{\gamma} \frac{B_y(r)}{w(r)(r-s)} dr; \qquad -1 < s < 1 \qquad (40)$$

where

$$\gamma = a/L \qquad (41)$$

and $w(s)$ is the characteristic function of the singular integral equation. The characteristic function depends on the behavior of $B_x(s)$ at the ends of the interval $(-1,1)$. It was found that the bounded solution for $B_x(s)$ does not satisfy the side condition (37). Hence $B_x(s)$ must be singular at -1 and 1, and

$$w(s) = (1 - s^2)^{-\frac{1}{2}} \tag{42}$$

Furthermore, the singular solution for $B_x(x)$ can be shown to satisfy (37) automatically.

Substituting $B_x(s)$ from (40) into (32), normalizing the interval of integration by setting

$$s = \gamma\zeta; \qquad r = \gamma\omega \tag{43}$$

and separating the Cauchy kernel, we obtain

$$(1 - \beta^2) \int_{-1}^{1} \frac{B_y(\omega)}{\omega - \zeta} d\omega - \beta^2 \int_{-1}^{1} k(\omega,\zeta) B_y(\omega) d\omega = \frac{\pi T}{C} ; \qquad -1 < \zeta < 1 \tag{44}$$

where

$$k(\omega,\zeta) = \frac{1}{\omega - \zeta} \left[\left(\frac{1 - \gamma^2\omega^2}{1 - \gamma^2\zeta^2} \right)^{\frac{1}{2}} - 1 \right] \tag{45}$$

is the bounded kernel. Since $B_y(x)$ has to be bounded at $-a$ and a, it must satisfy the following consistency condition [12, §118]:

$$\int_{-1}^{1} (1 - \gamma^2\omega^2)^{\frac{1}{2}} B_y(\omega) \int_{-1}^{1} \frac{d\zeta}{(1 - \zeta^2)^{\frac{1}{2}}(1 - \gamma^2\zeta^2)^{\frac{1}{2}}} d\omega = -\frac{\pi^2 T}{\beta^2 C} \tag{46}$$

which provides the additional relation for finding the unknown length a. It may be noted that $B_y(x)$ is an odd function of x due to the symmetry of the problem, and that (38) is satisfied automatically.

The interface tractions follow from (25-28) as

$$\sigma_{xy}(x,0) = -\frac{C}{\pi} \int_{-L}^{L} \frac{B_x(\xi)}{\xi - x} d\xi ; \qquad |x| > L \tag{47}$$

$$\sigma_{yy}(x,0) = T - C \left\{ \beta B_x(x) + \frac{1}{\pi} \int_{-a}^{a} \frac{B_y(\xi)}{\xi - x} d\xi \right\} ; \qquad |x| > a \tag{48}$$

Using the dimensionless variables s and r defined by (43) and $B_x(s)$ from (40), the interface shearing tractions become

$$\sigma_{xy}(s,0) = -\frac{\beta C}{\pi} \frac{\text{sgn } s}{(s^2 - 1)^{\frac{1}{2}}} \int_{-\gamma}^{\gamma} \frac{B_y(r)(1 - r^2)^{\frac{1}{2}}}{r - s} dr; \qquad |s| > 1 \tag{49}$$

and it is seen that they are square root singular at $|x| = L+$. Since $B_x(x)$ is square root singular at $|x| = L-$ and $B_y(x)$ is bounded everywhere, we can conclude from (48) that the normal tractions are square root singular at $|x| = L-$ in the contact zones, but are bounded at $|x| = L+$ in the adhesion zones. Referring to (17-20), it is seen that these conclusions are in agreement with the asymptotic analysis for $f = 0$.

The stress intensity factor K_2 for the shearing stresses can be defined in the conventional manner as

$$K_2 = \lim_{x \to L} \left\{ [2(x - L)]^{\frac{1}{2}} \sigma_{xy}(x,0) \right\} \tag{50}$$

and from (49) and (40) it follows that

$$K_2 = CL^{\frac{1}{2}} \lim_{s \to 1} \left\{ [2(1 - s)]^{\frac{1}{2}} B_x(s) \right\} \tag{51}$$

The crack extension force computed from the change in the free (total potential) energy by appropriately modifying the formula given by Bilby and Eshelby [13] is

$$G = \frac{\pi K_2^2}{4C} \tag{52}$$

In order to obtain more specific results, the integral equation must be solved numerically. Such a solution employing current methods for discretizing singular integral equations of the Cauchy type [14] led to the following conclusions [7]:

1. The contact zones for the interface crack loaded in tension are extremely small in comparison to the size of the crack. Thus it was found that for β varying from 0.4854 to 0.4239 the ratio a/L changed from $1 - 10^{-4}$ to $1 - 10^{-7}$. The small size of the contact zones makes the numerical solution of the integral equation (44) difficult and, even using double precision, it was not possible to consider a/L smaller than $1 - 10^{-7}$.

2. The stress intensity factor defined by (50) was $K_2 = 1.050 \text{ TL}^{\frac{1}{2}}$ for $\beta = 0.4854$ and $K_2 = 1.008 \text{ TL}^{\frac{1}{2}}$ for $\beta = 0.4239$. It was reasoned, however, that $K_2/\text{TL}^{\frac{1}{2}}$ is practically equal to 1 for $\beta \neq 0$ in the range in which it was not possible to solve the integral equation with sufficient accuracy. At $\beta = 0$, there is a discontinuous change in the singularity and stress intensity

factors, as a direct transition from adhesion to separation becomes possible for such material combinations, and the problem is essentially the same as for a crack in a homogeneous material.

3. Although the interface normal tractions $\sigma_{yy}(x,0)$ are finite in the adhesion zone ahead of the crack tip, they assume large positive values. Thus, $\sigma_{yy}(L+,0)/T = 23.36$ for $\beta = 0.4854$. Furthermore $\sigma_{yy}(L+,0)$ increases as β is decreased.

The formulation of the interface crack loaded in shear is obtained from (31) and (32) by setting $T = 0$. Although the problem can be approached in a manner similar to that used for the tension crack, its solution is complicated by lack of symmetry and the fact that one of the contact zones is even smaller than before [9]. The numerical solution of the integral equation yielded $a/L = 1 - 10^7$, but $b/L = 0.345$ for $\beta = 0.5$ and $S > 0$. Thus, while one of the contact zones is extremely small, the other is quite large. The stress intensity factors are defined as

$$K_2(\pm L) = \lim_{x \to \pm L} \left\{ [2(x \mp L)]^{\frac{1}{2}} \sigma_{xy}(x,0) \right\} \tag{53}$$

and, for $\beta = 0.5$, the results were $K_2(+L)/SL^{\frac{1}{2}} = 1.03$ and $K_2(-L)/SL^{\frac{1}{2}} = -0.45$. The fact that $K_2(-L) < 0$ is rather surprising. It is, however, in accordance with the conclusion reached in the asymptotic analysis that $C \geq 0$ for $\beta > 0$. This is seen by making the appropriate transformation required to transfer the coordinate system shown in Fig. 1 to the left tip of the crack in Fig. 3.

THE EFFECT OF FRICTION IN THE CONTACT ZONES

The normal tractions transmitted in the contact zones are very large, in fact singular at the crack tips. This suggests that friction may have a pronounced effect on the deformations. It is usually difficult to solve elasticity problems with friction because the stress state depends not only on the instantaneous values of the applied forces, but involves also the history of loading. Generally, contact problems with friction require an incremental formulation [15]. The interface crack with friction in the contact zones belongs to a class of problems, however, for which no incremental formulation is necessary if the loading is started from zero and increased monotonically. The reason for this is that slip takes place along the whole length of the contact zones, and the extent of the contact zones does not depend on the level reached by the monotonically increasing load.

The interface crack loaded in shear is perhaps more challenging than the case of tensile loading because it lacks symmetry, and we consider this problem. The governing integral equations again follow directly from (25-28) by noting that the climb dislocation density $B_y(x)$ vanishes in the contact zones:

$$S + C \left\{ \beta B_y(x) - \frac{1}{\pi} \int_{-L}^{L} \frac{B_x(\xi)}{\xi - x} d\xi \right\} = 0; \qquad - a < x < b \tag{54}$$

$$\beta B_x(x) + \frac{1}{\pi} \int_{-a}^{b} \frac{B_y(\xi)}{\xi - x} d\xi = 0; \qquad - a < x < b \tag{55}$$

$$S - \frac{C}{\pi} \int_{-L}^{L} \frac{B_x(\xi)}{\xi - x} d\xi = Cf \text{ sgn } x \left\{ \beta B_x(x) + \frac{1}{\pi} \int_{-a}^{b} \frac{B_y(\xi)}{\xi - x} d\xi \right\};$$

$$- L < x < - a \quad \text{and} \quad b < x < L \tag{56}$$

The first two of these equations simply reflect the requirement that the shearing and normal tractions vanish in the open part of the crack. The third equation or (56) may need some explanation. If we adopt the Coulomb law for dry friction, the relation between the shearing and normal tractions is

$$\left| \sigma_{xy}(x,0) \right| = f \left| \sigma_{yy}(x,0) \right|; \qquad f > 0 \tag{57}$$

in the contact zones that are in the process of slipping. Moreover, the sign of $\sigma_{xy}(x,0)$ must be consistent with the direction of slip at any given instant. The slip direction in the contact zones can be established, however, from the solution for no friction: If the applied shearing tractions start from zero and increase monotonically, they must be the same. It follows next from either the asymptotic analysis or the solution for the interface crack in a shear field [9] that, for $S > 0$ and $\beta > 0$, the slip is in such a direction that $\sigma_{xy}(x,0) < 0$ in the left contact zone $- L < x < - a$ and $\sigma_{xy}(x,0) > 0$ in the right contact zone $b < x < L$. Noting that $\sigma_{yy}(x,0) < 0$ in both contact zones, we can replace (57) with

$$\sigma_{xy}(x,0) = f \sigma_{yy}(x,0); \qquad - L < x < - a \tag{58}$$

$$\sigma_{xy}(x,0) = - f \sigma_{yy}(x,0); \qquad b < x < L \tag{59}$$

where $f > 0$. Assuming that the right contact zone does not extend to the left of the origin, or that $b > 0$, (58) and (59) can be replaced with the single condition

$$\sigma_{xy}(x,0) = - f \text{ sgn } x \sigma_{yy}(x,0) \tag{60}$$

and (56) follows from (25-28). We also note that the solution of the integral equations (54-56) must satisfy (37) and (38).

Equations (54) and (56) can be combined into a single integral equation valid in the interval $-L < x < L$:

$$\frac{1}{\pi} \int_{-L}^{L} \frac{B_x(\xi)}{\xi - x} \, d\xi = \frac{S}{C} + \beta [H(x + a) - H(x - b)] B_y(x)$$

$$- f \, \text{sgn} \, x \left\{ \beta B_x(x) + \frac{1}{\pi} \int_{-a}^{b} \frac{B_y(\xi)}{\xi - x} \, d\xi \right\} ; \qquad -L < x < L \tag{61}$$

Next, proceeding similarly to the solution for the tension crack, we arrive at the following integral equation for the density of climb dislocations:

$$- f \beta (1 - \beta^2) \, \text{sgn} \, x \, B_y(x) + \frac{1 - \beta^2}{\pi} \int_{-a}^{b} \frac{B_y(\xi)}{\xi - x} \, d\xi$$

$$= \frac{\beta^2}{\pi} \int_{-a}^{b} \frac{B_y(\xi)}{\xi - x} \left[\frac{w(x)}{w(\xi)} - 1 \right] d\xi - \frac{S f \beta^2}{C} \, \text{sgn} \, x$$

$$+ \frac{S\beta}{\pi C} w(x) \int_{-L}^{L} \frac{d\xi}{(\xi - x)w(\xi)} - D\beta (1 + f^2 \beta^2) w(x)$$

$$- \frac{f\beta}{\pi^2} w(x) \int_{-a}^{b} \frac{B_y(\xi')}{\xi' - x} \int_{-L}^{L} \frac{\text{sgn} \, \xi}{w(\xi)} \left[\frac{1}{\xi - x} - \frac{1}{\xi - \xi'} \right] d\xi \, d\xi' ;$$

$$- a < x < b \tag{62}$$

where

$$w(x) = (L^2 - x^2)^{-\lambda} \tag{63}$$

and λ is defined by (16). The constant D in (62) is determined from (37). Since $B_y(x)$ must be bounded at $-a$ and b, a consistency condition must be imposed. This condition together with (38) determines the two unknown parameters a and b.

The integral equation was put in an appropriate form for numerical calculations and discretized using the method of Krenk [16]. The resulting expressions are too lengthy for inclusion in this paper, and we merely give the dependence of the large contact zone on the friction coefficient. The results are shown in Fig. 4, and it is seen that friction increases the size of the larger contact zone. Additional details of this problem will be presented in a future publication.

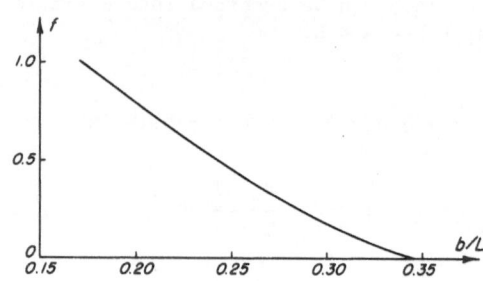

Fig. (4) - Effect of friction on the extent of the
large contact zone for shear loading

CONCLUSION

We have attempted to show how to overcome the contradictions arising from
the conventional formulation of interface cracks which assumes a direct transition from adhesion to separation and leads to oscillatory singularities.
The fact remains, however, that a large number of solutions involving oscillatory singularities with their contradictory features have been worked out
in the past and are now embedded in the literature devoted to fracture mechanics. The question is therefore what to do with them: Should the previous results be discarded and the solutions repeated by incorporating the
necessary contact zones at the crack tips? This is not an attractive pros-
pect because of the large number of problems that would have to be reformu-
lated and solved anew. Can the previous solutions be salvaged by reinterpret-
ing the results? Some guidance in this respect can be gained from the two
solutions worked out for frictionless contact zones [7,9].

The contact zones for an interface crack loaded in tension are extremely
small. The global nature of the stress field is consequently very close to
that of England's solution [2] which involves oscillatory singularities. For
instance at $x/L = 1.001$, $\sigma_{yy}(x,0)/T = 13.74$ from the solution with contact
zones and 13.59 from England's solution. The stress intensity factors proposed in the past for the interface crack bear no relation to the stress intensity factor defined by (50). However, it was found that the energy release
rate compares favorably with some previous results based on oscillatory singularities [7]. The situation is entirely different when the interface crack
is loaded in shear. One of the contact zones then is very large, and it affects strongly the global nature of the stress field [9]. In such case, no
useful results can be extracted from a solution involving oscillatory singularities.

It appears, therefore, that a fair value of the energy release rate can be
extracted from solutions involving oscillatory singularities, provided the

contact zones are small in comparison to the size of the crack. In turn, the sizes of the contact zones can be estimated by computing the zones in which overlapping of material is predicted by the oscillatory singularities.

ACKNOWLEDGEMENT

The authors are pleased to acknowledge the support by the National Science Foundation through the grants AER 75-00187 (J.D.) and ENG 77-25032 (M.C.).

REFERENCES

[1] Erdogan, F., J. Appl. Mech., Vol. 30, p. 232, 1963.

[2] England, A. H., J. Appl. Mech., Vol. 32, p. 400, 1965.

[3] Erdogan, F., J. Appl. Mech., Vol. 32, p. 403, 1965.

[4] Rice, J. R. and Sih, G. C., J. Appl. Mech., Vol. 32, p. 418, 1965.

[5] Malyshev, B. M. and Salganik, R. L., Int. J. Fracture Mech., Vol. 1, p. 114, 1965.

[6] Dundurs, J. and Comninou, M., J. Elasticity, in press.

[7] Comninou, M., J. Appl. Mech., Vol. 44, p. 631, 1977.

[8] Comninou, M., J. Appl. Mech., Vol. 44, p. 780, 1977.

[9] Comninou, M., J. Appl. Mech., in press.

[10] Palaniswamy, K. and Knauss, W. G., "On the problem of crack extension in brittle solids under general loading," Mechanics Today, Vol. 4, S. Nemat-Nasser, ed., Pergamon Press, Oxford, pp. 87-148, 1978.

[11] Comninou, M., Phil. Mag., Vol. 36, p. 1281, 1977.

[12] Muskhelishvili, N. I., Singular Integral Equations, P. Noordhoff, Groningen, 1953.

[13] Bilby, B. A. and Eshelby, J. D., "Dislocations and the theory of fracture," Fracture, Vol. 1, H. Liebowitz, ed., Academic Press, New York, pp. 99- 182, 1968.

[14] Erdogan, F. and Gupta, G., Q. Appl. Math., Vol. 29, p. 525, 1972.

[15] Goodman, L. E., J. Appl. Mech., Vol. 29, p. 515, 1962.

[16] Krenk, S., Q. Appl. Math., Vol. 33, p. 225, 1975.

Section III

FRACTURE ANALYSIS

FRACTURE MECHANICS OF COMPOSITE MATERIALS

G. C. Sih

Lehigh University, Bethlehem, Pennsylvania 18015 USA

INTRODUCTION

Fracture mechanics has now been widely accepted as a useful discipline for characterizing the toughness of materials that are homogeneous and isotropic at the macroscopic scale level. Metallic alloys fall into this category. The common experience supporting the classical concept[*] of fracture mechanics is that the instability of a structure or structural element is controlled by incipient fracture of a macrocrack. With the advent of high performance composites, there is concern whether the additional complexity of heterogeneity and anisotropy would preclude practical application of fracture mechanics to composites. This fundamental issue has thus far not received adequate attention and will be addressed in this paper.

The modes of failure in a composite can be manifold arising from broken fibers, flaws in the matrix, debonded interfaces or their combination. Crack initiation invariably starts from small, inherent defects in the constituents of the composites as in metals. The task of including all these failure mechanisms into an analytical model would be overwhelming and defeat the objective of fracture mechanics which is to provide a simple approach of characterizing the response of composites. The following questions are selected for discussion:

A. Is it possible to borrow the test procedures used for metals such as compliance measurement, critical stress intensity factor, etc., to characterize the fracture toughness of composites?

B. Can the basic equations of fracture mechanics be modified to include material nonhomogeneity and/or anisotropy?

C. To what extent is the path of crack propagation determined by the initial crack geometry, loading, and material orientation?

---

[*]The term "fracture mechanics" as used in this paper includes only unstable crack propagation. Initiation and subcritical growth are excluded.

D. What type of theoretical and experimental investigations are required for developing predictive techniques to forecast composite behavior?

FRACTURE MODES

The success of applying fracture mechanics to composites depends to a large extent on the analyst not to misinterpret the theory. Strictly speaking, the classical concept of fracture toughness is restricted to a single phase material [1] containing a dominant crack. The force necessary to drive this crack is controlled by the energy release rate, G_{1c} and can be related to the critical stress intensity factor, K_{1c}. For a homogeneous and isotropic material, the relationship is of the form [2]

$$G_{1c} = \frac{(1-\nu^2)K_{1c}^2}{E} \text{ (plane strain)} \tag{1}$$

where ν is the Poisson's ratio and E the Young's modulus. If the material is homogeneous and anisotropic, equation (1) can be modified to read as [3]

$$G_{1c} = K_{1c}^2 \left(\frac{b_{11}b_{22}}{2}\right)^{1/2} \left[\left(\frac{b_{22}}{b_{11}}\right)^{1/2} + \frac{2b_{12}+b_{22}}{2b_{11}}\right]^{1/2} \tag{2}$$

Since the assumption of self-similarity* was invoked in deriving equation (2), the crack has to coincide with one of the principal axes of material symmetry such that it extends straight ahead. In this case, the system must necessarily be orthotropic in nature and the elastic coefficients b_{ij} in equation (2) can be expressed in terms of the principal moduli of elasticity and the Poisson's ratios [4]:

$$b_{11} = \frac{1}{E_1}\left[1 - \left(\frac{E_1}{E_2}\right)\nu_{12}^2\right], \quad b_{22} = \frac{1}{E_2}(1-\nu_{23}^2)$$

$$b_{12} = -\frac{\nu_{12}}{E_1}(1+\nu_{23}), \quad b_{66} = \frac{1}{\mu_{12}} \tag{3}$$

in which $E_1 > E_2$. Another limitation of equation (2) is that the fractured surface should have a smooth appearance as opposed to a fibrous shaving brush appearance that is often encountered in the fracture of fiber-reinforced composites. Two of the commonly observed modes of fracture are shown in Figures 1

---

*Self-similarity can be observed if the applied loads and material properties are aligned symmetrically with reference to the crack such that it runs straight.

and 2. Figure 1 shows a matrix crack propagated past a number of unbroken fibers across the crack leaving the fibers to support the load. Frequently, the

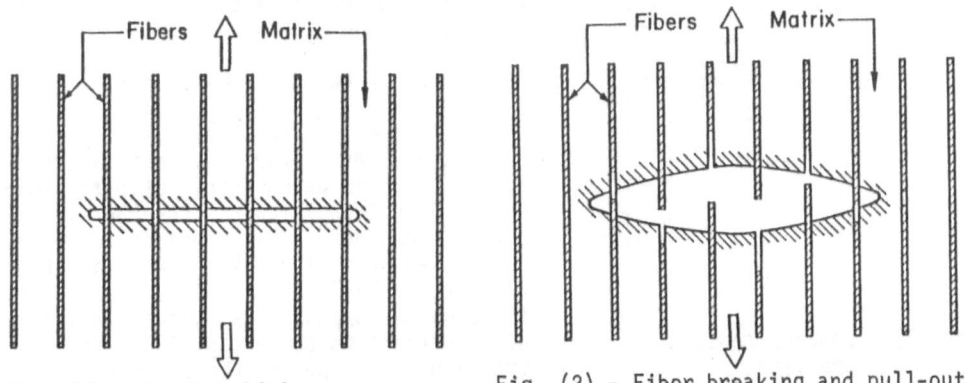

Fig. (1) - Crack bridging   Fig. (2) - Fiber breaking and pull-out

fibers contain many points of weakness and may break in the region ahead and behind the crack tip. The breaks obviously will not coincide with the crack surfaces and this causes a certain amount of fiber-matrix detachment leading to fiber pull-out as the crack opens. This mode of failure is illustrated in Figure 2. So far, there is no theory in fracture mechanics that can adequately analyze crack propagation perpendicular to the fibers unless the fibers and matrix fracture simultaneously along a smooth plane normal to the applied load. The problem, however, is tractable under some if still very restricted conditions. In particular, when crack propagation parallel to the fibers is the predominant failure mode, then the fracture toughness may be defined in terms of a few constants.

To reiterate, microcracking or transverse ply failures occur in all composites at relatively low stress levels, some more than others. This phenomenon can be equated to localized yielding in metallic structures. Continuum fracture mechanics makes no attempt to include all the failure modes in the composite. It adopts the simplistic approach of focusing attention on the onset of the last ligament of unstable fracture and is particularly effective for predicting brittle and catastrophic fractures. Graphite and E-glass fiber composites which distort very little before fracture fall into this category.

TESTING OF COMPOSITE MATERIALS

Techniques of fracture toughness measurements of metals have been well documented [5] and the possibility of using them to characterize the toughness of fiber reinforced composites will be explored. That is, the stress field near the crack tip in a composite can also be characterized by a single stress field parameter [3], K_1, the stress intensity factor. The value of K_1 at the load which triggers rapid crack growth, termed K_{1c}, may provide a useful measure of the resistance of composites to crack propagation. The standard K_{1c} or G_{1c} tests employ the compact tension specimen, Figure 3 and will be used to deter-

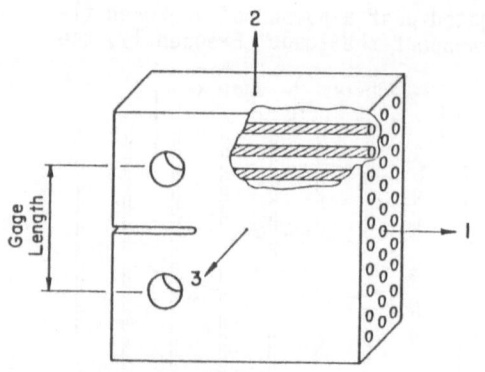

Fig. (3) - Compact tension
specimen

mine the fracture toughness values of glass
and graphite fiber reinforced composites
for different crack sizes under monotonically rising load.

A. Fiber Composite Materials

The composite materials tested include specimens of the resin matrix ERL-
2256/ZZL0820 in which unidirectional E-glass
and unidirectional Thornel 50-graphite fibers are embedded. All specimens were tested with an initial crack inserted in the direction parallel to the fiber. Tested are
the E-glass composite specimens consisting
of 10 percent to 60 percent fiber volume
fraction and graphite composites of 50 percent fiber volume fraction, Type A and B.

Typical values of the elastic constants for a unidirectional E-glass fiber composite are given in Table 1 for different percentages of the fiber volume fraction. Table 2 gives the values of the elastic constants for the graphite fiber
composite specimens.

TABLE 1 - PRINCIPAL ELASTIC CONSTANTS FOR
GLASS FIBER SPECIMENS

Percent of Volume Fraction	Elastic Constants				
	E_1 (psi × 10^6)	E_2 (psi × 10^6)	μ_{12} (psi × 10^6)	ν_{12}	ν_{23}
0	0.50	0.50	0.20	0.35	0.35
10	1.51	0.60	0.24	0.33	0.33
20	2.52	0.66	0.29	0.32	0.32
50	5.55	1.33	0.54	0.28	0.30
60	6.56	1.70	0.68	0.26	0.29

TABLE 2 - PRINCIPAL ELASTIC CONSTANTS FOR
GRAPHITE FIBER SPECIMENS

Percent of Volume Fraction	Elastic Constants				
	E_1 (psi × 10^6)	E_2 (psi × 10^6)	μ_{12}	ν_{12}	ν_{23}
0	0.50	0.50	0.20	0.35	0.35
50A	25.2	1.00	0.55	0.28	0.30
50B	25.2	1.00	0.55	0.28	0.30

B. Compliance Measurement on G_{1c}

This test is designed to determine the G_{1c} value of a cracked specimen by direct measurement of the displacement δ of a certain gage length for varying lengths of cracks and at different critical loads P. The ratio δ/P is the "compliance" of the specimen denoted by C and hence from [5]

$$G_{1c} = \frac{1}{2B} P^2 \frac{\partial C}{\partial a} \qquad (4)$$

where B is the specimen thickness. Cracks of increasing length were cut into the specimens and carefully measured. To prevent premature failure of the specimens at the longer crack lengths, it was necessary to progressively reduce the maximum load. The linear load-displacement curve gives the compliance C which varies as a function of the crack length a. With this information, the value of $\partial C/\partial a$ corresponding to crack instability is found and when substituted into equation (4) renders G_{1c}.

C. Critical Stress Intensity Factor K_{1c}

Fracture tests were run on the same experimental set up that was described previously for compliance measurement. The idea here is to measure critical values of the load P and crack length a at the point of unstable crack extension so that the stress-intensity factor as derived from the linear theory of elasticity [5]

$$K_{1c} = F(a/W) \frac{P\sqrt{a}}{BW} \qquad (5)$$

can be calculated. In the preceding equation, $F(a/W)^*$ is a function depending upon the ratio of the crack length to the width of the specimen W.

D. Relationship Between G_{1c} and K_{1c}

Since the precise relationship between the strain energy release rate G_1 and the stress-intensity factor K_1 of a heterogeneous system such as the fiber reinforced composite is not known and may not exist, it would be informative to investigate the possibility of using an orthotropic model based on equation (2).

Let $K_{1c}^{(1)}$ be the critical stress intensity factor found directly from the fracture test, equation (5) and hence $G_{1c}^{(1)}$ is computed indirectly from equa-

---

*Bowie and Freese [6] have shown that, for a wide range of geometries, the anisotropic stress intensity factor is closely approximated by its isotropic counterpart.

tion (2). Values of $G_{1c}^{(2)}$ are the energy release rate obtained directly from the compliance measurement while $K_{1c}^{(2)}$ is calculated using equation (2). All these are average values for their respective fiber volume fractions. The percentage of deviation compares the experimentally measured values of G_{1c} and K_{1c} using equations (4) and (5). The idea is to check whether equation (2) would hold for a fiber reinforced composite. The deviation is based on the value which was found more directly, that is,

$$\text{Percent deviation of } G_{1c} = \frac{G_{1c}^{(2)} - G_{1c}^{(1)}}{G_{1c}^{(2)}} \times 100 \tag{6}$$

for the energy release rate and

$$\text{Percent deviation of } K_{1c} = \frac{K_{1c}^{(1)} - K_{1c}^{(2)}}{K_{1c}^{(1)}} \times 100 \tag{7}$$

for the critical stress intensity factor.

The data recorded in Table 3 for glass fiber composites indicate that the orthotropic model is not accurate in the low-fiber concentration range.

TABLE 3 - FRACTURE TOUGHNESS VALUES FOR
GLASS FIBER COMPOSITES

Percent of Volume Fraction	Fracture		Compliance		Percent Deviation	
	$G_{1c}^{(1)}$ (1b-in/in²)	$K_{1c}^{(1)}$ (1b/in³/²)	$G_{1c}^{(2)}$ (1b-in/in²)	$K_{1c}^{(2)}$ (1b/in³/²)	G_{1c}	K_{1c}
0	0.75	661	0.77	674	+2.5	- 2
10	4.05	1808	2.98	1533	-36	+15
20	3.63	1710	2.47	1418	-47	+17
50	2.32	1932	2.31	1932	-0.5	0
60	1.97	1950	1.81	1879	-9	+ 4

The nonhomogeneity of the system appears to play a significant role in the fracture process for low density fibrous composites. Note, however, that when the volume fraction of fiber is increased, the orthotropic model is in close agreement with the experimental results.

Results for the graphite fiber composite shown in Table 4 suggest that the orthotropic model does not properly describe the fracture mode for this system. In fact, it was observed during the tests that the graphite composite resulted in a substantial amount of debonding or separation of the fibers from the matrix. The quality control on the graphite specimens is found to be very poor. Hence, a considerable amount of free surfaces are created through fiber separa-

TABLE 4 - FRACTURE TOUGHNESS VALUES FOR
GRAPHITE FIBER COMPOSITES

Percent of Volume Fraction	Fracture		Compliance		Percent Deviation	
	$G_{1c}^{(1)}$ (1b-in/in²)	$K_{1c}^{(1)}$ (1b/in^{3/2})	$G_{1c}^{(2)}$ (1b-in/in²)	$K_{1c}^{(2)}$ (1b/in^{3/2})	G_{1c}	K_{1c}
0	0.75	661	0.77	674	+2.5	- 2
50A	1.37	1356	0.84	1064	-63	+22
50B	0.40	732	0.24	571	-67	+22

tion, which is not accounted for in the analytical model.

EFFECT OF NONHOMOGENEITY

The homogeneous anisotropic elasticity crack model discussed earlier assumes that the crack extends into a material with the combined average properties of the fibers and matrix. On physical grounds, the production of free surface from a crack in the composite can only take place either in the matrix, the fiber or at the interface between the fiber and matrix, i.e., heterogeneity is an inherent feature of cracking in the multi-phase material. In what follows, the discrete nature of the fiber reinforced composite, which is modeled as a cracked isotropic layer sandwiched between the edges of two anisotropic solids. The isotropic layer represents the matrix, and the anisotropic solids are assumed to have the gross mechanical properties of the fiber and matrix. The results obtained from the homogeneous and nonhomogeneous models will be discussed and compared with existing experimental data on E-glass-fiber-epoxy-resin and graphite-fiber-epoxy-resin composites. The objective is to isolate those variables, such as loading angle, composite geometry, physical properties of fiber-matrix composition, etc., which have significant effect on the failure load of the composite. It is desired to have a single parameter that remains effectively constant while the foregoing variables are changed.

The strain energy density criterion [7] will be applied to investigate the fracture of fiber reinforced composites subjected to off-axis loadings*. Briefly, the criterion may be stated in terms of the strain energy density factor S:

Crack initiation is assumed to occur in the radial direction along which the local strain energy density possesses a stationary value.

The direction of crack propagation θ_o can be obtained by requiring that

$$\frac{\partial S}{\partial \theta} = 0 \text{ at } \theta = \theta_o \tag{8}$$

---

*In this case, the crack will not run straight and hence the G_{1c} or K_{1c} approach is no longer valid.

The critical value of S, S_c at θ_o is assumed to determine the onset of unstable crack propagation and is used as an intrinsic material parameter:

$$S_c = S(K_1, K_2) \text{ at } \theta = \theta_o \tag{9}$$

A. Homogeneous Anisotropic Model

Consider an infinite elastic solid whose axes of elastic symmetry coincide with those of the Cartesian coordinates x and y as shown in Figure 4. In

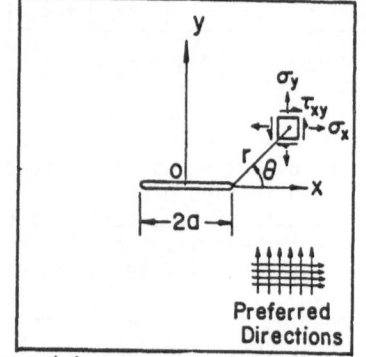

Fig. (4) - A crack in a homogeneous anisotropic elastic solid

an anisotropic material, the direction of crack propagation may not be known beforehand even though the applied load may be normal to the crack. For this reason, the strain energy density criterion is at least in principle applicable, while the classical concept K_1 or G_1 cannot be used in general except for cases where both load and material symmetry prevail.

The stress field in the neighborhood of the crack tip in Figure 4 depends on the elastic constants b_{ij} (i,j = 1,2) in equations (3) and is different from the corresponding crack problem in isotropic elasticity. Refer to [3] for details. The strain energy density factor for the anisotropic material has been derived previously [8] and takes the form

$$S = A_{11}K_1^2 + 2A_{12}K_1K_2 + A_{22}K_2^2 \tag{10}$$

where A_{ij} are complicated functions of b_{ij} and the angle θ in Figure 4. The stress intensity factors K_1 and K_2 in equation (10) are defined such that they correspond to the isotropic case [9]:

$$K_1 = \sigma\sqrt{\pi a} \sin^2\beta$$

$$\tag{11}$$

$$K_2 = \sigma\sqrt{\pi a} \sin\beta \cos\beta$$

in which β is the angle between the uniaxial load and the crack plane, Figure 5.

B. Nonhomogeneous Isotropic Model

The discrete nature of the fiber composite suggests the model of a cracked isotropic layer of matrix material sandwiched between the edges of two semi-infinite orthotropic solids whose gross mechanical properties coincide with those of the composite. In Figure 6, 2h is the layer spacing and 2a the crack

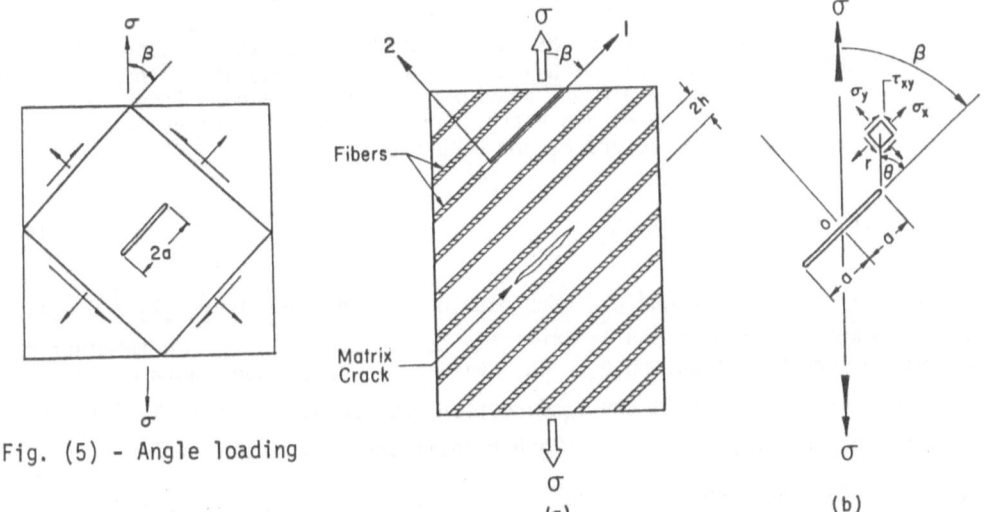

Fig. (5) - Angle loading

Fig. (6) - Off-axis loading of a matrix crack:
(a) crack parallel to fibers, (b) idealized line crack

length. Assuming that the unidirectional composite is transversely isotropic in the 2-3 plane, the orthotropic elastic constants are given by

$$E_1 = E_{f_1}V_f + E_mV_m, \quad E_2 = \left[\frac{E_{f_2}+E_m+(E_{f_2}-E_m)V_f}{E_{f_2}+E_m-(E_{f_2}-E_m)V_f}\right]E_m$$

$$\nu_{12} = \nu_fV_f + \nu_mV_m, \quad \nu_{23} = \nu_fV_f + \left[\frac{1+\nu_m-\nu_{12}E_m/E_1}{1-\nu_m^2+\nu_m\nu_{12}E_m/E_1}\right]\nu_mV_m \tag{12}$$

$$\mu_{12} = \left[\frac{\mu_f+\mu_m+(\mu_f-\mu_m)V_f}{\mu_f+\mu_m-(\mu_f-\mu_m)V_f}\right]\mu_m$$

where the directions 1, 2 and 3 are defined in Figure 3, and

V_f = fiber volume fraction, $\quad\quad\quad\quad$ ν_f = fiber Poisson's ratio
V_m = matrix volume fraction = $1-V_f$, $\quad\quad$ ν_m = matrix Poisson's ratio
E_{f_1} = fiber longitudinal Young's modulus, $\quad$ μ_f = fiber shear modulus $\quad\quad$ (13)
E_{f_2} = fiber transverse Young's modulus, $\quad\quad$ μ_m =, matrix shear modulus

and

E_m = matrix Young's modulus

The criterion of fracture instability assumes that there is an interior element of the matrix material which initiates crack propagation. This is based on the energy density in the element reaching a critical value. This energy quantity, referred to as the strain energy density factor S, can be computed from the stress intensity factors K_1 and K_2 as follows:

$$S = a_{11}K_1^2 + 2a_{12}K_1K_2 + a_{22}K_2^2 \qquad (14)$$

The coefficients a_{11}, a_{12} and a_{22} have been defined previously in [7] and will not be repeated. As mentioned earlier, the condition, $\partial S/\partial \theta = 0$, determines the direction of crack propagation θ_0. The critical value S_c, once determined at the failure of an element in the θ_0 direction, can be used to calculate the failure stress of the composite for any other positions of loading.

The foregoing analysis requires the determination of two functions $\Phi(1)$ and $\Psi(1)$ in the crack tip stress intensity factor expressions

$$K_1 = \Phi(1)\ \sigma\sqrt{\pi a}\ \sin^2\beta$$

$$\qquad (15)$$

$$K_2 = \Psi(1)\ \sigma\sqrt{\pi a}\ \sin\beta\ \cos\beta$$

where the functions $\Phi(1)$ and $\Psi(1)$, depending on the geometry and elastic properties of the composite, are found numerically by solving the Fredholm integral equations given in [10].

C. Comparison of Results

Figures 7 and 8 represent graphical plots of the critical stress as a function of β for the two types of composites mentioned previously. One of the curves refer to the homogeneous model and the other to the nonhomogeneous model. Figure 7 refers to glass fiber composite where the critical crack length is 0.578. Experimental data points [11] are also given for comparison purposes. For $45° < \beta \leq 90°$, the theoretical curves agree with experimental data very well. A slight deviation is observed for $0° < \beta < 45°$. In the absence of experimental data, Figure 8 contains only theoretical curves. The assumed stress at failure $(\sigma_c)_\beta$ is normalized by division by the value at $\beta = 90°$, $(\sigma_c)_{\beta=\pi/2}$. Once the value of $(\sigma_c)_{\beta=\pi/2}$ is determined experimentally, the failure stress at any angle β can be obtained from the figures.

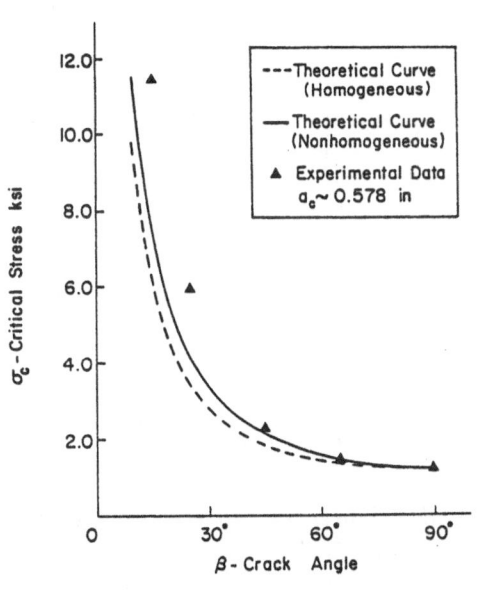

Fig. (7) - Critical stress versus β for glass fiber composite

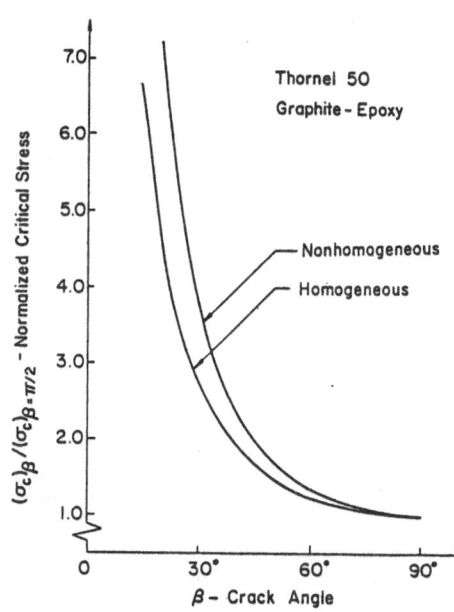

Fig. (8) - Normalized critical stress versus β for graphite fiber composite

The calculated strain energy density factors for the two foregoing models agree reasonably well for composites with low fiber volume fractions and differ significantly for increasing fiber volume fractions. It appears that the condition of homogeneity could be better preserved for composites with low fiber volume fractions. The nonhomogeneous model, however, is considered to be more reliable for situations where the angle between the fiber and load is not too small, so that crack propagation occurs essentially in the matrix material. For small angles of loading, the failure mode of fiber breaking may be important.

ANGLE-PLY LAMINATES

Following the fracture study of unidirectional fibrous composites, it is natural to inquire into the fracture behavior of angle-ply laminates. Figure 9(a) illustrates a four-layer laminate with length $2a^*$ and width $2b$. It is extended uniformly by an amount $\pm u^*$ along the x-axis. Each layer is considered to be a plate element made of unidirectional fibrous composites. These plates are stacked and bonded together with various angles between the fibers in one plate to the next identified by the angles $\pm\beta$ (angle-ply) as illustrated in Figure 10. Upon loading, the entire assembly will act as a single unit in that the load will be transferred from layer to layer through the interfaces. Moreover, the laminate is said to be balanced when the layers are stacked symmetrically with reference to the mid-plane of the laminate as shown in Figures 9(a) and 10.

It has been observed experimentally [12] that balanced laminates, when subjected to uniaxial loading, failed by a combination of thru-lamina and interlami-

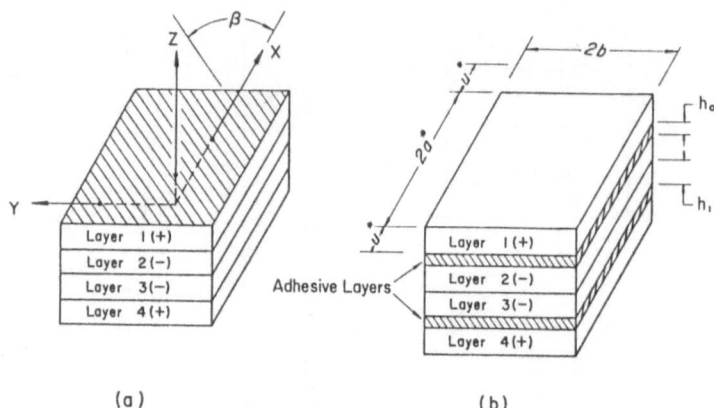

Fig. (9) - Four-layered laminate: (a) idealized model
(b) with adhesive layers

nar (delamination) cracking. These two modes of failure tend to trade off with
one another as the fiber angles are varied.

A. Analytical Model

     The interlaminar cracking between the plies, or delamination, at low fi-
ber angles has been shown [12] to take place within a thin layer of the fibrous
material. Instead of taking this layer as an idealized interface of zero thickness* as in Figure 9(a), it would be more realistic to introduce an additional
layer of material representing the adhesive between the plies and delamination
occurs by breaking the adhesive layer or the material nearby. Figure 9(b) shows
the analytical model of the four-layer laminate containing additional adhesive
layers of thickness h_0 which is small in comparison with the plate element of
thickness h_1. The plate element is assumed to possess three mutually perpendicular directions of material symmetry and hence will be analyzed by the homogeneous,
orthotropic theory of elasticity while the adhesive layer is isotropic in nature.

B. Thru-Laminar Cracking

     For laminates with fiber angles $45° \leq |\beta| \leq 90°$, the dominant fracture mode
is that of cracking through the individual laminae with little or no delamination. Failure of the balanced four-layered glass fiber composite is attributed
to crack propagation initiated from crack-like imperfections inherent in the material. It is assumed that the entire laminate fractures when any one lamina

---

*By definition, delamination cannot occur at the interface where the theoretical
analysis assumes a perfect bond by requiring the stresses and displacements to
be continuous. Hence, only the material next to the interface can possibly
fail. The problem of adhesive and cohesive failure is a subject in itself and
will not be addressed in detail.

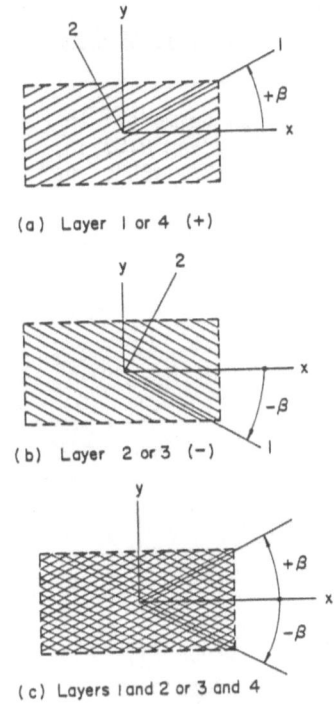

(a) Layer 1 or 4 (+)

(b) Layer 2 or 3 (−)

(c) Layers 1 and 2 or 3 and 4

Fig. (10) - Fiber orientations:
(a) layer 1 or 4 (+),
(b) layer 2 or 3 (−),
(c) layers 1 and 2 or
3 and 4

reaches the point of unstable rapid crack propagation. Thus, the strain energy density criterion of fracture will be applied to a single lamina as in the case of the unidirectional composites.

Since the crack is generally aligned along the fibers making an angle with the tensile axis, the initial segment of crack extension will not be collinear with the main crack and is assumed to take place in a layer of epoxy resin with $E = 4.5 \times 10^5$ psi and $\nu = 0.35$. The layer height 2h is very small in comparison with the crack length 2a so that the stress intensity factors can be approximated by

$$K_1 = 0.29 \ \sigma_2 \sqrt{\pi a}$$

$$K_2 = 0.17 \ \tau_{12} \sqrt{\pi a} \tag{15}$$

which correspond to a glass fiber composite with a fiber volume fraction of 56.5%. The fiber reinforced material surrounding the cracked layer is characterized by the average properties

$$E_1 = 5 \times 10^6 \text{ psi}, \ E_2 = 1.67 \times 10^6 \text{ psi}$$

$$\nu_{12} = 0.05, \ \mu_{12} = 7.04 \times 10^5 \text{ psi} \tag{16}$$

The stresses σ_2 and τ_{12} are referred to the lamina principal axes as indicated in Figure 10. Their expressions are complicated functions of the laminate properties [13] and will not be outlined here.

Once K_1 and K_2 are known, the strain energy density factor S can be found:

$$S = (0.084 \ a_{11}\sigma_2^2 + 0.049 \ a_{12}\sigma_2\tau_{12} + 0.029 \ a_{22}\tau_{12}^2)a \tag{17}$$

where the coefficients a_{ij} (i,j = 1,2) are the same as those for an isotropic material since in this case the crack is assumed to lie in the epoxy resin. Differentiating S with respect to θ embedded in a_{ij} and setting the result equal to zero, a relation between the direction of crack initiation, measured by the fracture angle θ_o, and the fiber angle β is obtained. The results are presented graphically in Figure 11 and when they are used in equation (17), S becomes a material constant S_c.

A comparison between theory and experiment will now be made. In [12], balanced four-layer Scotchply laminate specimens 10 in. long and 0.05 in. thick (0.01 in. per ply) with fiber orientations of ±15°, ±30°, ±45°, ±60°, and ±75°

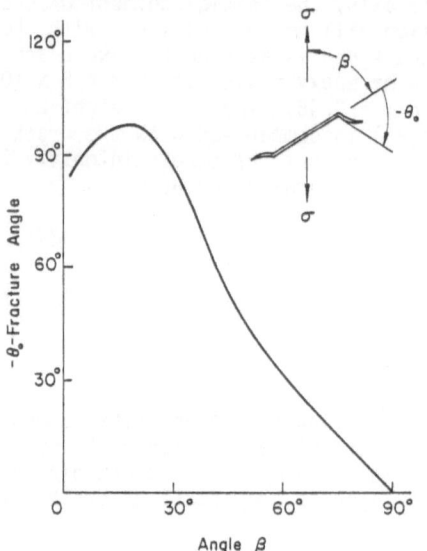

Fig. (11) - Crack growth direction versus
fiber angle for angle-ply
laminates

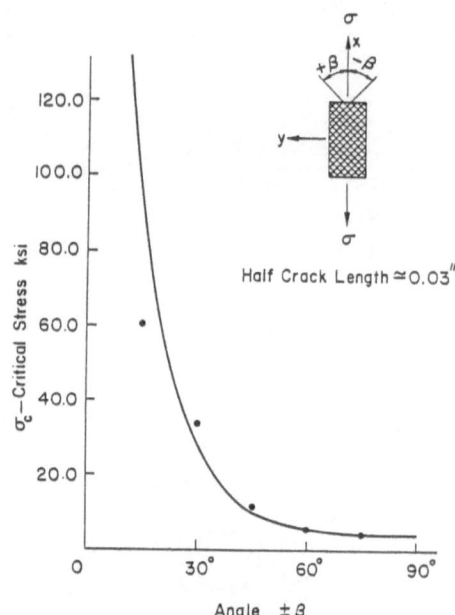

Fig. (12) - Variations of critical
stress with fiber orien-
tation

were tested by loading in the zero direction. The tensile loads at incipient fracture were measured and they are given as round dots in Figure 12 plotted as a function of the fiber angle β. The half crack length in these specimens is estimated to be approximately 0.03 in. The theoretical results represented by the solid curve is obtained from equation (17) by holding S_c at $\beta=90°$ to be constant. Note that the agreement is good for $30° \leq |\beta| \leq 90°$. At $\beta=\pm15°$, the theoretical prediction is somewhat higher than the experimental result. This is because a substantial amount of delamination has occurred at this fiber orientation which is not accounted for in the thru-laminar crack model.

C. Delamination

At low fiber orientations $15° \leq |\beta| \leq 30°$, the specimens exhibited inter-laminar cracking (i.e., cracking between the "+" and "-" layers in Figures 9 prior to complete thru-laminar failure. The region of interlaminar cracking or delamination is triangular in shape as illustrated in Figure 13(a). This scissoring effect was not evident in the high angle fiber orientations. Experiments for angles below $\beta=\pm15°$ are not available. It is anticipated that the failure mode tends to that of the unidirectional composite for small values of β.

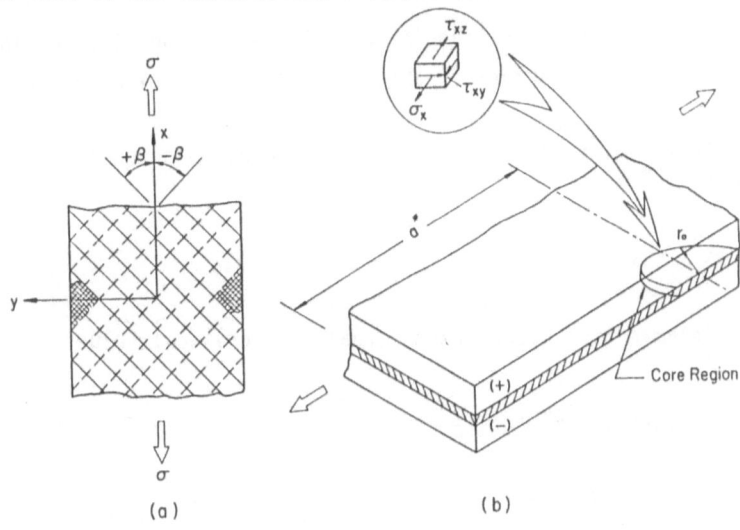

Fig. (13) - Interlaminar cracking: (a) plane view, (b) core region

Delamination failure is, in fact, the propagation of a surface flaw between layers. Since the stress analysis for a surface flaw is overwhelmingly difficult, the S-criterion will be applied to investigate initiation of possible failure without assuming the presence of an initial flaw. In Figure 13(b), reference is made to a core region of radius r_o within which delamination may occur. To this end, a strain energy density factor S is computed

$$S = r_o\left(\frac{dW}{dV}\right) = \frac{r_o}{2\Delta}\left[(\overline{Q}_{22}\overline{Q}_{66}-\overline{Q}_{26}^2)\sigma_x^2 + (\overline{Q}_{11}\overline{Q}_{22}-\overline{Q}_{12}^2)\tau_{xy}^2 + 2(\overline{Q}_{12}\overline{Q}_{26}-\overline{Q}_{16}\overline{Q}_{22})\sigma_x\tau_{xy}\right.$$

$$\left. + \frac{\Delta}{\mu^*}\tau_{xz}^2\right] \tag{18}$$

in which Δ stands for

$$\Delta = \overline{Q}_{11}\overline{Q}_{22}\overline{Q}_{66} + 2\overline{Q}_{12}\overline{Q}_{16}\overline{Q}_{26} - \overline{Q}_{16}^2\overline{Q}_{22} - \overline{Q}_{11}\overline{Q}_{26}^2 - \overline{Q}_{12}^2\overline{Q}_{66} \tag{19}$$

The stiffness coefficients $\overline{Q}_{ij}$ are referred to the x and y axes can be related to Q_{ij} referred to the principal axes 1 and 2 through the angle β as follows:

$$\bar{Q}_{11} = Q_{11} \cos^4\beta + 2(Q_{12}+2Q_{66}) \sin^2\beta \cos^2\beta + Q_{22} \sin^4\beta \tag{20}$$

Similar expressions can be written for $\bar{Q}_{12}$, $\bar{Q}_{16}$ etc. The stresses in equation (18) depend only on the variable y and hence the S-criterion may be stated as

1. Interlaminar cracking occurs at a location determined by the stationary value of S in equation (18), i.e.,

$$\frac{\partial S}{\partial y} = 0 \text{ at } y = y_0 \tag{21}$$

2. Delamination occurs when S reaches a critical value

$$S_c = S(\sigma_x, \tau_{xy}, \tau_{xz}) \text{ for } y = y_0 \tag{22}$$

Figure 14 gives a plot of the normalized S-value in equation (18) against the dimensionless distance y/b along the plane of symmetry where delamination may occur. Indeed, S possesses a minimum value for each β. The radius of the core region on which failure is assumed to occur coinciding with S_{min} increases for small β reaching a maximum at $\beta \approx 35°$ and then decreases as β increases. Experimental data on delamination is lacking. It must be cautioned that the measured values of the critical load or σ_c for $\beta = \pm 15°$ and 30° in [12] at which delamination was observed also include the effect of thru-laminar cracking which is left out in this portion of the analysis.

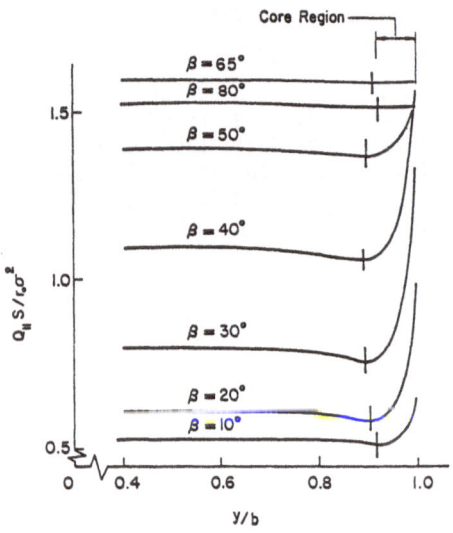

Fig. (14) - Variations of S along the y-direction

IMPACT RESPONSE OF COMPOSITES

There is limited information on the failure of composites due to impact. Procedures for applying fracture mechanics to analyze the dynamics of composites involve a consideration of the modes of failure.

For the purpose of illustration, imagine a multi-layered composite system. One of the layers with thickness 2h contains a crack of length 2a and material properties μ_1, ν_1, ρ_1. The material properties of the outer layers are averaged and enter into the analysis through the constants μ_2, ν_2, ρ_2 as indicated in Figure 15. Here, μ_j is the shear modu-

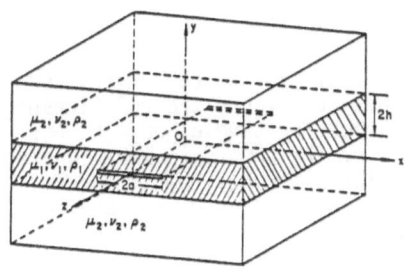

lus, ν_j the Poisson's ratio and ρ_j the mass density of the material. Suppose that the composite is initially at rest and the crack is suddenly opened by a normal stress of magnitude σ_0. This gives rise to the elevation of stress and energy intensity in the immediate vicinity of the crack and the possibility of triggering catastrophic fracture. Such a situation is particularly vulnerable to those high performance composites that are brittle in nature.

Fig. (15) - Crack in a layered composite

Without going into details, the amplification of the local stress field can be measured by the time dependent dynamic stress intensity factor [14]

$$K_1(t) = \frac{\sigma_0\sqrt{\pi a}}{2\pi i} \int_{Br} \frac{\Phi^*(1,p)}{p} e^{-pt} dp \qquad (23)$$

where Br denotes the Bromwich path of integration. The numerical values of $\Phi^*(1,p)$ can be found in [14]. Figure 16 displays the variations of $K_1(t)/\sigma_0\sqrt{\pi a}$

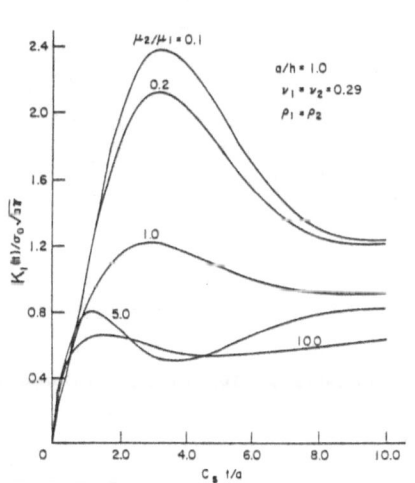

with the dimensionless time variable $c_s t/a$ normalized against the shear wave speed $c_s = \sqrt{\mu_1/\rho_1}$ and the half crack length a. It is assumed that $\rho_1 = \rho_2$, $\nu_1 = \nu_2 = 0.29$ and μ_2/μ_1 takes the values 0.1, 0.2, 1.0, 5.0 and 10.0. Note that $K_1(t)$ tends to rise very quickly at first. This corresponds to the arrival of the reflected waves from the composite interfaces to the crack tip region. All the curves reach a peak and then decrease in amplitude and tend to the static solution for sufficiently long time. In comparison with the homogeneous case of $\mu_1 = \mu_2$, the maximum $K_1(t)$ values are higher for $\mu_2 < \mu_1$ and lower for $\mu_2 > \mu_1$. Hence, the difference in

Fig. (16) - Dynamic stress intensity factor as a function of time for a/h = 1.0 and different ratios of μ_2/μ_1

the shear moduli of the adjoining materials is seen to have a significant influence on the dynamic load transfer to a defect in the composite. The ratio a/h can also play an important role.

These are some of the sensitive parameters that need to be controlled in order to optimize the overall mechanical performance of the composite under impact.

CONCLUSIONS AND FUTURE CONSIDERATIONS

Both theoretical and experimental investigations of fibrous composites containing flaws were presented to determine under what conditions the relationship, equation (2), based on conventional fracture mechanics for single phase homogeneous materials would apply if the discrete nature of the composite were ignored by using the averaged material properties of the fibrous composite. The critical stress intensity factor and energy release rate were measured and analyzed. The results were particularly informative, indicating good agreement with fracture mechanics predictions over the range where the composite responded as a unit. Large discrepancies were encountered when significant separation of the material phases of the composite occur. These discrepancies in energy release rate result because fracture mechanics based on the assumption of homogeneity does not take into account the additional energy necessary to create the free surfaces which result from internal debonding of the composite. Thus, there is a range of conditions over which the assumption of homogeneity can be applied to fibrous composites without substantial error. Beyond this range, energy is transferred to surface energy as the system starts to decompose, and this phenomenon must be accounted for in the analytical model.

Thru-lamina and interlaminar (delamination) cracking were analyzed for a four-layered laminate. These two modes of failure were found to trade off with one another as the fiber angles in the laminate are varied. For the glass fiber composite, a large amount of delamination is observed at low fiber angles within the range $15° < |\beta| < 30°$ and thru-lamina cracking predominates at the higher fiber angles. The two modes of failure tend to interact with each other and their individual contribution cannot be easily sorted out.

An important area for future considerations is the characterization of the conditions under which a composite will respond until fracture without substantial decomposition. In-depth understanding of the response of composites to impact loads is also of paramount importance to the construction of aircrafts and space vehicles.

The analytical modelling of the mechanical interface between the fiber and matrix is another area that needs attention. The usual assumption of continuous stress and displacement across a zero-thickness interface or a layered interface with homogeneous elastic properties may not be realistic since the properties of the fiber-matrix interfaces are known to have a significant influence on the load carrying capacity of the composite. The conditions of adhesive and cohesive failure should be identified and analyzed in terms of the basic properties of the two components of the composite, namely the fiber and matrix.

Aside from cost-effective considerations, it is important to understand the various failure mechanisms that occur in the composite, not only descriptively but quantitatively as well. After all, failure prediction is a prerequisite to the establishment of simplified design procedures. Fracture mechanics should not be solely identified with the assumptions of homogeneity and isotropy as commonly used for metals. With the proper selection of failure criteria, failure in the presence of heterogeneity and anisotropy can be analyzed in accordance with the concept of fracture mechanics which has been illustrated in many of the examples discussed earlier.

REFERENCES

[1]   Linear Fracture Mechanics, edited by G. C. Sih, R. P. Wei and F. Erdogan, Envo Publishing Company, Inc., Bethlehem, Pa., p. 85, 1975.

[2]   Sih, G. C., "Mechanics of Fracture: Linear Response", Numerical Methods in Fracture Mechanics, edited by A. R. Luxmoore and D. R. J. Owen, Swansea, p. 155, 1978.

[3]   Sih, G. C. and Liebowitz, H., "Mathematical Theories of Brittle Fracture", Fracture, edited by H. Liebowitz, Vol. 2, Academic Press, New York, p. 67, 1968.

[4]   Sih, G. C., Hilton, P. D., Badaliance, R., Shenberger, P. S. and Villarreal, G., "Fracture Mechanics of Fibrous Composites", ASTM Special Technical Publication No. 521, p. 98, 1973.

[5]   Plane Strain Crack Toughness Testing of High Strength Metallic Materials, edited by W. F. Brown, Jr. and J. E. Srawley, ASTM Special Technical Publication No. 410, 1966.

[6]   Bowie, O. L., "Solutions of Plane Crack Problems by Mapping Technique", Methods of Analysis and Solutions of Crack Problems, Vol. I. Mechanics of Fracture, edited by G. C. Sih, Noordhoff International Publishing, Leyden, p. 1, 1973.

[7]   Sih, G. C., "A Special Theory of Crack Propagation", Methods of Analysis and Solutions to Crack Problems, edited by G. C. Sih, Noordhoff International Publishing, Leyden, p. XXI, 1972.

[8]   Sih, G. C., Chen, E. P., Huang, S. L. and McQuillan, E. J., "Material Characterization on the Fracture of Filament-Reinforced Composites", Journal of Composite Materials, Vol. 9, p. 167, 1975.

[9]   Sih, G. C., Paris, P. C. and Erdogan, F., "Crack-Tip, Stress Intensity Factors for Plane Extension and Plate Bending Problems", Journal of Applied Mechanics, Vol. 29, p. 306, 1962.

[10]   Sih, G. C. and Chen, E. P., "Fracture Analysis of Unidirectional and Angle-Ply Composites", Technical Report NADC-TR-73-1, IFSM 73-26, Institute of Fracture and Solid Mechanics, Lehigh University, 1973.

[11]   Wu, E. M. and Reuter, R. C., "Crack Extension in Fiber-Glass Reinforced Plastics", University of Illinois TAM Report No. 275, 1965.

[12]   Lauraitis, K., "Failure Modes and Strength of Angle-Ply Laminates", University of Illinois TAM Report No. 345, 1971.

130

[13] Sih, G. C. and Chen, E. P., "Fracture Analysis of Angle-Ply Composites", Proceedings of the 12th Annual Society of Engineering Science, p. 615, 1975.

[14] Sih, G. C. and Chen, E. P., "Impact Response of a Layered Composite with a Crack", IFSM Technical Report, Institute of Fracture and Solid Mechanics, Lehigh University, (in press).

INTERACTION OF CRACKS IN FIBROUS MEDIA

G. Vanin

Institute of Mechanics
Academy of Sciences of the Ukrainian SSR

Fracture of fibrous materials in prolonged and cyclic loading and particularly in the case of weak adhesion between components is manifested in accumulation of local damage which consists of microcracks in region with highest stress concentration and, in particular, near the fiber-matrix border. Subsequently, some of the cracks formed from microcracks that are already in the matrix. This process is initiated at low levels of applied stress long before the fracture of the material as a whole occurs. Changes in the material properties, local redistribution of stresses, interaction and coalescence of microcracks are known to take place prior to material separation.

In this study, the effect of cracks and imperfections on the contact surfaces of fiber-matrix on the properties and brittle fracture of the linearly-reinforced medium is investigated [1,2].

1. An infinite elastic fibrous medium with hexagonal structure is analyzed with perfect contact of components in the field of mean stresses and strains. Averaging of the latter is made within the limits or within the volume of a unit cell shown in Figure 1a. For this structure, the relations of the elasticity law between the mean strains and stresses contain 5 independent constants as in the case of a transversely isotropic body. When a microcrack appears along the entire length of a fiber in the unit cell shown in Figure 1b, this is equivalent to a transformation of the matrix with the elastic parameters grouped as

$$
\left\|
\begin{array}{cc}
\dfrac{1}{G^{\circ}_{12}} & 0 \\[2ex]
0 & \dfrac{1}{G^{\circ}_{12}}
\end{array}
\right\|
\rightarrow
\left\|
\begin{array}{cc}
\dfrac{1}{G_{12}} & \dfrac{\mu_{12,13}}{G_{13}} \\[2ex]
\dfrac{\mu_{13,12}}{G_{12}} & \dfrac{1}{G_{13}}
\end{array}
\right\|
\tag{1}
$$

for the case of longitudinal shear and

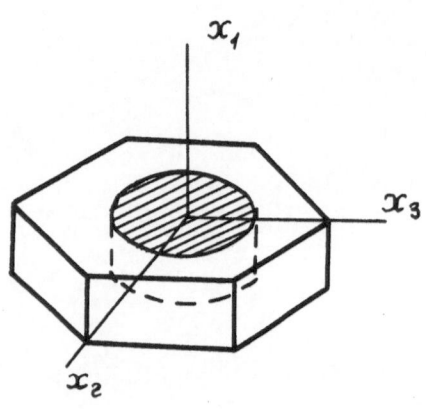

Fig. (1) - (a) Unit cell without a
   crack

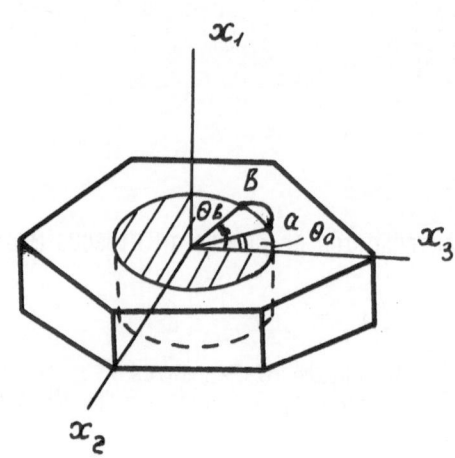

Fig. (1) - (b) Unit cell with a
   crack

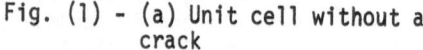

$$
\left\|\begin{array}{cccc}
\dfrac{1}{E^o_{11}} & -\dfrac{v^o_{12}}{E^o_{22}} & -\dfrac{v^o_{12}}{E^o_{22}} & 0 \\[2mm]
-\dfrac{v^o_{21}}{E^o_{11}} & \dfrac{1}{E^o_{22}} & -\dfrac{v^o_{23}}{E^o_{22}} & 0 \\[2mm]
-\dfrac{v^o_{21}}{E^o_{11}} & -\dfrac{v^o_{23}}{E^o_{22}} & \dfrac{1}{E_{22}} & 0 \\[2mm]
0 & 0 & 0 & \dfrac{1}{G_{23}}
\end{array}\right\|
\ \rightarrow\
\left\|\begin{array}{cccc}
\dfrac{1}{E_{11}} & -\dfrac{v_{12}}{E_{22}} & -\dfrac{v_{13}}{E_{33}} & \dfrac{\eta_{1,23}}{G_{23}} \\[2mm]
-\dfrac{v_{21}}{E_{11}} & \dfrac{1}{E_{22}} & -\dfrac{v_{23}}{E_{33}} & \dfrac{\eta_{2,23}}{G_{23}} \\[2mm]
-\dfrac{v_{31}}{E_{11}} & -\dfrac{v_{32}}{E_{22}} & \dfrac{1}{E_{33}} & \dfrac{\eta_{3,23}}{G_{23}} \\[2mm]
\dfrac{\eta_{23,1}}{E_{11}} & \dfrac{\eta_{23,2}}{E_{22}} & \dfrac{\eta_{23,3}}{E_{33}} & \dfrac{1}{G_{23}}
\end{array}\right\|
\qquad (2)
$$

for the case of transverse shear and three-axial tension. The reciprocity relations among the parameters are [3]

$$
\frac{v^o_{12}}{E^o_{22}} = \frac{v^o_{21}}{E^o_{11}}, \quad \frac{1}{G^o_{23}} = \frac{1+v^o_{23}}{2E^o_{22}},
$$

and

$$
\frac{\mu_{12,13}}{G_{13}} = \frac{\mu_{13,12}}{G_{12}}, \quad \frac{v_{12}}{E_{22}} = \frac{v_{21}}{E_{11}}, \quad \frac{v_{13}}{E_{33}} = \frac{v_{31}}{E_{11}}, \quad \frac{v_{23}}{E_{33}} = \frac{v_{32}}{E_{22}}
$$

$$\frac{n_{1,23}}{G_{23}} = \frac{n_{23,1}}{E_{11}}, \quad \frac{n_{2,23}}{G_{23}} = \frac{n_{23,2}}{E_{22}}, \quad \frac{n_{3,23}}{G_{23}} = \frac{n_{23,3}}{E_{33}}$$

The formation of microcracks does not change the form of the basic equations relating the mean strains and stresses, i.e.,

$$<\varepsilon_{ik}> = Z_{iksn} <\sigma_{sn}>, \tag{3}$$

There results, however,

(a) the change of the value of initial parameters

$$Z^{\circ}_{iksn} \rightarrow Z_{iksn};$$

(b) the alteration of the symmetry of the medium, leading to the appearance of 17 parameters instead of 5 constants, and

(c) the appearance of unstable states, corresponding to crack growth associated with certain relations of the stresses and crack dimensions.

The development of brittle fracture of the fibrous materials is reduced to transformation of the Z-matrix, the elements of which are analytical functions of the parameters describing the details of the microstructure. The explicit dependence of the Z-matrix on microstructure is derived from a special procedure [1,2].

2. For the longitudinal shear of a medium with a crack in the form of the arc of a circle (Figure 1b) the following asymptotic relations are derived:

$$\frac{G_{12}}{G_s} = \frac{4(1+\xi\cos\theta+\eta G_s/G_a)^2 - \xi^2\sin^4\theta}{L(\xi,\theta)+\xi^2\sin^4\theta+4\xi(1+G_s/G_a)\sin^2\theta\cos2\alpha} + \cdots$$

$$\frac{G_{13}}{G_s} = \frac{4(1+\xi\cos\theta+\eta G_s/G_a)^2 - \xi^2\sin^4\theta}{L(\xi,\theta)-\xi^2\sin^4\theta-4\xi(1+G_s/G_a)\sin^2\theta\cos2\alpha} + \cdots \tag{4}$$

$$\mu_{12,13} = \frac{4\xi(1+G_s/G_a)\sin^2\theta\sin2\alpha}{L(\xi,\theta)-\xi^2\sin^4\theta-4\xi(1+G_s/G_a)\sin^2\theta\cos2\alpha} + \cdots$$

where ξ and $\eta=1-\xi$ are volumetric content of the fiber and matrix while $2\theta=\theta_b-\theta_a$ and $2\alpha=\theta_b+\theta_a$. The indices "a" and "s" refer to quantities associated with the fiber and matrix,

$$L(\xi,\theta) - 4[1 - \xi^2\cos\theta + 2(1+\xi^2\cos\theta)G_s/G_a + (1-\xi^2)(G_s/G_a)^2]$$

The above relations indicate that effective parameters are monotonic functions of the angle θ, and

$$\lim_{\theta\to 0} (G_{ik}-G^\circ_{ik}) = 0, \quad \lim_{\theta\to 0} \frac{\partial}{\partial\theta} (G_{ik}-G^\circ_{ik}) = 0,$$

Consequently, for sufficiently small cracks, the shear moduli do not change appreciably.

The condition of unstable state of the medium with cracks is a consequence of the Griffith's criterion [4,5]

$$\frac{\partial}{\partial\theta} (4\lambda\gamma\theta) = \frac{F}{2} (<\sigma_{12}>^2 \frac{\partial}{\partial\theta} \frac{1}{G_{12}} + <\sigma_{13}>^2 \frac{\partial}{\partial\theta} \frac{1}{G_{13}} + <\sigma_{12}><\sigma_{13}> \frac{\partial}{\partial\theta} \frac{\mu_{12,13}}{G_{13}})$$

in which λ is the fiber radius, F the cross-sectional area of the unit, γ a constant expressing the resistance of the medium to crack growth. The last equation may be conveniently expressed in the form

$$\frac{<\sigma_{12}>^2}{a^2} + \frac{<\sigma_{13}>^2}{b^2} + \frac{<\sigma_{12}><\sigma_{13}>}{\kappa^2} = 1, \tag{5}$$

where

$$a^2 = \frac{8\lambda\gamma}{F} \frac{\partial}{\partial\theta} (\frac{1}{G_{12}}), \quad b^2 = \frac{8\lambda\gamma}{F} \frac{\partial}{\partial\theta} (\frac{1}{G_{13}}), \quad \kappa^2 = \frac{4\lambda\gamma}{F} \frac{\partial}{\partial\theta} (\frac{\mu_{12,13}}{G_{13}})$$

In the case of symmetrically distributed cracks, with $\theta_a=\theta_b$, $\alpha=0$, $\theta=\theta_b$ and $\mu_{12,13}=0$, numerical analysis was made to determine the gross mechanical properties of the glass-reinforced plastic with $\nu_a = 0.2$, $E_a = 7.10^{10}N/m^2$, $\nu_s = 0.382$ and $E_s = 0.315 \times 10^{10}N/m^2$ for hexagonal and tetragonal microstructures. The asymptotic relations simplify to the form

$$\frac{G_{12}}{G_s} = \frac{2(1+\xi\cos\theta+\eta G_s/G_a)-\xi\sin^2\theta}{2(1-\xi\cos\theta)+2(1+\xi)G_s/G_a+\xi\sin^2\theta} + \cdots$$

$$\frac{G_{13}}{G_s} = \frac{2(1+\xi\cos\theta+\eta G_s/G_a)+\xi\sin^2\theta}{2(1-\xi\cos\theta)+2(1+\xi)G_s/G_a-\xi\sin^2\theta} + \cdots \tag{6}$$

Curves 1, 3 and 2,4 in Figures 2a, 3a and 4a illustrate the relation between shear moduli G_{12} and G_{13} for $\xi = 0.5$, 0.6 and 0.7 of the hexagonal (solid lines)

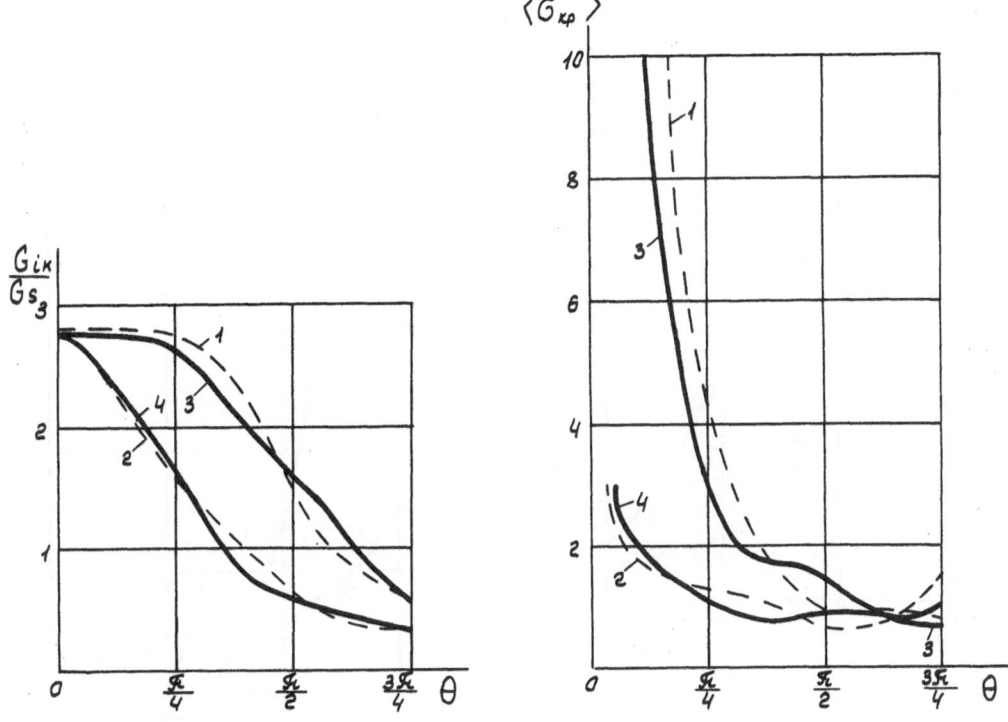

Fig. (2) - (a) Dependence of longitudinal shear moduli on the angle θ
for ξ = 0.5 and (b) stress intensity variations with θ for
ξ = 0.5

and tetragonal structures and the angle θ. The corresponding curves in Figures 2b, 3b and 4b characterize changes in stress with increase of θ for the same structures. It follows from the curves that

(1) Tetragonal structure of equal G ensures a higher hardness of the material.

(2) The effect of the type of fiber packing upon the shear modulus is insignificant at ξ < 0.6.

(3) The stable state of the crack for tetragonal structure occurs at the angle 2θ=π, for hexagonal one at 2π/3. It increases with the rise of hexagonal microstructure. This is due to the increased localization of cracks at shear.

3. For materials with symmetrically arranged defects such as the structure considered here, it is possible to derive approximate relations for the determining of other mechanical properties. The asymptotic relation for the trans-

136

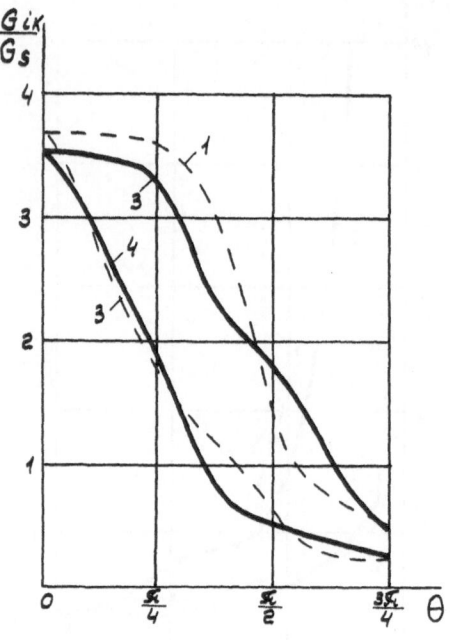

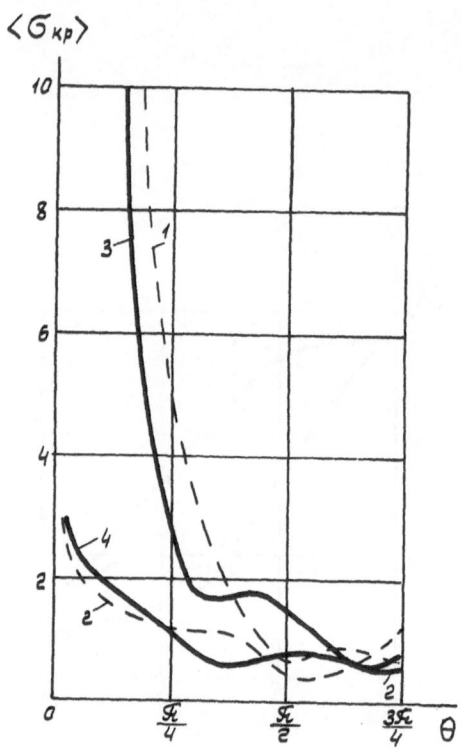

Fig. (3) - (a) Dependence of longitudinal shear moduli on the angle θ
for ξ = 0.6 and (b) stress intensity variations with θ for
ξ = 0.6

verse shear modulus as a first approximation is

$$\frac{G_{23}}{G_s} = \frac{\eta D_s + \xi D_a + \xi Nf}{\eta D_s + \xi [D_a + Nf] G_s / G_a} + \ldots \tag{7}$$

where

$$D_s = \frac{1-g}{1+\kappa_s} (1 + \kappa_s G_s / G_a), \quad f = (\tfrac{1}{2} + 2\beta^2) \sin^2\theta,$$

$$D_a = 1 + \frac{\Omega_c}{\mu(0)} + \frac{1}{1-g} \frac{f}{G} [\frac{\Omega_0}{2} \mu''(0) - \mu'(0) - 1] - \frac{\Omega_c}{6 (0)} \mu'''(0) + \frac{\mu^{iv}(0)}{24 (0)};$$

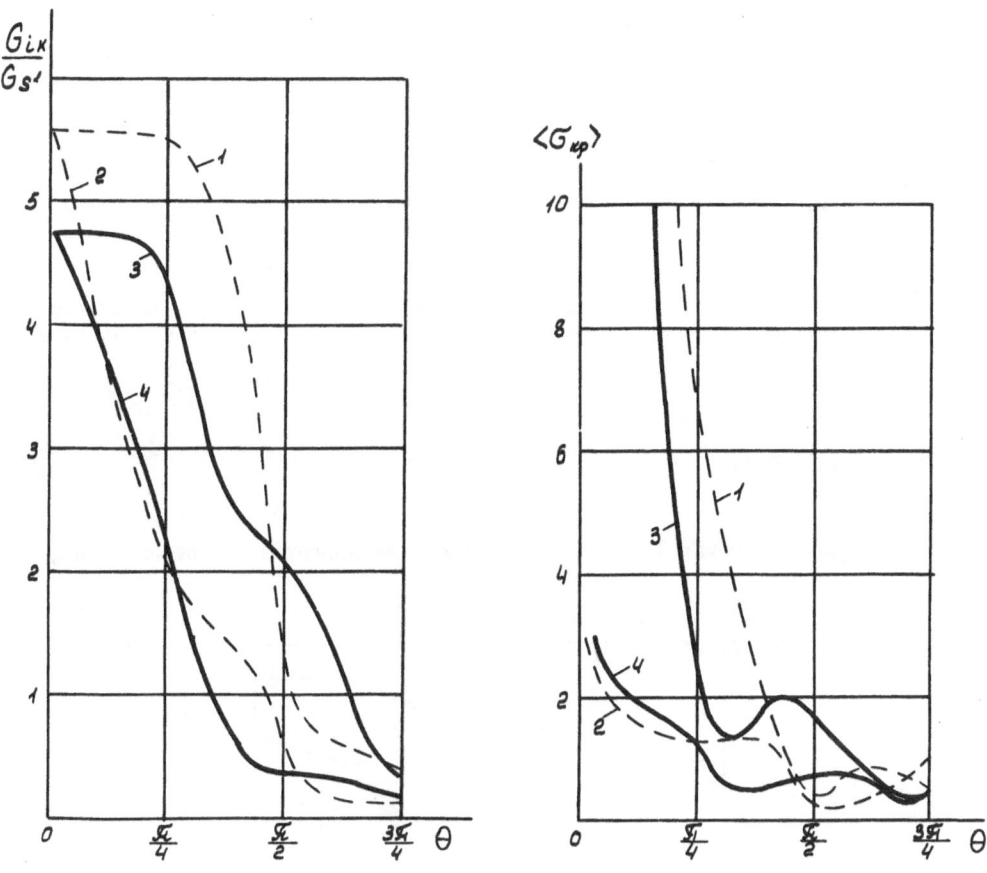

Fig. (4) - (a) Dependence of longitudinal shear moduli on the angle θ
for ξ = 0.7 and (b) stress intensity variations with θ for
ξ = 0.7

$$N = 4\cos^2\theta - 1 + 2 \frac{\mu'(o)}{\mu(o)} \cos\theta + \frac{\mu'(o)}{2\mu(o)} + \frac{1}{1-g} \frac{f}{G} [\mu'(o) + 2\mu(o)\cos\theta]$$

$$+ \frac{1}{2f} [\frac{\mu^{iv}(o)}{12\mu(o)} - \frac{\mu'''(o)}{3\mu(o)} \Omega_e]$$

Further,

$$g = e^{2\pi\beta} = -\frac{\kappa_s + G_s/G_a}{1 + \kappa_a G_s/G_a}; \quad \Omega_o + i\Omega_e = (1+i)\cos\theta + 2(1-i)\beta\sin\theta;$$

$$\mu(z) = \frac{1}{z - e^{-i\theta}} - \left(\frac{z - e^{-i\theta}}{z - e^{i\theta}}\right)^{\frac{1}{2} - i\beta} = \mu(o) + z\mu'(o) + \ldots$$

$$G = \frac{1 + \mu(o)\Omega_o}{1 - g}$$

In Figure 5, curve 3 gives the approximate alteration of the transverse modulus for various angles of crack opening. Stresses at the crack tip have the singularity

$$\frac{1}{\sqrt{\rho}} \left\{ \begin{matrix} \sin \\ \cos \end{matrix} \right\} (\Delta \ell n \rho), \tag{8}$$

where Δ = const. Curves 1 and 2 characterize the approximate change of G_{12} and G_{13}.

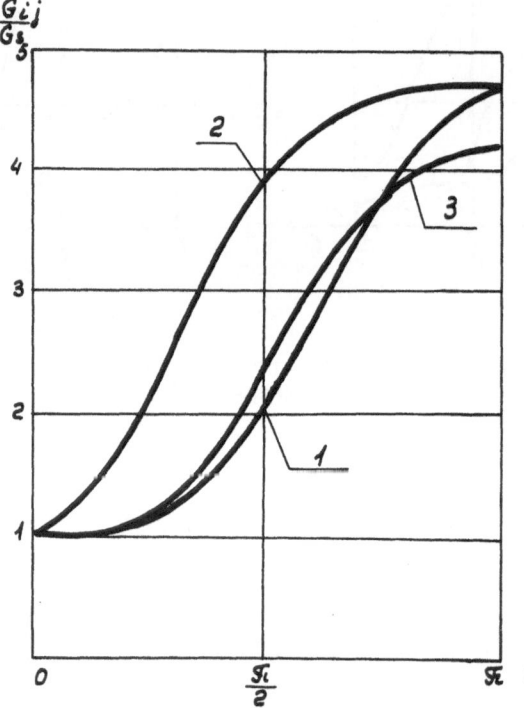

Fig. (5) - Dependence of shear moduli on θ for $\xi = 0.7$

Approximate relations for determining the elastic constants for transverse tension are obtained in the form

$$\frac{1}{E_{12}} = \frac{\nu_{21}}{E_{11}} + \frac{1+\kappa_s}{2G_s} \frac{(\kappa_s+G_s/G_a)\eta+[2(\kappa_a-1)Q+L]\xi G_s/G_a}{(1+\kappa_s)(4Q+L)\xi+4\eta(\kappa_s+G_s/G_a)} + \dots$$

$$(9)$$

$$\frac{\nu_{23}}{E_{22}} = -\frac{\nu_{21}\nu_{31}}{E_{11}} + \frac{1}{2G_s} \frac{4\nu_s(\kappa_s+G_s/G_a)\eta-[2(\kappa_s-1)Q-L]\xi G_s/G_a}{(1+\kappa_s)(4Q+L)\xi+4\eta(\kappa_s+G_s/G_a)} + \dots$$

Here

$$Q = 1 + \frac{2(1-g)f-(1-g)^2 G-2[1+\mu(o)\Omega_o][\kappa_s+G_s/G_a+(\kappa_s+1)G_s/G_a]}{2(1-g)(1+\kappa_a)GG_s/G_a+4\kappa_s+4G_s/G_a};$$

$$L = R-P + S \frac{\kappa_s+G_s/G_a+(\kappa_a+1)G_s/G_a}{(1-g)(1+\kappa_a)GG_s/G_a+2\kappa_s+2G_s/G_a};$$

$$R = 2f(4\cos^2\theta - 1 + 2\frac{\mu'(o)}{\mu(o)}\cos\theta + \frac{\mu''(o)}{2\mu(o)} - \Omega_e\frac{\mu'''(o)}{3\mu(o)} + \frac{\mu^{iv}(o)}{12\mu(o)};$$

$$P = \frac{2}{\mu(o)}(\Omega_e + \mu(o) + \frac{\Omega_e}{6}\mu'''(o) - \frac{\mu^{iv}(o)}{24};$$

$$S = 2f[\mu'(o) + 2\mu''(o)\cos\theta] + \Omega_e\mu''(o) - 2\mu'(o) - 2f$$

The modulus of elasticity for transverse tension of the medium along the longitudinal axis is

$$\frac{1}{E_{33}} = \frac{\nu_{31}^2}{E_{11}} + \frac{1+\kappa_s}{2G_s} \frac{(\kappa_s+G_s/G_a)\eta-[2(\kappa_a-1)B+C]\xi G_s/G_a}{4(\kappa_s+G_s/G_a)\eta-(4B+C)(1+\kappa_s)\xi}$$

$$(10)$$

where

$$B = f - \frac{1}{2}(1+\kappa_a G_s/G_a)\frac{2f+(1-g)G}{2(1+\kappa_a G_s/G_a)-(1+\kappa_a)GG_s/G_a}$$

$$+ [1 + \Omega_o\mu''(o)]\frac{\kappa_s+G_s/G_a-(1+\kappa_a)G_s/G_a}{(1-g)(1+\kappa_a)GG_s/G_a+2\kappa_s+2G_s/G_a};$$

$$C = P-R + S \frac{\kappa_s + G_s/G_a - (1+\kappa_a)fG_s/G_a}{(1-g)(1+\kappa_s)GG_s/G_a + 2\kappa_s + 2G_s/G_a}$$

The results of calculations of some of the mechanical properties of glass-reinforced plastics with cracks are presented by curves in Figures 6 and 7. Curves 1 and 2 (Figure 6) and 1 (Figure 7) represent change in E_{22} and E_{33}, as well as ν_{23} relative to the increase of crack opening 2 . Stresses at the crack tips have a singularity of the form in equation (8).

The tensile modulus in longitudinal direction is given by the relation

$$E_{11} = \xi E_a + \eta E_s + \frac{8\xi\eta(\nu_a - \nu_s)^2 dG_s}{(1+\kappa_s)\xi d + 2(1+\kappa_a G_s/G_a)\eta - (1+\kappa_a)\eta dG_s/G_a} \qquad (11)$$

where

$$d = \frac{1}{1-g} [1 - g(\cos\theta + 2\beta\sin\theta)],$$

The transverse effects are accounted for by

$$\nu_{21} = \nu_s + \xi(\nu_a - \nu_s)$$

$$- \frac{\xi\eta(\nu_a - \nu_s)\{[q + 2f\mu(o)(3\cos\theta - 2\beta\sin\theta) - P]G_s/G_a - (1-g)(\kappa_s - 1)d\}}{(1-g)[(1+\kappa_s)\xi d + 2\eta(1+\kappa_a G_s/G_a) - \eta(1+\kappa_a)dG_s/G_a]}$$

$$\nu_{31} = \nu_s + \xi(\nu_a - \nu_s)$$

$$(12)$$

$$- \frac{\xi\eta(\nu_a - \nu_s)\{[q + 2f\mu(o)(3\cos\theta - 2\beta\sin\theta) - P]G_s/G_a - (1-g)(\kappa_s - 1)d\}}{(1-g)[(1+\kappa_s)\xi d + 2\eta(1+\kappa_a G_s/G_a) - \eta(1+\kappa_a)dG_s/G_a]}$$

where

$$q = (\kappa_a - 1)[1 + \Omega_o\mu(o)]; \quad p = 2\mu'(o) - \Omega_o\mu'''(o) - 2f$$

The change of E_{11} and ν_{21}, ν_{31} with increasing θ is illustrated by curve 3 in Figure 6 and curves 2 and 3 in Figure 7. Stresses have a singularity as in equation (8) at the crack tip. Longitudinal cracks have a weak effect on the values of the elastic constants under longitudinal tension. For the medium with symmetric cracks, the number of essentially independent constants is 9. Other types of microcracks of the composite structure are also analyzed by the above mentioned method. These are three-dimensional defects which can be represented by a set of interacting two-dimensional cracks with variable dimensions.

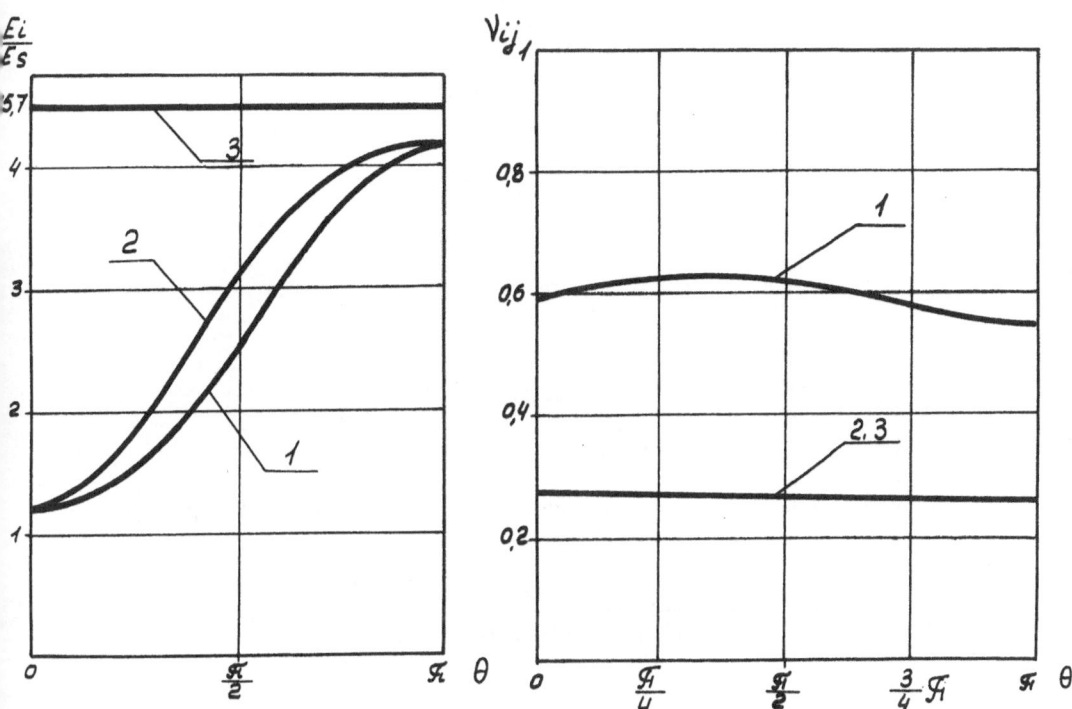

Fig. (6) - Dependence of longitudinal normal moduli on θ for ξ = 0.7

Fig. (7) - Variations of Poisson's ratios with θ for ξ = 0.7

4. A discussion is made on the relation between the stress intensity at the ends of microcracks and the mean stresses near a macrocrack in the plate made of linearly reinforced glass-fiber plastic. It is assumed that the material has a hexagonal structure, Figure 1a, with a symmetric crack in each unit cell occurring at the border of contact between the fiber and the matrix. We assume, further, that the macrocrack has a form of a flattened ellipse with semi-axis a and is oriented perpendicular to the fiber, Figure 8. When the plate is in uniform tension with stresses $<\sigma^\circ_{11}>$ at infinity along the direction of fiber packing the mean stress concentration at the tip of the macrocrack is [6,7]

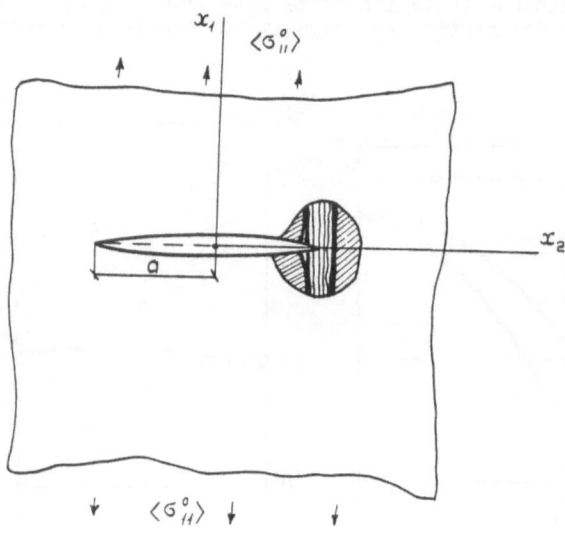

Fig. (8) - A possible form of local failure near the crack tip region

$$\langle\sigma_{11}\rangle = \langle\sigma_{11}^{o}\rangle \sqrt{\frac{a}{2R}} \, Re[\frac{s_1 s_2}{s_1 - s_2} (\frac{s_2}{\sqrt{\cos\phi + s_2 \sin\phi}} - \frac{s_1}{\sqrt{\cos\phi + s_1 \sin\phi}})],$$

(13)

$$\langle\sigma_{12}\rangle = \langle\sigma_{11}^{o}\rangle \sqrt{\frac{a}{2R}} \, Re[\frac{s_1 s_2}{s_1 - s_2} (\frac{1}{\sqrt{\cos\phi + s_2 \sin\phi}} - \frac{1}{\sqrt{\cos\phi + s_1 \sin\phi}})],$$

where R is the macroscopic distance from the crack tip, ϕ is the inclination angle of the area, and

$$s_1 s_2 = - \sqrt{\frac{E_{11}}{E_{22}}}, \quad s_1 + s_2 = i\sqrt{2(\sqrt{\frac{E_{11}}{E_{22}}} - \nu_{21} + \frac{E_{11}}{2G_{12}})}$$

Two cases of plate fracture are discussed:

(1) The plate material has a high strength of adhesion between its components. There are no microcracks at the interfaces.

(2) The material contains damaged regions at the contact areas, caused by preloading or by weak adhesion (cohesion) between the matrix and fibers.

The mean stress opening the crack in the first case is

$$<\sigma_{11}>^{\circ}_{B} = <\sigma^{\circ}_{11}> \sqrt{\frac{a}{2R}} \sqrt{\frac{E^{\circ}_{11}}{E^{\circ}_{22}}} = \frac{K_1}{\sqrt{2r}} \sqrt{\frac{E^{\circ}_{11}}{E^{\circ}_{22}}}$$

The constant K_1 is determined essentially by the resistance of fibers to the transverse growth of the crack. In the second case, the critical stresses according to the same fracture criterion will be higher due to a reduction of the elastic modulus E_{22} and the condition $E^{\circ}_{11} \approx E_{11}$

$$<\sigma_{11}>_{B} = <\sigma_{11}>^{\circ}_{B} \sqrt{\frac{E^{\circ}_{22}}{E_{22}}} > <\sigma_{11}>^{\circ}_{B}$$

For that material, a probable type of fracture is the breaking of fibers when the driving force are the tangential stresses $<\sigma_{12}>$. True stresses near the crack tip at longitudinal shear

$$\sigma_{12} = <\sigma_{12}> \sqrt{\frac{A\sin\theta}{\rho}} \left(\frac{\sin\frac{\theta}{2}}{1+\xi\cos\theta-0.5\xi\sin^2\theta+\eta G_s/G_a} + \ldots \right) \tag{14}$$

are increasing near the macrocrack in accordance with equation (13) and

$$<\sigma_{12}> = <\sigma^{\circ}_{11}> \sqrt{\frac{a}{2R}} \sqrt{\frac{A\sin\theta}{\rho}} \, Re \, \frac{\sqrt{S_1 S_2}}{\sqrt{S_1}+\sqrt{S_2}} \left(\frac{\sin\frac{\theta}{2}}{1+\xi\cos\theta-0.5\xi\sin^2\theta+\eta G_s/G_a} + \ldots \right) \tag{15}$$

where ρ is the distance from crack tip along the circular border of the fiber. From equation (15), it is evident that the intensity of tangential stresses at the microcrack tip is related to the mean tensile stress through the product of the intensity of mean stresses and the intensity of microstresses. The first factor characterizes the effect of the change of mean stresses, that is determined by the characteristics of the medium and by geometrical features of macrodefect, while the second factor characterizes the features of the microstructure.

It is evident that the parameters controlling the contact between the fiber and matrix can lead to a wide range in both the mechanical properties and the process of fracture of fibrous media. Optimal conditions of contact depend on the nature of loading and fracture of the material governed by the effective fiber length and other parameters.

REFERENCES

[1] Vanin, G. A., "A new method of considering interactions in the theory of composite systems", Dokladi Akademii Nauk, Ser. A., No. 4, pp. 321-324, 1976.

[2] Vanin, G. A., "On the theory of fibrous media with imperfections", Prik-ladnaya mechanika, No. 10, pp. 14-22, 1977.

[3] Lechnitsky, S. G., "The theory of elasticity of an anisotropic body", Moscow, 1977.

[4] Barenblatt, G. I., "Mathematical theory on cracks formed in brittle frac-ture", PMTF, No. 4, p. 68, 1961.

[5] Cherepanov, G. P., "Mechanics of brittle fracture", Moscow, 1974.

[6] Savin, G. N., "Stress concentration near holes", Moscow, 1951.

[7] Lechnitsky, S. G., "Anisotropic plates", Moscow, 1957.

OBSERVATIONS OF THREE DIMENSIONAL GEOMETRIC EFFECTS IN CRACK GROWTH AND
IMPLICATIONS FOR COMPOSITE MATERIAL STRUCTURES

C. W. Smith

Department of Engineering Science and Mechanics
Virginia Polytechnic Institute and State University

ABSTRACT

After reviewing experimental techniques employed by the author to
measure stress intensity factor distributions and flaw geometries in
cracked photoelastic models, the results obtained from applying the
techniques to a three dimensional cracked body of current technological
interest will be presented. Finally implications of these results in
applying fracture mechanics to polyphase (composite) materials will be
noted.

INTRODUCTION

Substantial progress has been made in recent years in the generation of stress intensity solutions for a broad class of two and three
dimensional problems [1], [2], [3], [4]. However. except for a very
few cases, such as [5], [6], full field solutions to three dimensional
cracked body problems still present formidable mathematical difficulties to those seeking the closed form result for cracked bodies of
finite geometry. Problems such as the surface flaw are still receiving
intensive study, [7], [8] and, with improved computational facilities,
many numerical solutions are being proposed [9]. For such problems,
however, rigorous convergence proofs are often difficult to obtain and
independent computer code verification becomes desirable, if not manda-
tory.
Nearly a decade ago, the author and his associates began an effort
to adapt an idea proposed by Irwin [10] for measuring stress intensity
factors (SIFs) photoelastically to the "frozen stress" analysis of
three dimensional cracked bodies. These studies evolved from an initial
effort [11] into a current approach described in [12], [13] and [14].
The technique capitalizes upon the diphase optical and mechanical properties of transparent polymeric materials and effects a marriage between
a simple programmed least squares data analysis and the photoelastic
data above "critical" temperature for estimating SIF distributions in
three dimensional problems.
Very recently, as the result of a cooperative program [15], [16]
with researchers in the laboratory for Thermal Power Engineering at the
Delft University of Technology, the author found that cracks grown in
polymeric photoelastic models at critical temperature in complex three
dimensional geometries under monotonically increasing loads exhibited
shapes virtually identical to those obtained from tension-tension fatigue
tests on geometrically similar steel models, that flaw growth was non-self-
similar, and crack shapes were more complex than the simple curve shapes
assumed by numerical analysts in modeling the problem with finite element
techniques. Moreover, the studies implied that arbitrarily assumed simple
flaw shapes generate SIF distributions with significantly higher gradients

along the flaw borders than are found with "natural" cracks of more complex shapes implying that natural flaw shapes tend to minimize SIF gradients along the flaw border for such problems. The SIF distribution along natural flaws will not be uniform, however, due to varying constraint conditions along the flaw border. Since it is well known [17], [18] that constraint variations exist even in "two dimensional" problems, and that such variations constitute three dimensional effects, the implication here then is that, for fatigue crack growth, a spectrum of values of threshold SIF's lie along the flaw border which must be reached for flaw growth to continue.

It may be conjectured that such complex geometric effects as those described above in single phase materials may also influence stress intensity distributions and crack growth phenomena for cracks embedded in the matrix of a polyphase (composite) material.

After reviewing the techniques employed by the author to measure SIF distributions in photoelastic models and basic concepts regarding self similar and non-self similar fatigue crack growth, the concepts described above will be experimentally quantified for a geometrically complex 3D cracked body problem of current technological interest. Finally, implications of these results in applying fracture mechanics to composite materials will be identified.

METHOD OF ANALYSIS

Before discussing the use of the method, a brief review of its foundations would not seem inappropriate.

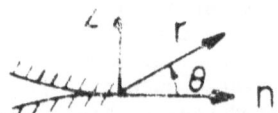

A) Analytical Background – A Brief Review

For the case of Mode I loading, one begins with equations of the form:

$$\sigma_{ij} = \frac{K_I}{r^{1/2}} f_{ij}(\theta) + \sigma_{ij}^o \qquad (1)$$

for the stresses in a plane mutually orthogonal to the flaw surface and the flaw border referred to a set of local rectangular cartesian coordinates as pictured in Figure 1, where the terms containing K_I, the SIF, are identical to Irwin's Equations for the plane case and σ_{ij}^o represent the contribution of the regular stresses to the stress field in the measurement zone. The σ_{ij}^o are normally taken to be constant for a given point along the flaw border, but may vary from point to point. Observing that stress fringes tend to spread approximately normal to the flaw surface (Figure 2), Eqs. 1 are evaluated along $\theta = \pi/2$ (Figure 1) and

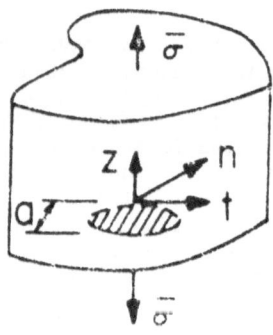

Figure 1 – Problem Geometry and Notation for Mode I Loading

$$\tau_{max} = 1/2 \ [(\sigma_{nn} - \sigma_{zz})^2 + 4\sigma_{nz}^2]^{1/2} \qquad (2)$$

which, when truncated to the same order as Equations (1), leads to the two parameter Equation:

$$\tau_{max} = \frac{A}{r^{1/2}} + B \text{ where } \begin{array}{l} A = K_I/\sqrt{8\pi} \\ B = f(\sigma_{ij}^{0}) \end{array} \quad (3)$$

which can be rearranged into the normalized form

$$\frac{K_{AP}}{q(\pi a)^{1/2}} = \frac{K_I}{q(\pi a)^{1/2}} + \frac{f(\sigma_{ij}^{0})(8)^{1/2}}{q} \left\{\frac{r}{a}\right\}^{1/2} \quad (4)$$

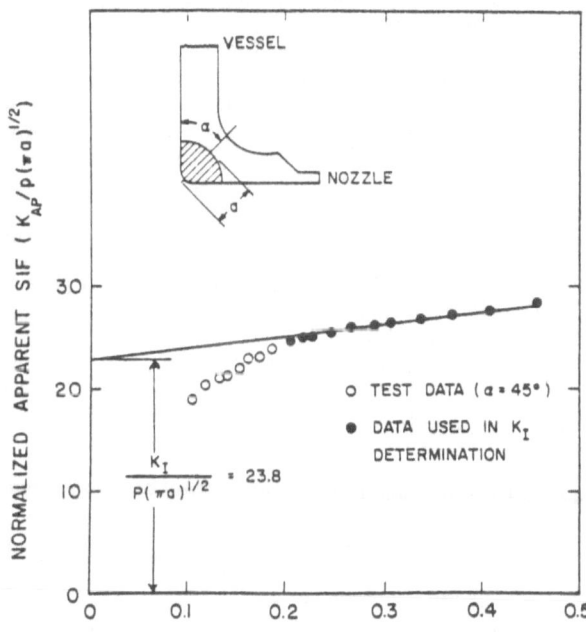

Figure 2 - Mode I Stress Fringe Pattern

where $K_{AP} = \tau_{max}(8\pi r)^{1/2}$ and, from the Stress-Optic Law, $\tau_{max} = Nf/2t'$ where N is the stress fringe order, f the material fringe value and t' the slice thickness in the t direction, q is the remote loading parameter (such as uniform stress, pressure, tec.) and a the characteristic flaw depth. Equation (4) prescribes that, within the zone dominated by Equations (1) with σ_{ij}^{0} as described above, a linear relation exists between the normalized apparent stress intensity factor and the square root of the normalized distance from the crack tip. Thus, one need only locate the linear zone in a set of photoelastic data and extrapolate across a very near field non-linear zone [19][20] to the crack tip in order to obtain the SIF. An example of this approach using data from nozzle tests described in the sequel is given in Figure 3.

By following similar arguments, but not specifying $\theta = \pi/2$, equations for the mixed mode case can als be developed [14].

B) Frozen Stress Method Applied to Cracked Bodies

The frozen stress method was introduced by Oppel [21] in 1937. It capitalized upon the observation that certain transparent materials exhibit both birefringence and diphase mechanical behavior.

Figure 3 - Estimating SIF Value from Test Data - Typical Result

Such materials respond to load in an anelastic manner when loaded at
room temperature but above a certain temperature, called "critical",
the anelastic effect vanishes and the material becomes linearly elastic
and incompressible i.e. (Poisson's Ratio → 0.5). All loads are applied
above critical temperature and bodies are then cooled under load, "freezing" in both the deformation and fringe patterns obtained above critical
temperature so that they remain in the model at room temperature even
after unloading and slicing. Above "critical", the material modulus is
typically 1% of its room temperature value and the material fringe
value is typically 4% of its room temperature value.

Cracks to be studied are grown from starter cracks. To insert a
starter crack, a sharp blade is fixed in contact with and normal to the
surface of the body at room temperature (or below) at the desired initiation locus and is struck, producing a small crack under the blade. This
crack grows when loaded above critical temperature in a principal plane
and takes the shape which apparently tends to minimize the SIF gradient
along the flaw border. When the crack reaches its desired size, the load
is reduced to stop growth and cooling is carried out "freezing in" stress
fringes and deformation fields due to the reduced load. It has been found
that cracks produced in this way exhibit geometric similarity with those
produced in A508 reactor vessel steel models under tension-tension fatigue
loading [22] provided all of the flaw growth occurs in the same plane.

In applying the method, cracks of progressively larger size are
grown and "frozen" into otherwise identical photoelastic models from which
slices are removed mutually orthogonal to the flaw border and flaw surface at intervals along each flaw border. Photoelastic data are then
obtained along $\theta = \pi/2$ (Figure 1) for the Mode I case using a crossed
circular polariscope with white light, reading tint of passage and
utilizing the Tardy Method at about 15X. These data are reduced through
a simple least squares computer program, yielding estimates of the SIF
along the flaw borders.

In summary, the method possesses the potential for simultaneously
providing natural flaw shapes resulting from stable fatigue crack growth
together with SIF distributions along such flaw borders for complex three
dimensional cracked body problems. It is limited to cases where body
geometry and elastic behavior controls the flaw shape and to incompressible
materials. The effect of the latter has been assessed approximately by
the author and his colleagues [18] and, in three dimensional problems is
judged to increase the SIF a maximum of the order of the experimental
error (i.e. 5%).

EXPERIMENTAL RESULTS

As a vehicle for presenting quantified observations on three dimensional effects in cracked body problems, the author has selected results
from a study conducted on scale models of a boiling water reactor vessel
containing two diametrically opposite nozzles (Figure 4) each with a
flaw at the vessel-nozzle juncture oriented normal to the hoop stress
direction. Observed flaw shapes are pictured in Figure 5 along with
quarter elliptic shapes of the same semi-axis values. The corresponding
normalized Mode I SIF distributions are given in Figure 6. From Figures
5 and 6, it is clear that flaw growth is not self-similar, and that the SIF
distribution changes with increasing flaw depth. These observations indicate
that simple flaw shapes used in numerical analyses of this problem such

Figure 4 - Photo of BWR Model with Closeup of Nozzle

as quarter circular or quarter elliptical ones could only be valid over a narrow a/T range of flaw sizes. The same is true of the SIF distributions. It is also noted that, for flaws deeper than a/T ≈ 0.30, the smallest crack growth rate occurs in the central part of the flaw where the SIF is the greatest.

DISCUSSION OF RESULTS

A) Preliminary Observations

In seeking to compare experimental results with approximate analytical solutions for problems of this class, it should be noted that the analytical flaw shapes are normally simple curves while the natural flaw shapes are generally of a

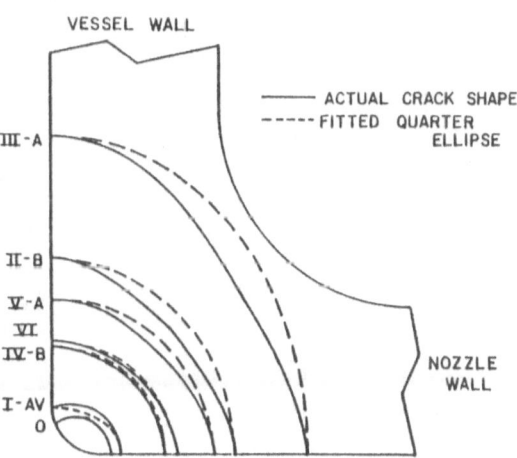

Figure 5 - Assumed Analytical and Experimental Flaw Shapes

150

more complex shape. In the cooperative study with Delft University mentioned earlier, the results
of which are described in Refs.
[22] and [24], flat A508 reactor
vessel steel plates, each containing a prenotched central nozzle
were subjected to tension-tension
fatigue tests in order to grow
fatigue cracks from the nozzle-
plate corner. Measurements of the
flaw depth were then made along
the plate and along the nozzle
and these values (a_p and a_n) were
used as semi-axes for a quarter
elliptic shape assumed for the
flaw in a finite element model.
When the author attempted to
duplicate this geometry photoelastically, he observed a reduced flaw growth rate near the
center of the flaw similar to

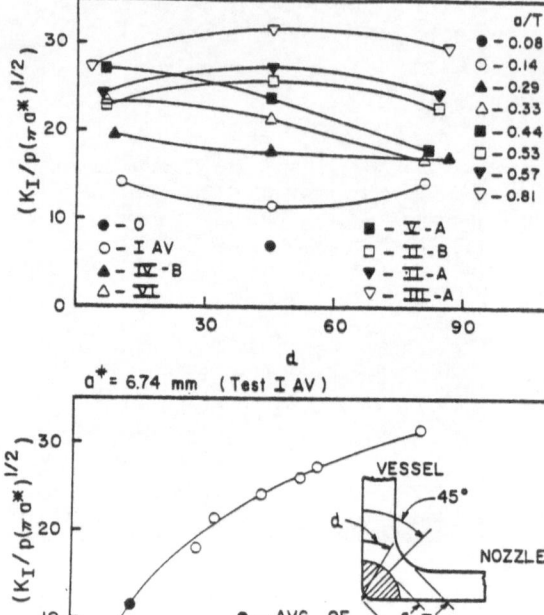

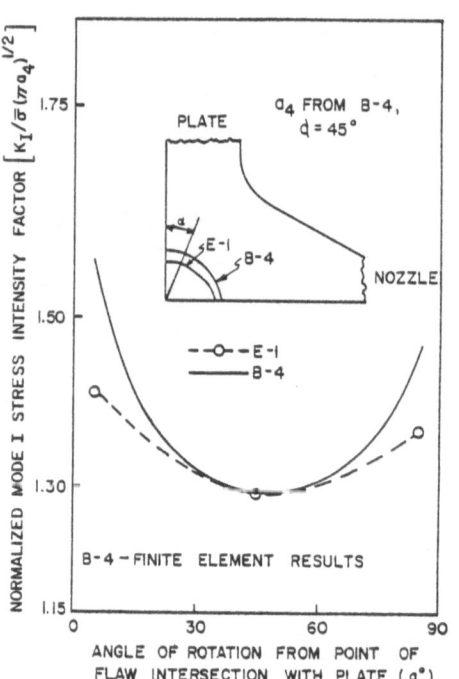

Figure 6 – Normalized SIF Distributions
Along Experimental Flaw Shapes

that displayed by flaws II-B
and V-A in Figure 5. Moreover,
the natural flaw shapes produced
SIF distributions which exhibited
lower SIF gradients along the
flaw borders than the assumed
simple curves in the finite element model. A typical result
is shown in Figure 7.

B) Analytical Comparisons

Several analytical studies
employing approximate methods
have been proposed for thin
walled pressure vessels containing cracked nozzles. These
studies have indicated SIF distributions which are concave
upward [23, 24], nearly constant

Figure 7 – Comparison of Experimental
Results with Theory

[25] and concave downward [26] depending upon flaw shapes and sizes. The most frequently used solution in the U.S.A. is due to Gilman and Rashid [27] and employs quarter circular crack shapes centered on the point of intersection of the extension of the inner boundaries of the vessel and the nozzle. They employed a compliance type of approach which led to a single average SIF for a given crack size. Figure 8 shows a comparison of the current results at α = 45° with the Gilman-Rashid results for a similar nozzle shape. Differences between theory and experimental results for shallow cracks may be due to the fact that experimental and analytical flaw shapes for the very shallow flaws were quite different for equal a/T values (Figure 9).

C) Crack Growth Phenomena

It was noted under Experimental Results that the observed flaw growth for the case under study was significantly non-self similar. This may not be surprising if one conjectures that there are at least three different crack growth regimes in the problem:

i) Very shallow flaws (Flaw depth of order of the inner fillet radius) Flaw growth here may be conjectured to depend primarily upon the uncracked body stress field and the fillet radius.

ii) Moderate depth flaws (a/T ≈ 0.3 to 0.6) Here the flaw growth may depend less upon inner fillet radius and more upon transverse constraint and its interaction with the uncracked stress field.

iii) Deep flaws (a/T > 0.6) Here we conjecture the outer boundary shape to exert a significant influence along with the effects noted in (ii).

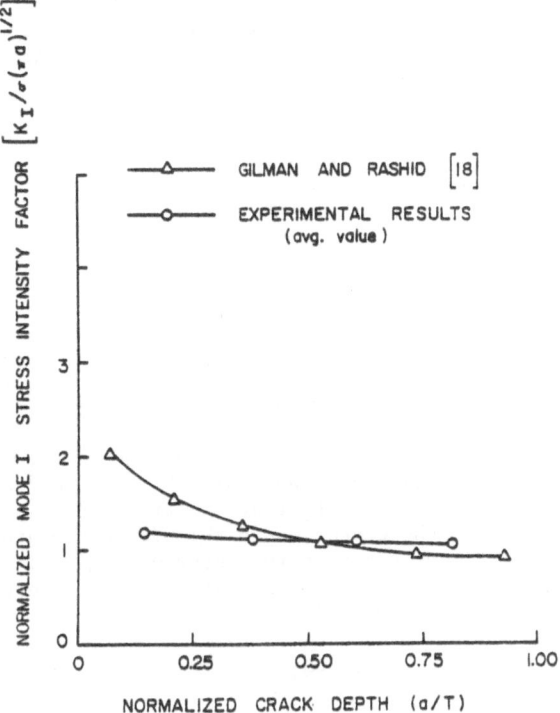

Figure 8 – Influence of Assumed Flaw Shape on SIF Gradient

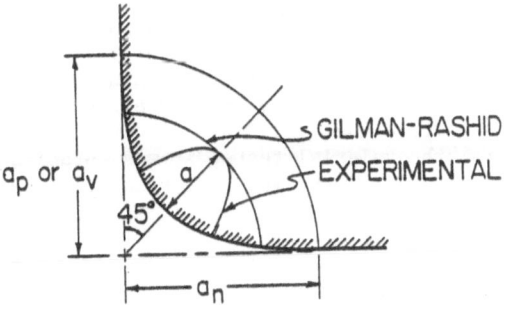

Figure 9 – Shallow Flaw Shapes and Flaw Notation

Unfortunately, most of our knowledge about crack growth is empirical, and is based upon two dimensional observations. The simplest and most common expression for crack growth rate is:

$$\frac{da}{dN} = C(\Delta K)^n \tag{5}$$

where

  a = crack depth
  N = Number of stress cycles
  $\Delta K = K_{max} - K_{min}$
  C,n = Material Parameters

and C,n are determined by conducting fatigue tests (normally two dimensional) on geometries for which the functional form of the SIF is known. Once C,n are computed, they are used in computing ΔK from measurements of a,N in an equation such as Equation 5.

Equation 5 implies that the crack growth rate will increase with increasing ΔK. However, the experimental results cited in the previous section indicate that the growth rate along the flaw border is slowest near the center where the SIF values are the highest, implying that C,n are not constant for three dimensional problems.

If one assumes that $\Delta K \propto$ COD, the crack opening displacement, it can be shown [27] that

$$\frac{da}{dN} = A' \left\{\frac{\Delta K}{E}\right\}^2 \tag{6}$$

It is known [18] that, whenever a crack is introduced into a finite body a state of constraint approximating plane strain exists near the crack tip. However, this constraint may vary as one moves away from the crack tip, or towards an unrestrained boundary. It is suggested that, in the present case, local constraints elevate the effective value of the elastic modulus E* so as to reduce the crack growth rate in the central part of the crack for moderate to deep flaws in the preceding example. In a recent study, [29], the author and his associates used a full field elasto-plastic mathematical model [30] which coalesced locally with the asymptotic solutions of Hutchinson [31] and Rice and Rosengren [32] to predict near field non-linear behavior utilizing test data from uniaxial stress-strain tests on photoelastic models above critical temperature and comparing the results with experimental data taken from the near field of stress frozen cracked photoelastic bodies. From the uniaxial tests, it was established that the strain hardening exponent n in the stress-strain relation:

$$\sigma = C''\varepsilon^n \tag{7}$$

was n = 1.15. However, from the frozen stress photoelastic data on cracked bodies it was found that n = 1.79. In the absence of strain hardening phenomena in photoelastic materials, this result may be interpreted as indicating an increase in effective modulus (E).

---

*An interesting and very different approach which models changes in elastic constants in damaged material has been described by Tamuzh [28].

A second phenomenon which might be present in fatigue crack growth is that of fatigue crack closure [24]. However, it is expected that the effect of crack closure would tend to lower the effect of ΔK, expecially near the boundaries and this is the opposite of the observed behavior [22][24]. Consequently, it is conjectured that the influence of fatigue crack closure is small, or negligible in the present case.

IMPLICATIONS FOR COMPOSITE MATERIAL STRUCTURES

In reviewing efforts to apply fracture mechanics to composite materials in the U.S.A. some time ago, the author noted [33] two basic approaches to the problem. The first consisted of empirical two parameter methods based heavily upon experimental data, such as those of Waddups et al [34] and Whitney and Nusmier [35]. The second approach, based directly upon linear elastic fracture mechanics, included comprehensive analytical treatments which were developed by Wu [36] and by Sih and Chen [37]. These theories implied that the crack was in the matrix and was usually oriented parallel to the reinforcing fibers. When so placed in thin, unidirectionally reinforced specimens, excellent correlation between theory and experiments with Scotch ply 1002 specimens was observed. Most current experimental studies dealing with stress raiser and fatigue effects continue to deal with thin laminates consisting of three to six layers of reinforced lamina. In general, the direction of crack growth from a notch in a unidirectional composite depends upon the relative properties of the reinforcing and matrix materials. Very recent studies [38] using graphite-epoxy reinforced laminates ($E_R/E_M \approx 10$ to 100) suggest that, indeed, if a stress raiser is introduced into a thin, unidirectionally reinforced composite laminate, fatigue cracks tend to develop in the matrix parallel to the fibers regardless of load orientation (Figure 10a). Moreover, when a stress raiser is placed in a multi-layered off-angle laminate, the same phenomenon tends to occur, with the result that the cracks grow in different directions in different layers, producing delaminations and a diffuse "damage" zone ahead of the stress raiser, Figure 10b.

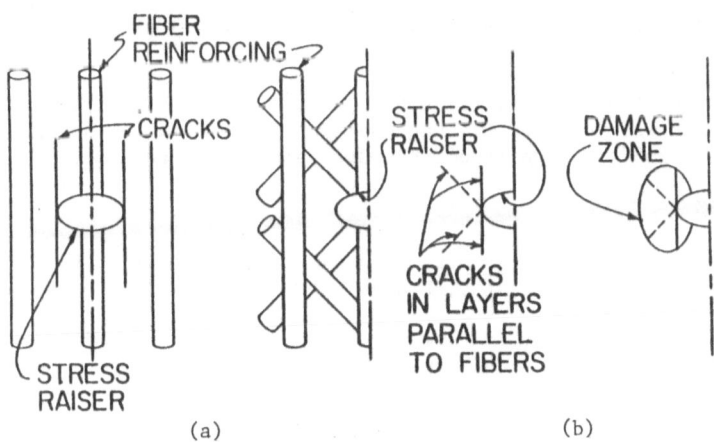

Figure 10 - Cracks in Composites

For some values of E_R/E_M. some fiber breakage may also occur.

In structural applications of composite materials, laminates of many layers are generally involved, and when through thickness stress raisers occur, they can be expected, under fatigue loads, to develop damage zones of the type depicted in Figure 10b. This process appears to generate strong Mode II-III effects on the individual cracks created by interlaminar shear. As of this writing, the way is not open for integrating the simultaneous effects of matrix cracking and delamination into a linear elastic fracture criterion. Instead, current approaches in the USA consist of empirical approaches identified earlier as two parameter methods. For example, see references [39] and [40].

One might conjecture that the three dimensional effect illustrated with the nozzle corner crack problem would be most significant in a relatively thick or many layered composite structure. Due to the complex nature of the "damage" zone, the influence of this effect in the composite laminate is not clear, but it might be conjectured that, if constraint builds up in the central region of the laminate ahead of the cracks, then conditions for delamination might be enhanced on the interior of the laminate structure.

SUMMARY

After reviewing the foundations of a frozen stress photoelastic method for simultaneously obtaining crack shapes and SIF distributions for three dimensional cracked body problems exhibiting non-self-similar stable flaw growth, and noting its limitations, its application was illustrated by applying it to the currently important nozzle corner crack problem from the reactor pressure vessel field. A three dimensional effect in the form of an increase in apparent elastic modulus was conjectured in order to explain the observed behavior. Finally the role which this effect might play in a multi-layered composite structure is briefly considered.

ACKNOWLEDGEMENTS

The author wishes to acknowledge the laboratory assistance of W. H. Peters and T. L. Fleishman, discussions with W. W. Stinchcomb and K. L. Reifsnider and the collective support of the Delft University of Technology, under Contract No. VPI-808923, Union Carbide Corporation under Sub-Contract No. 7015 and the National Science Foundation under Grant No. Eng. 76-20824 for parts of the work described herein.

REFERENCES

[1] Sih, G. C. Handbook of Stress-Intensity Factors, Institute of Fracture and Solid mechanics, Lehigh University, 1973.

[2] Tada, H., Paris, P. and Irwin, G. R., The Stress Analysis of Cracks Handbook, Del Research Corp., 1973.

[3] Rooke, D. P. and Cartwright, D. J., Compendium of Stress Intensity Factors, Pendragon House, Palo Alto, Calif., 1976.

[4] Smith, D. G. and Mullinix, B. R., <u>Fracture</u> <u>Mechanics</u> <u>Design</u> <u>Handbook</u>, TR-RL-77-5 U.S. Army MIssle Command, Redstone Arsenal, Alabama, 1976.

[5] Sneddon, I. N., "The Distribution of Stress in the Neighborhood of a Crack in an Elastic Solid" <u>Proceedings</u> <u>of</u> <u>the</u> <u>Royal</u> <u>Society</u>, Series A, Vol. 187, pp. 229-260, 1946.

[6] Green, A. E. and Sneddon, I. N., "The Distribution of Stress in the Neighborhood of a Flat Elliptical Crack in an Elastic Solid", <u>Proceedings</u> <u>of</u> <u>the</u> <u>Cambridge</u> <u>Philosophical</u> <u>Society</u>, Vol. 46, pp. 159-163, 1950.

[7] Swedlow, J. L., Ed., "The Surface Crack; Physical Problems and Computational Solutions" Symposium <u>Proceedings</u> <u>ASME</u> <u>Committee</u> <u>for</u> <u>Computing</u> <u>in</u> <u>Applied</u> <u>Mechanics</u> <u>of</u> <u>AMD</u>, Nov. 1972.

[8] Anon. <u>Proceedings</u> <u>of</u> <u>a</u> <u>Symposium</u> <u>on</u> <u>Part</u> <u>Through</u> <u>Crack</u> <u>Life</u> <u>Prediction</u> (In Press) ASTM STP, Oct. 1977.

[9] Rybicki, E. F., and Benzley, S. E., Eds., <u>Computational</u> <u>Fracture</u> <u>Mechanics</u>, Symposium Proceedings ASME Computer Technology Committee of Pressure Vessels and Piping Division, June 1975.

[10] Irwin, G. R., Discussion, <u>Proceedings</u> <u>of</u> <u>the</u> <u>Society</u> <u>for</u> <u>Experimental</u> <u>Stress</u> <u>Analysis</u>, V16n1, pp. 42-96, 1958.

[11] Smith, D. G. and Smith, C. W., "A Photoelastic Investigation of Closute and Other Effects Upon Local Bending Stresses in Cracked Plates" <u>International</u> <u>Journal</u> <u>of</u> <u>Fracture</u> <u>Mechanics</u>, V6n3, pp. 305-318, Sept. 1970.

[12] Jolles, M., McGowan, J. J., and Smith C. W., "Use of Hybrid, Computer Assisted, Photoelastic Technique for Stress Intensity Determination in Three Dimensional Problems", Ref. 9, pp. 63-82, 1975.

[13] Smith, C. W., "Stress Intensity Estimates by a Computer Assisted Photoelastic Method" (In Press), <u>Proceedings</u> <u>of</u> <u>the</u> <u>International</u> <u>Conference</u> <u>on</u> <u>Fracture</u> <u>Mechanics</u> <u>and</u> <u>Technology</u>, March 1977.

[14] Smith, C. W., Jolles, M. and Peters, W. H., "Stress Intensities for Cracks Emanating from Pin Loaded Holes" <u>Flaw</u> <u>Growth</u> <u>and</u> <u>Fracture</u>, ASTM-STP 631, pp. 190-201, 1977.

[15] Smith, C. W., Jolles, M. and Peters, W. H., "An Experimental Study of the Plate-Nozzle Tensile Test for Cracked Reactor Vessel Nozzles", VPI-E-76-26, July 1977, (Condensed in <u>Proceedings</u> <u>of</u> <u>15th</u> <u>Midwest</u> <u>Mechanics</u> <u>Conference</u>, March 1977).

[16] Smith, C. W., Jolles, M. I. and Peters, W. H., "Geometric Influences Upon Stress Intensity Distributions Along Reactor Vessel Nozzle Cracks", <u>Transactions</u> <u>of</u> <u>the</u> <u>4th</u> <u>International</u> <u>Conference</u> <u>on</u> <u>Structural</u> <u>Mechanics</u> <u>in</u> <u>Reactor</u> <u>Technology</u>, Paper No. G413, Aug. 1977.

[17] Irwin, G. R., "Measurement Challenges in Fracture Mechanics"
William Murray Lecture SESA Fall Meeting, Indianapolis, INd.,
Oct. 1973.

[18] Smith, C. W., McGowan, J. J. and Jolles, M. I., Effects of
Artificial Cracks and Poisson's Ratio Upon Photoelastic Stress
Intensity Determination" Experimental Mechanics, V16n5, pp. 188-
193, May 1976.

[19] Smith, C. W., "Stress Intensity Estimates by a Computer Assisted
Photoelastic Method" (Invited Paper) Proceedings of the International Conference on Fracture Mechanics and Technology, Vol. 1,
pp. 591-606, Noordhoff Internatonal, 1977.

[20] Schroedl, M. A. and Smith, C. W., "A Study of Near and Far Field
Effects in Photoelastic Stress Intensity Determination", Engineering Fracture Mechanics, Vol. 7, pp. 341-355, 1975.

[21] Oppel, G., "Photoelastic Investigation of Three Dimensional Stress
and Strain Conditions", NACA TM 824 (Translation by J. Vanier),
1937.

[22] Smith, C. W. and Peters, W. H., "Prediction of Flaw Shapes and
Stress Intensity Distributions in 3D Problems by the Frozen Stress
Method" (In Press), Proceedings of the Sixth International Confer-
ence on Experimental Stress Analysis, Munich, 1978.

[23] Reynen, J., "On the Use of Finite Elements in the Fracture Analysis
of Pressure Vessel Components", ASME Paper No. 75-PVP-20, June 1975.

[24] Broekhoven, M. J. G., "Fatigue and Fracture Behavior of Cracks at
Nozzle Corners", Comparison of Theoretical Predictions with Experimental Data", Proceedings of the Third International Conference
on Pressure Vessel Technology (Part II) Materials and Fabrication,
pp. 839-852, April 1977.

[25] Hellen, T. K. and Dowling, A. H., "Three Dimensional Crack Analysis
Applied to an LWR Nozzle-Cylinder Intersection", International
Journal of Pressure Vessels and Piping, Vol. 3, pp. 57-74, 1975.

[26] Schmitt, W., Bartholome, G., Grostad, A. and Miksch, M., "Calculation of Stress Intensity Factors for Cracks in Nozzles", International Journal of Fracture, Vol. 12, No.3 June 1976, pp. 381 390.

[27] Hahn, G. T., Sarrate, M., and Rosenfeld, A. R., "Experiments on
the Nature of the Fatigue Crack Plastic Zone", Proceedings of
the Air Force Conference on Fatigue and Fracture of Aircraft Structures and Materials, AFFDL TR-70-194, pp. 425-450, Sept. 1970.

[28] Tamuzh, V., "Calculation of Constants for Damaged Material", Mechanics of Polymers, No. 5, 1977, pp. 838-845.

[29] Smith, C. W., McGowan, J. J. and Peters, W. H., "A Study of Crack Tip Non-Linearities in Frozen Stress Fields" (In Press), _Journal of Experimental Mechanics_, 1978.

[30] McGowan, J. J. and Smith, C. W., "A Plane Strain Analysis of the Blunted Crack Tip Using Small Deformation Plasticity Theory", _Advances in Engineering Science_, Vol. 2, p. 585, Nov. 1976.

[31] Hutchinson, J. W., "Plastic Stress and Strain Fields at a Crack Tip", _Journal of Mechanics and Physics of Solids_, Vol. 16, p. 337, 1968.

[32] Rice, J. R. and Rosengren, G. F., "Plane Strain Deformation Near a Crack Tip in a Power Hardening Material", _Journal of Mechanics and Physics of Solids_, Vol. 16, p. 1, 1968.

[33] Smith, C. W., "Limitations of Fracture Mechanics as Applied to Composites", _Inelastic Behavior of Composite Materials_, C. T. Herakovich, Ed., AMD V13, pp. 157-176, Dec. 1975.

[34] Waddups, M. E., Eisenmann, J. R. and Kaminski, B. E., "Macroscopic Fracture Mechanics of Composite Materials", _Journal of Composite Materials_, Vol. 5, Oct. 1971, pp. 446-454.

[35] Whitney, J. M. and Nusmier, R. J., "Stress Fracture Criteria for Laminated Composites Containing Stress Concentrations", _Journal of Composite Materials_, Vol. 8, July 1974, pp. 253-265.

[36] Wu, E. M., "Strength and Fracture of Composites", _Fatigue and Fracture_, Vol. 5, _Composite Materials_, L. J. Broutman, Ed., Academic Press, New York, 1974, pp. 191-247.

[37] Sih, G. C. and Chen, E. P., "Fracture Analysis of Unidirectional Composites", _Journal of Composite Materials_, Vol. 7, April 1973, pp. 230-244.

[38] Stinchcomb, W. W. and Reifsnider, K. L., "Fatigue Damage Mechanisms in Composite Materials", ASTM Symposium of Fatigue Damage in Composite Materials", May 1978.

[39] Smith, D. G. and Mullinix, B. R., "Fracture Mechanics Design Handbook for Composite Materials", _U.S. Army Missile Research and Development Command_, Redstone Arsenal, Ala., TR-T-78-6, Sept. 1977, 127 pp.

[40] Kim, R. Y. and Whitney, J. M., "Fracture of Composite Laminates by Three Point Bend" (In Press) _Experimental Mechanics_, 1979.

A DELAMINATION FAILURE MODE OF COMPOSITE BARS IN COMPRESSION

Yu. M. Tarnopolskiy

Institute of Polymer Mechanics
L.S.S.R. Academy of Sciences

INTRODUCTION

Delamination of filament wound or layup construction composite structures is a common failure mode. At first glance, delamination may appear in different forms. For example, unwinding due to pressure [1] or centrifugal force [2], edge effects in tension-compression of multidirectional materials [3], a premature peel of the skin of metal-composite bars of various cross sections in compression [4], buckling of surface layers in compression and bending [5] of laminate bars, the formation of a "Chinese" lantern in compression of unidirectional tubular bars as a result of delamination along fibers. It should be noted that the danger of delamination increases with the increase of thickness of the composite subjected to initial tensile stresses [6].

The occurrence of delamination is not uncommon in the case of laminate structures. Such mode of failure is well known in wood materials [7]. The peeling of surface layers for GFRP in compression was described in [8] and elsewhere. However, analytical description of the delamination process is most difficult. We succeeded in predicting the onset of delamination in the case of unwinding by accounting for the spiral arrangement of circuits, the low composite shear and transverse tension strengths in the analytical model.

The energy approach proposed by Griffith [9] who relates the critical compressive load to the specific surface energy γ at failure will be used to characterize the composite "delamination strength". Although it is difficult to evaluate γ experimentally, the proposed approach can determine the lowest level of critical stress and gain an understanding on how to increase the "delamination strength".

In the paper, the energy approach is employed to predict the magnitude of loads at which the onset of delamination of different types of structures occurs. For example, delamination of a short aluminum faced CFRP bar of various profile in compression and bending, and ring type specimens of various layups under external pressure.

UNWINDING

One example of delamination failure mode is the so-called unwinding of composite rings under external pressure. Unwinding has been detected in annular specimens and shells made by continuous winding [1]. Filament wound structures are characterized by low shear strength. There are two causes of interlaminar shear stresses τ. The first is associated with a spiral circuit arrangement as shown by the design calculations in [10] where the magnitude of τ'' in a resin interlayer is given by

$$\tau'' = \frac{\sigma' h'}{2\pi R} \tag{1}$$

where σ' is the compressive stress in the reinforcing layer of thickness h' and R is the radius of the layer. Hereafter, the single prime refers to the reinforcing layer while the double prime to the interlayer. The shear stresses are not damaging to the cured resins at normal temperature. Another situation can arise during manufacturing. At that stage, the matrix shear strength is low and the applied stress may initiate significant interlaminar creep and change the initial tension in the winding [10]. Another cause involves the existence of loose ends or notches on the outer surface both of which can act as tangential stress concentrators. Calculation of the stresses in wound rings follows that of the problem of plane bond failure, Figure 1. The solution is

$$\tau'' = \sigma' \sqrt{\frac{G''}{E'} \frac{h'}{h''}} \tag{2}$$

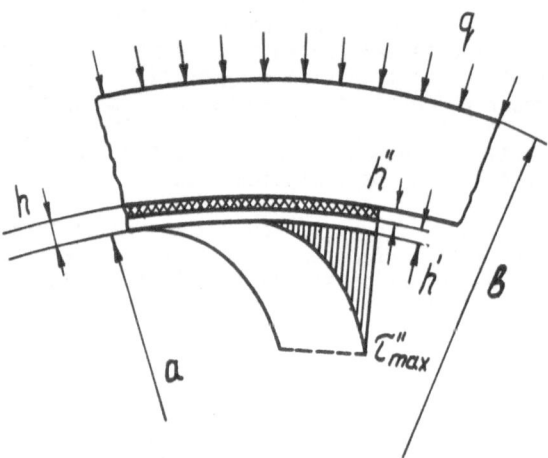

Fig. (1) - Failure due to unwinding

where E' is Young's modulus of the reinforcing layer, G'' the matrix shear modulus and σ' is obtained from the solution of Lame's problem given by

$$\sigma' = \sigma_\theta \frac{h}{h'}; \quad \sigma_\theta = q f_q(\beta) \tag{3}$$

in which

$$f_q(\beta) = \beta/(C^{\beta+1} - C^{1-\beta}); \quad C = \frac{a}{b}; \quad \beta = \sqrt{E_\theta/E_r}$$

In the case of thin rings $f_q(\beta) = \frac{b}{b-a}$, and a and b are the inner and outer radii of the ring. E_θ and E_r are elastic moduli of the composite in the tangential and radial directions. Equation (2) refers to a plane bond. The effect of bond curvature is not significant. This can be shown by a more detailed analysis.

Consider the simple strength criterion $|\tau_{max}| < [\tau]$. It is possible to evaluate the external bursting pressure q_p of unwinding

$$q_p = \sqrt{\frac{E' \, h''}{G'' \, h'}} \frac{[\tau]h'}{f_q(\beta)h} = \sqrt{\frac{E_\theta}{G_{r\theta}}} \frac{[\tau]}{f_q(\beta)} \tag{4}$$

Since the matrix shear strength is independent of the sign of the tangential stress, unwinding can occur either due to the application of external and/or internal pressure. If the properties of the resin interlayer are elasto-plastic with proportional limit ψ^*, yield strength $[\tau]$ and ultimate shear strain ψ, then q_p is determined from the following relation:

$$q_p = \sqrt{\frac{E_\theta}{G_{r\theta}}} \frac{[\tau]}{f_q(\beta)} [1 + \frac{E'}{E_\theta} \sqrt{2(\frac{\psi}{\psi^*} - 1)}] \tag{5}$$

where $G_{r\theta}$ is the interlaminar shear modulus. The relationship for the case of internal pressure has a similar form. It follows from equation (5) that in order to preclude the danger of unwinding, it is necessary to increase $[\tau]$ and ψ. By comparing the bursting pressure q_p with respect to the strength criterion, it is possible to define the shear characteristics of the matrix such that unwinding can be avoided.

FLYWHEEL

The delamination failure mode can also occur in a composite superflywheel. Besides the well-known delamination development in thick-wall flywheels due to initial stresses, the danger of unwinding is observed in the stationary stage and in the spinning up-down stage should the composite transverse tension strength n_r^+ [11,12] be inadequate. The critical angular rotation ω_{cr} for a flat disk and a rim flywheel can be determined from the centrifugal stress expression in equation (2) by substituting σ' for the solution of the rotation of an ortho-tropic disk [13]:

$$\sigma' = \sigma_\theta \frac{h}{h^+}; \quad \sigma_\theta = \rho\omega_{cr}^2 b^2(3+\nu_{r\theta})f_\omega(\beta)$$

In the case of anisotropic materials the circumferential stresses in the surface layer are the most important. At the outer radius, f_ω is equal to

$$f_\omega(\beta) = \begin{cases} \dfrac{1}{\beta^2-g}\left(3 + \beta\,\dfrac{c^{2\beta}-2c^{\beta+3}+1}{c^{2\beta}-1}\right) + \dfrac{1}{3+\nu_{r\theta}}; & \beta \neq 3 \\[4mm] \dfrac{c^6}{c^6-1}\,\ell nC + \dfrac{3-\nu_{r\theta}}{3+\nu_{r\theta}}; & \beta=3 \end{cases}$$

where ρ is the density and $\nu_{r\theta}$ is Poisson's ratio. Hence

$$\omega^2 \geq \frac{n_{r\theta}E_\theta}{\rho(3+\nu_{r\theta})b^2f_\omega(\beta)}\sqrt{\frac{h''}{h^+}\frac{1}{E'G''}} \qquad (6)$$

The number of revolutions is given in Table 1.

TABLE 1 - CRITICAL NUMBER OF REVOLUTIONS FOR DIFFERENT COMPOSITES

Materials	$n_{r\theta}$ kgf/mm²	ω_{cr} rpm		
		b = 0.1m	b = 0.5m	b = 1.0m
GFRP	4.50	40,000	8,000	4,000
BFRP	6.30	64,000	12,800	6,400
CFRP	6.75	63,000	12,500	6,300
Organic Plastic	2.75	40,000	8,000	4,000

A more rigorous solution of the rotational disk problem shows that a peak of radial stress σ_r develops near the outer end of the band. This tends to increase the chance of delamination due to $\tau_{r\theta}$. The problem will be discussed in a later contribution.

For fiberglass, carbon, boron and organic composites, it is possible to determine the range of superflywheel thicknesses, Figure 2, over which the low

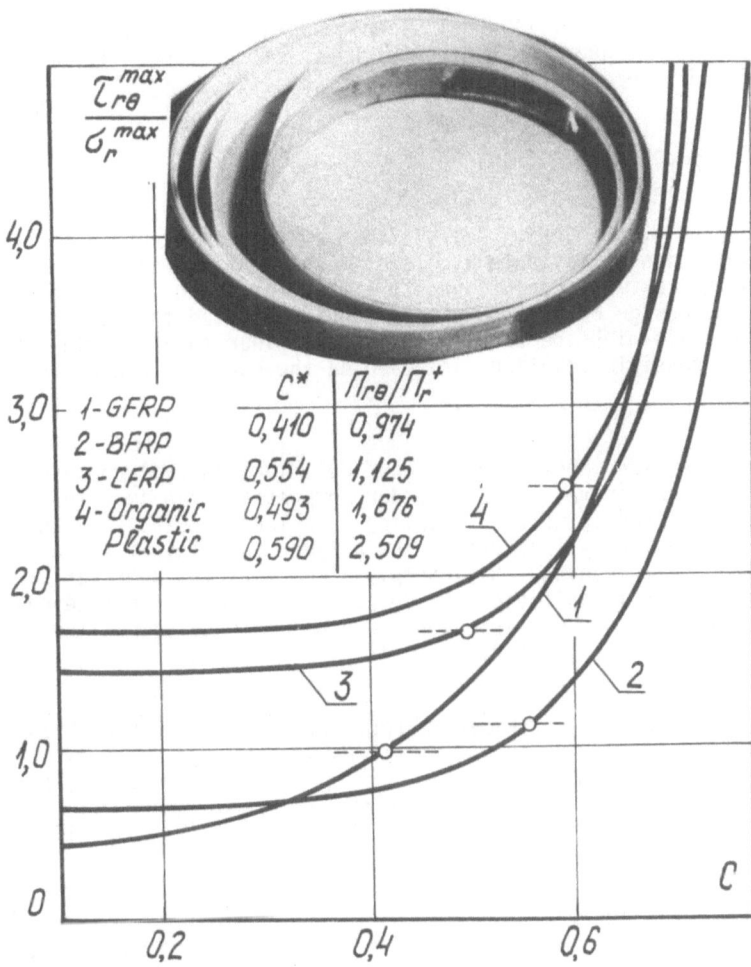

Fig. (2) - Maximum tangential and radial stress versus relative thickness. $\Pi_{r\theta}$ is the interlaminar shear strength; Π_r^+ the transverse tension strength; A is the failure caused by unwinding

transverse tension strength ($C<C^*$) is observed and the region within which unwinding due to $\tau_{r\theta}$ will most likely occur for different ratios of $\Pi_{r\theta}/\Pi_r^+$. If we assume that $\tau_{r\theta}^{max}/\sigma_\theta^{max} \geq 2\Pi_{r\theta}/\Pi_\theta^+$, then unwinding is the more likely failure mode in the rim flywheels rather than failure induced by the normal stress where

n_θ^+ is the fiber tensile strength in the circumferential direction. This finding has led to the idea of employing a chordwise winding pattern [2] for increasing ω_{cr}.

During the stage of flywheel spinning up-and-down, tangential stresses are developed and act over the entire thickness of the ring. The maximum shear stress $\tau_{r\theta}$ developed at the inner radius [14] is

$$\tau_{r\theta}^{max} = \frac{1}{4} \rho\dot{\omega}b^2 (c^2 - \frac{1}{c^2}) \tag{7}$$

Undoubtedly, the design calculations of failure by unwinding involve the proper selection of failsafe accelerations $\dot{\omega}$. At the same time, the failure by unwinding at maximal energy capacity attributes to the advantageous feature of the present design of flywheels. It should be noted that unwinding in the spinning down stage might be used in the interlaminar shear testing of materials. One of the possible solutions is based on the use of disks of varying thickness.

PEELING

Let us consider another example of delamination failure mode. The separation of surface layers in a zone under compression can lead to subsequent failure of a catastrophic nature. Peeling is a local instability phenomenon that can be accompanied by fracture of the polymer interlayer. From the viewpoint of linear failure mechanics, the interply plane exhibits poor resistance to crack propagation. It is sufficient to compare the specific surface energy at failure, γ, of the homogeneous metals and fibrous or laminate composites [4,15,16]. The results are shown in Table 2.

TABLE 2 - SPECIFIC SURFACE ENERGY OF VARIOUS MATERIALS

Materials	γ kgf/cm
Homogeneous metals	10 -100
GFRP { unidirectional laminated	0.8 -1.4 0.2 -0.5
CFRP { unidirectional laminated	0.08-0.10 0.03-0.05
Al-CFRP	0.05-0.08

The peeling of the inner ply of a GFRP ring is shown in Figure 3. The peeling occurs at a stress lower than the composite compression strength n_θ^-. At a thickness h, governed by the layup, the critical stress is [9]:

$$\frac{\sigma_{cr}}{E_\theta} = 0.916 \left[\left(\frac{h}{R}\right)^2 + \kappa\left(\frac{h}{R}\right)^{-1} \right]^{1/2} \tag{8}$$

where $\kappa = 4.77 \dfrac{\gamma}{E_\theta R}$ and R is the inner radius of the ring.

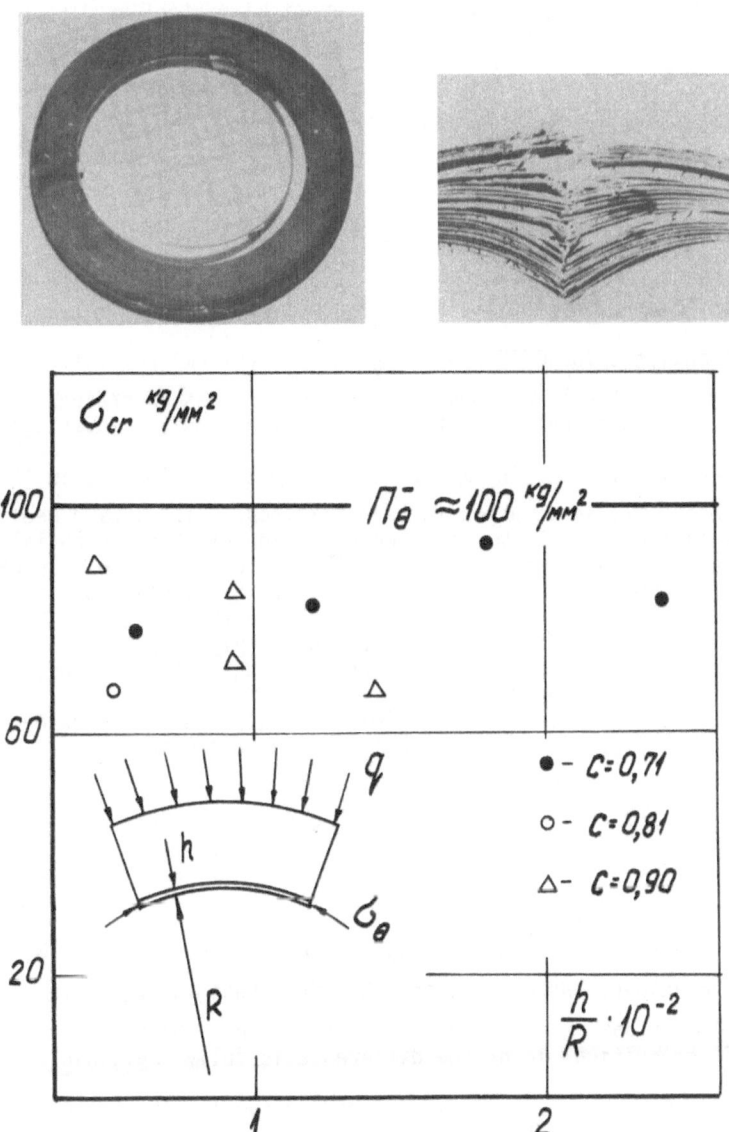

Fig. (3) - A peel type failure at external pressure

Delamination by peeling may occur in two ways. In the case of a spiral circuit arrangement, delamination may initiate unwinding even in the absence of

stress concentrators on the ring surface. In the case of the annular layup
(for example, a longitudinal-circumferential winding), the subsequent layer-by-
layer peeling may occur. Additional reinforcement with steel needles [17] in
the radial direction can reduce the possibility of peeling, Figure 3a.

In three-point bending, the critical stress σ_{cr}, at which peeling in a compressive zone prevails, is given by [5]:

$$\frac{\sigma_{cr}}{E_x} = 4\pi^2 \left(\frac{h}{\ell}\right)^2 \left(1 + \sqrt{1 + \frac{3\gamma\ell^4}{4\pi^4 E_x h^5}}\right) \tag{9}$$

where ℓ is the half-span between grips and the thickness h over which delamination occurs is governed by the stacking sequence.

A numerical analysis for GFRP having $E_x = 5.10^5$ kgf/cm² and different span
lengths has shown that the lowest bound of critical stress according to equation
(9) exceeds the bending strength η_σ of the types of composites investigated, i.e.,
$\sigma_{cr}^{min} < \eta_\sigma$ for very low γ. Peeling may be caused by the difference in the elastic
moduli of the peeled ply as compared to beam stiffness. The inward peeling de-
velops from the surface of a beam. The incorporation of circumferential circuits
near the external surface introduces another type of peel failure. In experiments performed on GFRP with a layup 10:1 such that the elastic modulus of the
outer layers under compression is significantly lower than that of a beam in bending, the failure is preceded by peeling of the outer ply with a blast. The calculated critical stresses in this case are lower than the material bending strength.

The energy approach was used in describing the delamination development of
metal-composite bars of various cross sections. In this case

$$\sigma_{cr} = \frac{\pi^2}{3} E \left(\frac{h}{\ell_0}\right)^2 + 2\sqrt{\frac{E\gamma}{h}} \tag{10}$$

where ℓ_0 is the peeled length.

Experiments [4] have shown that the calculated values of σ_{cr} are 15% lower
than the measured values. Here, h is the skin thickness and ℓ_0 is measured from
a cinematogram of the bar in the loading process. Since the criterion yields the
lowest bound of the critical load, the difference is fully explainable.

The energy approach, however, does not predict the length of the peeled region ℓ_0. The drawback of the theory is associated with the disregard of such
factors as the loading rate, skin plasticity, etc. Therefore, for an approximate
lower bound estimate with $\ell_0 >> h$, we may use

$$\sigma_{cr} = 2\sqrt{\frac{E\gamma}{h}}$$

Here, γ plays the leading role in the load carrying capacity of the metal-composite bar. It should be noted that the range of appropriate increase in γ is restricted by two factors. The possibility of passing over to another failure mechanism, i.e., from compression of the inner ply and the possibility of a peel type failure not on the metal-composite but on composite-composite interface. Peeling of this kind was observed in the experiments [4]. A thorough knowledge of the peeling mechanism is needed. One of the ways of increasing γ involves the application of a magnetic field during the molding process.

AXIAL COMPRESSION

Another interesting type of the delamination failure [18] occurs in thin-wall tubes made by axial or off-axial winding at a small angle due to an increase in the relative length L/D_o. The failed specimen assumes the shape of a "Chinese lantern", Figure 4a. The dependence of the critical load σ_z (related to cross

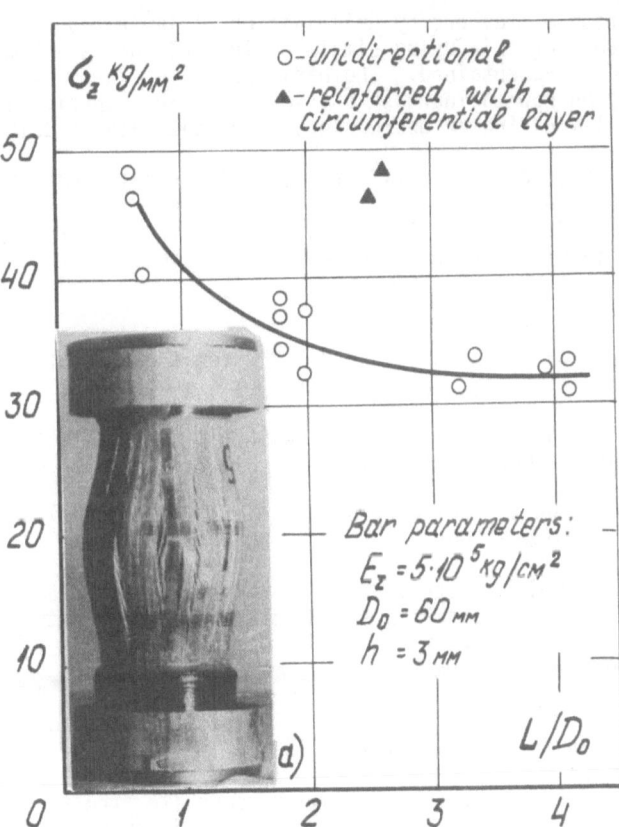

Fig. (4) - Breaking stress in compression versus relative length of a cylindrical bar. D_o is outside diameter, a -- is failure of a bar resembling a "Chinese lantern"

sectional area) on the relative length L/D_o (D_o is the outside diameter) is shown in Figure 4. Only for very short bars the critical load is equal to compressive strength $\bar{n}_z$ of unidirectional composites. In that case, another failure mechanism sets in. It may be assumed that the initial delamination is caused by the circumferential strain ε_θ due to Poisson's effect. The allowable tensile strains perpendicular to fibers $|\varepsilon_\theta|$ are low. Independent experiments have shown that $|\varepsilon_\theta| \leq 0.10\%$. The development of axial microcracks at the level of strains (experimentally detected by measuring the light transmission) leads to the splitting of the tube into a number of strips followed by buckling. The incorporation of circumferential overwraps prevents delamination. A single overwrap of 0.2 mm thickness causes a sharp increase in $|\varepsilon_\theta|$.

CONCLUSION

The description of failure by delamination of laminates is based on bars of a straight and annular axis using simple analytical models. Numerical estimates of critical loads were obtained. The basic parameters were singled out such that an understanding on how to increase the load carrying capacity of laminates has been gained by eliminating or decreasing the possibility of delamination.

The present paper does not claim to be an exhaustive contribution. It represents a survey of the works by the author and his coauthors published in the Journal of Polymer Mechanics. Attention is drawn to the very important feature of delamination in filament wound laminate structures.

REFERENCES

[1] Tarnopolskiy, Yu. M. and Rose, A. V., "Design features of reinforced plastic parts", Riga, "Zinātne", p. 274, 1969.

[2] Portnov, G. G. and Kulakov, V. L., "Failure of composite flywheels by unwinding", Polymer Mechanics, No. 4, pp. 615-621, 1978.

[3] Pagano, N. J., "On the calculation of interlaminar normal stress in composite laminates", J. Compos. Mater., Vol. 8, pp. 65-81, 1974.

[4] Tarnopolskiy, Yu. M., Khitrov, V. V., Shemshurin, M. V. and Vasil'evskiy, V. M., "Delamination danger in short metal-faced composite columns under axial compression", Polymer Mechanics, No. 1, pp. 27-33, 1978.

[5] Tarnopolskiy, Yu. M., Zhigun, I. G. and Pol'akov, V. A., "Analysis of the distribution of tangential stresses in composite beams in three-point bending", Polymer Mechanics, No. 1, pp. 56-62, 1977.

[6] Tarnopolskiy, Yu. M., "Thick-wall wound composite structures", Proceedings of ICCM-75, Vol. 1, AIME, New York, pp. 221-248, 1976.

[7] Rzhanitsin, A. R., "Theory of constituent rods of engineering structures", Moscow, p. 192, 1948.

[8] Lavenetz, B., Proc. 19th Conf. SPI Reinforced Plastics Division, 1964, Sec. 14-D; U.S. Navy N. Obs. 86347. Cited from: Advanced Composite Materials, Moscow, pp. 136-137, 1970.

[9] Kachanov, L. M., "Failure of composite materials due to stratification", Polymer Mechanics, No. 5, pp. 918-922, 1976.

[10] Beil', A. I., "Refined models of composite winding mechanics", Thesis of Cand. of Sci., Riga, 1977.

[11] Portnov, G. G. and Kulakov, V. L., "The effect of initial thermal stresses on the energy capacity of filament wound composite flywheels", Polymer Mechanics (in press), 1978.

[12] Dick, W. E., "Design and manufacturing considerations for composite flywheels", Proc. of the 1975 Flywheel Technology Symposium. Lawrence Livermore Laboratory, November 1975.

[13] Portnov, G. G. and Kulakov, V. L., "Investigation of energy capacity of filament wound composite flywheels", Polymer Mechanics, No. 1, pp. 73-81, 1978.

[14] Phillips, J. and Schrock, M., "Note on shear stresses in accelerating disks of variable thickness", Int. J. Mech. Sci., Vol. 13, No. 5, pp. 445-449, 1971.

[15] Kelly, A., "Strong solids", Clarendon Press, Oxford, 1973. Russian translation by S. T. Mileiko, Moscow, "Mir", 1976.

[16] Bugakov, I. I., "Fracture work in GFRP laminates at the interface", Strength Problems, No. 4, pp. 49-52, 1978.

[17] Zhmud', N. P., Petrov, V. Yu and Shalygin, V. N., "Laminated fiberglass reinforced plastic rings with an additional steel needle reinforcement in the radial direction", Polymer Mechanics, No. 2, pp. 226-231, 1978.

[18] Khitrov, V. V., Shemshurin, M. V. and Katarzhnov, Yu. I., "The effect of the reinforcing angle on load carrying capacity of filament wound columns in compression", Polymer Mechanics (in press), 1978.

ANALYSIS OF CRACKS IN COMPOSITE STRUCTURES SUBJECTED BY THERMAL LOADING

K. Herrmann
Lehrstuhl für Mechanik, FB 10, Gesamthochschule Paderborn, W-Germany

H. Braun
Institut für Technische Mechanik und Festigkeitslehre,
Universität Karlsruhe, W-Germany

ABSTRACT

A quasistatic approach of a thermal crack propagation in a self-stressed unidirectional low-fibre concentration composite with circular fibers in a hexagonal array is given by means of an approximate analytical solution based on the application of complex variable technique and on the method of integral equations as well as using the finite element method. Two different cracked unit cells of a two-phase composite are used in order to obtain numerical results concerning crack edge displacement, crack extension force as well as opening mode stress intensity factors for several material combinations. In case of a circular unit cell could be shown that the analytical solution of the crack-thermal stress problem is a sufficient approximation except for a small interval in the neighborhood of the fibre-matrix interface. In addition, the phenomenon of crack arrest inside the fibre was investigated.

INTRODUCTION

The study of crack propagation in inhomogeneous media due to the action of external body forces or of well-defined macroscopic self-stress fields or a combination of both of these effects is useful for the assessment of the strength of composites. A summary of the present state of the art concerning the fracture of composite materials was given by Corten [1], Sih and coworkers [2], Broutman [3], Erdogan [4], Cooper and Piggott [5]. The field of composite materials fracture research contains as a special discipline the investigation of thermal fracture of inhomogeneous materials which was already performed in the past experimentally and theoretically by several authors [6-14]. Therein from the standpoint of fracture mechanics, the thermal shock resistance of a fibre-reinforced composite depends on the position and shape of the interfacial area between the individual phases of the composite structure, on the shape of the free surface, and in addition on the propagation of an existing crack. At present, there exist some investigations concerning the thermal shock resistance of special materials and a number of appropriate parameters were defined [15, 16]. However, a complete mathematical description of the static or the dynamic state of the most structural composites is beyond reach. Therefore, in order to determine the mechanical properties of composites a number of effective continuum models of compound materials has been developed which are known in the literature as the effective modulus theories, the effective stiffness theories, mixture theories, continuum theories with microstructure. A review of those theories was given by Hegemier [17]. Therein the latter approach models a heterogeneous composite as a continuum theory with microstructure where the governing equations are completely determined from a knowledge of the geometry and constitutive relations of the composite microcomponents. Thus in the past the unit cell of fibre-reinforced composites has been investigated by many research workers in order to gain microstructural informations about the mechanical properties of the compound material.

In this paper, by use of two cracked representative elements of a unidirectional low-fibre concentration composite with circular fibers in a hexagonal array (cf. Fig. 1),

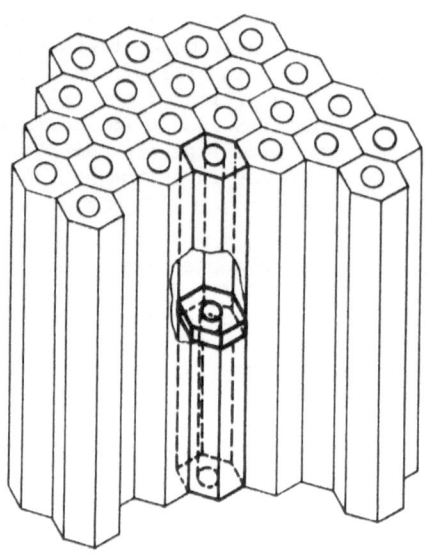

Fig. 1 - Fibre reinforced composite structure

a quasi-static approach of a thermal crack propagation caused by thermal loading is carried out. Fibre and matrix of those unit cells (composite circular or hexagonal cylinders of infinite length with a concentric or an eccentric fibre-matrix interface) consist of homogeneous, isotropic or anisotropic, and linearly elastic materials, where the thermoelastic material properties vary discontinuously at the fibre-matrix interface from the values E_f, ν_f, α_f of the fibre to the values E_m, ν_m, α_m of the matrix. Besides, the conditions of perfect contact are assumed at the interface. Further, at time $t=t^*$ the cracked two-phase solid, having for $t<t^*$ the temperature $T=T_o$ (temperature of the unstressed initial state) is subjected to a thermal shock producing the following temperature distribution in the cross section (cf. Fig. 2-3 for notation)

$$T(x,y) = \begin{cases} T_f \text{ in } A_f \\ T_m \text{ in } A_m \end{cases} \tag{1}$$

in which $T_f \neq T_m$ or $T_f = T_m \neq T_o$ and both temperatures are constant.

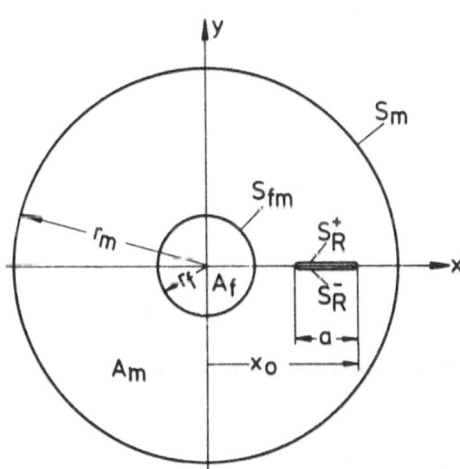

Fig. 2 - Cross section of a circular unit cell

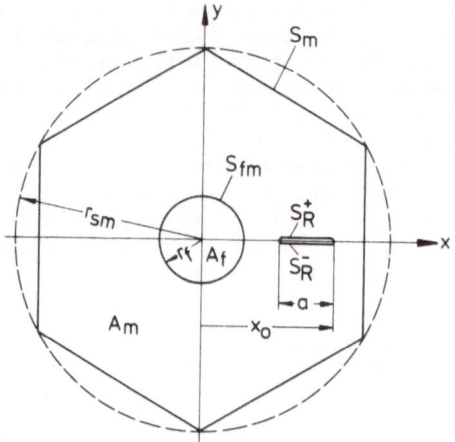

Fig. 3 - Cross section of a hexagonal unit cell

ANALYSIS

Boundary-Value Problem of the Cracked Inhomogeneous Solid

A. Analytical solution

Using the linear theory of quasi-static thermoelasticity for a plane strain state, and assuming temperature independence of the thermo-elastic material constants E_j, ν_j, α_j (j=f,m) as well as heat insulation of the interface and of the solid with respect to its environment, respectively, the thermal stress field existing in the cracked inhomogeneous solids with a concentric or an eccentric interface, respectively, can be decomposed into two parts:

1) a regular stress field in the uncracked inhomogeneous solid

2) a corrective stress field with two singularities of magnitude $\rho^{-1/2}$ at the crack tips where ρ is a local polar coordinate with respect to the crack tip.

Making use of the stress function method and of the complex variable technique the regular stress field for the circular unit cell can be obtained from the solution of a boundary-value problem of the bipotential theory[18]. Owing to the complicated shape of the boundary of the cracked inhomogeneous solids, a closed solution of the given crack-thermal stress problem is not available. But by consideration of certain assumptions concerning the crack length and the distances of the crack tips with respect to the boundaries S_m and S_{fm}, respectively, the corrective stress field in both cases (concentric and eccentric interface) can be obtained in an approximate manner from the solution of the following mixed boundary-value problem

$$\sigma^c_{yy}(\xi,0)=-\sigma^m_{yy}(\xi,0) \qquad ; \qquad (y=0, |\xi| \lessgtr \tfrac{a}{2}) \tag{2}$$

$$\sigma^c_{xy}(\xi,0)=0 \qquad ; \qquad (y=0, \forall\, \xi) \tag{3}$$

$$u^c_y(\xi,0)=0 \qquad ; \qquad (y=0, |\xi| > \tfrac{a}{2}) \tag{4}$$

with $\xi = x - x^*$ where x^* means the center of the crack.

The solution of the mixed boundary-value problem (2)-(4) is given in reference [12] and has been obtained by means of an analytical method based on the application of complex variable technique and on the method of integral equations. Therein the opening mode stress intensity factor at both crack tips of the internal crack is given by the formula

$$K_I = 4\sqrt{2\pi} \ \mathrm{Re} \ \{\lim_{z \to \pm \frac{a}{2}} (\sqrt{z \mp \tfrac{a}{2}}\ \Psi'(z))\} \tag{5}$$

The upper signs in formula (5) correspond to the right, the lower to the left crack tip. The complex potential $\Psi(z)$ represents the solution of the mixed boundary-value problem (2)-(4) and reads in case of a circular unit cell with a concentric fibre-matrix interface [13]:

$$\Psi(z) = \frac{Br_f}{2} \{ \frac{1}{r_m} \sqrt{z^2 - \tfrac{a^2}{4}} - (\frac{1}{r_m} + \frac{r_m}{(x^*)^2 - z^2})\, z + \frac{r_m z^2}{x^*((x^*)^2 - z^2)} +$$

$$+ \frac{r_m \sqrt{z^2 - \tfrac{a^2}{4}}}{(x^*+z)\sqrt{(x^*)^2 - \tfrac{a^2}{4}}} \} \tag{6}$$

with the definition of the bithermoelastic constant B

$$B = -\frac{1}{2} \frac{(1+\nu_f)\alpha_f T_f - (1+\nu_m)\alpha_m T_m}{\frac{r_f}{r_m}[\lambda_f(1-\kappa_f)-\lambda_m(1-\kappa_m)] - \frac{r_m}{r_f}[\lambda_f(1-\kappa_f)+\lambda_m(1+\kappa_m)]} \tag{7}$$

where the abbreviations read

$$\lambda_j = \frac{1-v_j^2}{E_j} \quad , \quad \kappa_j = \frac{v_j}{1-v_j} \quad ; \quad (j=f,m) \tag{8}$$

Fig. 4 shows the opening mode stress intensity factor K_I for a Carbon fibre reinforced Aluminium matrix with a concentric fibre-matrix interface by using a radius ratio $r_f/r_m = 0.1$. Therein the calculation was performed for a crack having originally zero length and starts at the point $\tilde{x}_o = x_o/r_f = 5.5$, $\tilde{y}_o = 0$ inside of the matrix and runs along the symmetry line of the specimen towards the fibre. Further, the stress intensity factors for the left and the right crack tip were calculated by consideration of the temperature dependence of the elastic and thermal material constants (cf. table 1) as functions of crack length $\tilde{a} = a/r_f$ and matrix temperature T_m. Negative matrix temperatures were chosen, because according to the assumption $T_f = 0$, a cooling of the matrix only leads to tensile stresses along the crack line.

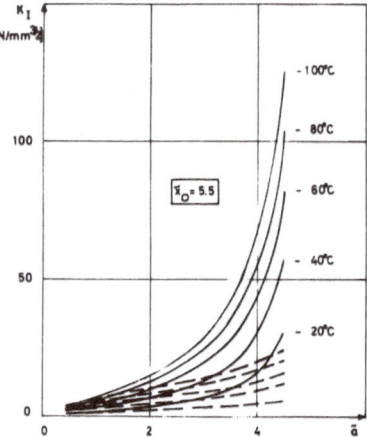

Fibre : Carbon

$E_f = 235440 \, N/mm^2$

$\alpha_f = 3 \cdot 10^{-6} \, K^{-1}$

$v_f = 0,27$

Matrix: Aluminium

T_m [°C]	E_m [N/mm^2]	α_m [$10^{-6} \cdot K^{-1}$]	v_m
- 20	72692	21.08	0.32
- 40	73281	20.05	0.32
- 60	73669	19.18	0.32
- 80	74458	18.24	0.32
-100	76047	17.15	0.32

Fig. 4 –
Opening mode stress intensity
factor in dependence on crack
length and with the matrix
temperature T_m as a parameter (concentric interface)
———————— left crack tip
– – – – – right crack tip

Table 1 –
Material properties of the composite
structure Carbon fibre/Aluminium
matrix in dependence on temperature

In addition, two remarks should be added concerning the approximate analytical solution of our present problem. Firstly, assuming the validity of the DUGDALE model, the length of the plastic zones in the vicinity of the crack tips can be determined. Therein the corrective stress field

$$\sigma_{yy}^{c}(\xi,0)=2\{\Psi'(z)+\bar{\Psi}'(\bar{z})\} \tag{9}$$

calculated by means of the complex potential given in formula (6) for a crack with the effective crack length a has to be superposed on an additional stress field resulting from a stress distribution σ_y (yield stress) acting on a small interval $\ell \leq |\xi| \leq (\ell+d)$ of the effective crack where ℓ means one half of the length of the true crack and d is the length of the plastic zone. Extended calculations also under consideration of strain hardening and softening in the plastic zones were performed and numerical results can be found in the references [13,14].

Secondly, if the matrix material of the composite structure consists of a linearly viscoelastic material (for instance Aluminium at higher temperatures) the corresponding stress intensity factors can be obtained by application of the correspondence principle of the linear thermo-viscoelasticity. According to the evaluation of formula (5) by using equation (6) in case of a crack situated in the matrix of an elastic fibre-reinforced viscoelastic material the stress intensity factors K_I^v have the following form

$$K_I^v=\tilde{B}(t)r_f\sqrt{2\pi a} \ \{r_m^{-1}+ \frac{r_m(x^* \mp \frac{a}{2})}{((x^*)^2 - \frac{a^2}{4})^{3/2}}\} \tag{10}$$

where again the upper sign in formula (10) corresponds to the right, the lower to the left crack tip. The factor $\tilde{B}(t)$ results from the bithermo-elastic constant B given in formula (7) by application of the Laplace transform

$$\bar{f}(x,s)=\int_0^\infty f(x,t)e^{-st}dt \tag{11}$$

with the transform parameter s>0 and by introduction of the relationships

$$\bar{\mu}_m(s)= \frac{1}{2} \frac{Q_m(s)}{P_m(s)} \quad ; \quad \bar{\nu}_m(s)= \frac{\bar{K}_m-2\bar{\mu}_m}{2(\bar{K}_m+\bar{\mu}_m)} \quad ; \quad \bar{K}_m(s)= \frac{H_m(s)}{F_m(s)} \tag{12}$$

instead of the elastic material properties μ_m, ν_m, K_m (bulk modulus), where the functions P_m, Q_m, F_m, H_m are polynomials in the variable s. Finally, applying the inverse Laplace transform to the solution of the associated thermoelastic problem in the image space gives the extensive expression for the quantity $\tilde{B}(t)$ which can be found in reference [19].

B. Solution using finite element method

Moreover, calculations by use of the finite element method were performed in order to obtain a limit of validity for the analytical solution
mentioned above. In addition, by preservation of the temperature distribution (1), a crack having originally zero length starts at the point $x=x_o$,
$y=o$ inside of the matrix and runs along the symmetry line of the specimen
towards the fibre in order to get through the interface. The phenomenon of
crack arrest inside the fibre was investigated.

The resulting boundary-value problems for the two unit cells shown in
Fig. 2-3 were solved with the aid of the standard finite element computer program SAP IV. The finite element mesh-work for one-half of the
symmetric specimens consists of about 350 to 650 triangular elements,
focused essentially along the prospective crack line with smallest elements of the size $a_{max}/53$ and by using linear displacement functions. The
figures 5-6 show the finite element mesh-works for one-half of the cracked
circular as well as hexagonal unit cells.

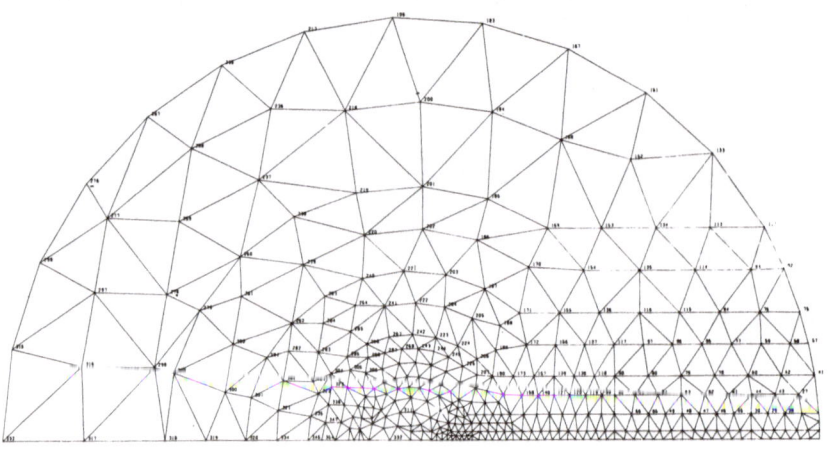

Fig. 5 - Finite element mesh-work for one-half
of a circular unit cell

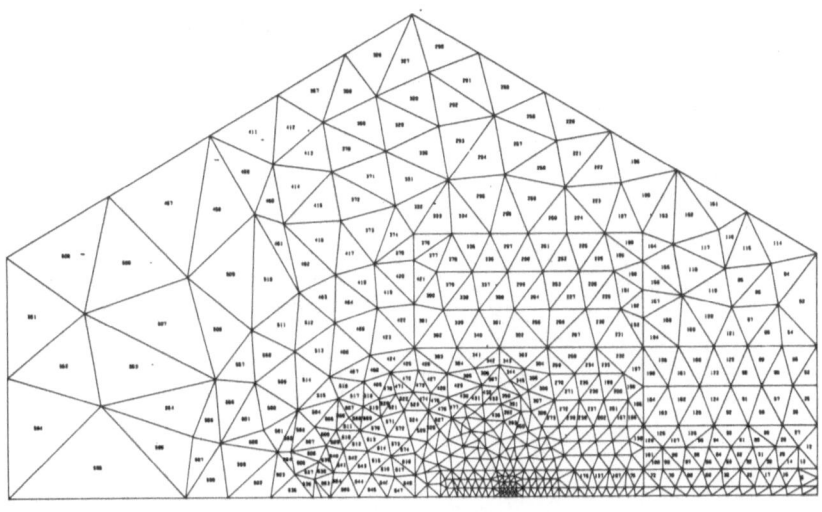

Fig. 6 - Finite element mesh-work for one-half
of a hexagonal unit cell

The determination of the crack extension force and of the opening mode
stress intensity factor was made applying a procedure given by Hayes [20]
which is based on a method originated by Bueckner [21]. The numerical
calculations of the finite element data, carried out on the computer
UNIVAC 1108 of the University of Karlsruhe, were performed for several
material combinations of which the temperature dependent material properties are given in the tables 2-6. Besides, in all cases the analysis was
based on a plane strain state.

The figures 7-8 show the result of the finite element calculations for a
cracked circular unit cell (cf. Fig. 2) using the material combination
Carbon fibre/Aluminium matrix where the mechanical properties are given
in table 1. The numerical calculations were performed for a radius ratio
$r_f/r_m = 0.1$ using a unit length of one millimeter. Further, a comparison
was made between the values of the crack extension force as well as of
the opening mode stress intensity factor for the left crack tip obtained
by the approximate analytical solution (cf. Fig. 4) on the one hand and
by the finite element method on the other hand. Therein the dotted curves
in the figures 7-8 show the adjoint analytical solution of the present
crack-thermal stress problem. Both graphs were given for a negative
matrix temperature $T_m = -100^\circ C$. It can be seen that the accordance of the
adjoint curves is quite good with the exception of a small interval in
the vicinity of the fibre-matrix interface S_{fm} which is represented in
the graphs by the vertical dotted line.

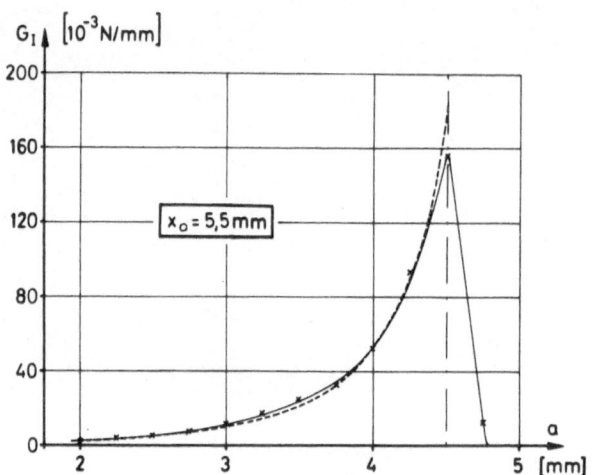

Fig. 7 - Crack extension force as function of crack length for a starting point $x_o = 5.5$ mm of a matrix crack and for a matrix temperature $T_m = -100°C$ (circular unit cell)

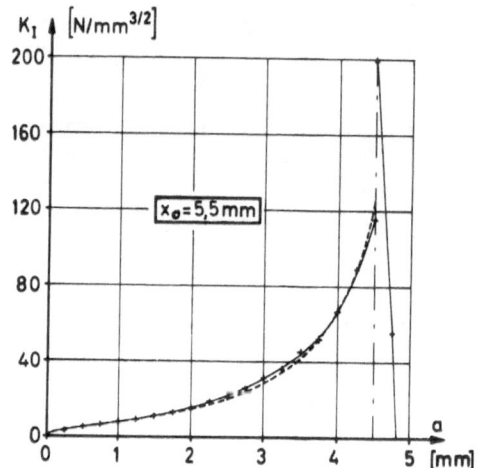

Fig. 8 - Opening mode stress intensity factor as function of crack length for a starting point $x_o = 5.5$ mm of a matrix crack and for a matrix temperature $T_m = -100°C$ (circular unit cell)

Furthermore, it can be seen that the crack extension force G_I and the stress intensity factor K_I, respectively, increase strongly with increasing crack length. By traversing the fibre-matrix interface S_{fm}, the K_I-value jumps discontinuously to a higher level because of the discontinuity of the material properties. In contrast to this behavior the crack extension force G_I always remains continuous which is clear for physical reasons. After the penetration of the interface the leading tip of the extending crack is in a region of hydrostatic pressure and with further extension the values for G_I and K_I rapidly tend towards zero; that means the crack will be arrested inside of the fibre. Further results can be found in reference [22].

The figures 9-10 give a comparison of numerical values for the crack extension force as well as the opening mode stress intensity factor for a cracked circular unit cell (material combination Carbon fibre/Aluminium matrix) obtained by the finite element method for a crack starting at the point $x_o = 4.0$ mm, $y_o = 0$ inside of the matrix and running towards the fibre-matrix interface. Therein the temperature distributions used are different.

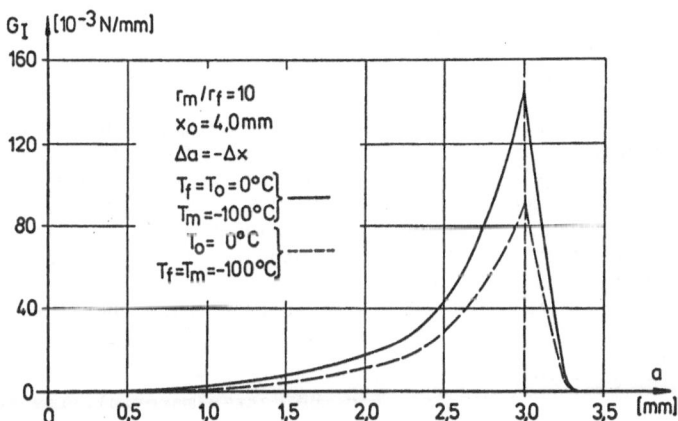

Fig. 9 - Crack extension force as function of crack length
for two different temperature distributions and for
a starting point $x_o = 4.0$ mm of a matrix crack
(circular unit cell)

182

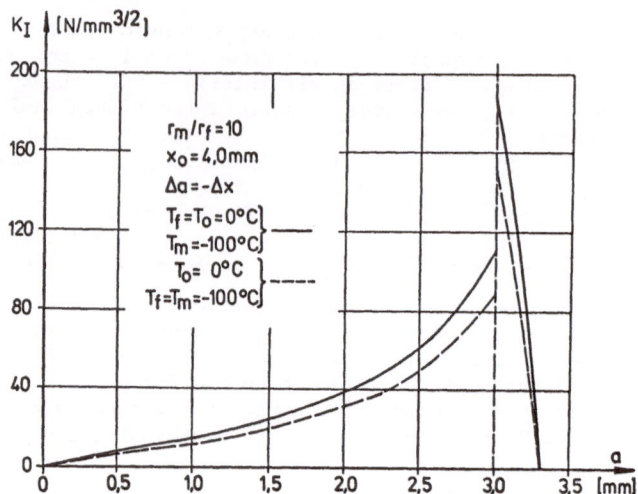

Fig. 10 - Opening mode stress intensity factor as function of crack
length for two different temperature distributions and for
a starting point x_o = 4.0 mm of a matrix crack
(circular unit cell)

It can be seen that in case of a discontinuous temperature distribution
$T_f \neq T_m$ the maximum value of the crack extension force G_I for instance is
higher by about 60 percent in comparison with the corresponding value in
case of a constant temperature distribution $T_f = T_m$. However, the numerical values for G_I and K_I show a similar behavior as in the figures 7-8.
Noteworthy is the fact that the stress intensity factor K_I for both temperature distributions is always vanishing at the same crack length
a_o = 3,3 mm whatever its magnitude is after traversing the fibre-matrix
interface S_{fm}.

The Fig. 11-13 describe the behavior of a crack starting at the fibre-
matrix interface S_{fm} and running into the matrix material. The study of
such kind of cracks is of special interest because in the uncracked unit
cell of a composite structure subjected by thermal loading there arises
the maximum value of the tension stress in the vicinity of the fibre-matrix
interface. Three different material combinations (Carbon/Al, C-fibre I/Al,
C-fibre II/Al) were investigated. The corresponding material properties of
those composite structures are given in the tables 1-4.

Carbon Fibre I							
T [°C]	$E_{xx}=E_{yy}$	E_{zz} [N/mm^2]	G_{xy}	$\alpha_x=\alpha_y$	α_z [10^{-6}K^{-1}]	v_{xy}	$v_{xz}=v_{yz}$
-273	25000	500 000	10593	4.5	0.0	0.18	0.26
0	25000	500000	10 593	6.9	-0.5	0.18	0.26
320	25000	500000	10593	8.7	-0.2	0.18	0.26
500	25000	500000	10593	9.2	0.2	0.18	0.26
700	25000	500000	10593	9.4	0.6	0.18	0.26

Table 2

Aluminium			
T [°C]	E [N/mm^2]	α [10^{-6}K^{-1}]	v
-100	74070	18.11	0.332
0	70390	22.50	0.339
100	66710	24.60	0.346
200	62290	25.83	0.352
300	57390	27.78	0.357
400	51800	30.23	0.362
500	43410	33.16	0.367
600	38750	36.48	0.372

Table 3

Carbon Fibre II							
T [°C]	$E_{xx}{=}E_{yy}$	E_{zz} [N/mm^2]	G_{xy}	$\alpha_x{=}\alpha_y$	α_z [10^{-6}K^{-1}]	ν_{xy}	$\nu_{xz}{=}\nu_{yz}$
-273	20 000	225 000	7042	4.5	0.0	0.42	0.3
0	20 000	225 000	7042	6.9	-0.5	0.42	0.3
320	20 000	225 000	7042	8.7	-0.2	0.42	0.3
500	20 000	225 000	7042	9.2	0.2	0.42	0.3
700	20 000	225 000	7042	9.4	0.6	0.42	0.3

Table 4

Table 2-4 – Material properties of the composite structures
C-fibre I/Aluminium matrix and C-fibre II/Aluminium matrix,
respectively, in dependence on temperature

Therein the C-fibres I and II consist of homogeneous, anisotropic and linearly elastic materials. Therefore, by existence of a plane strain state and by consideration of certain restrictions concerning the plane and the axis of symmetry of the anisotropic material the following constitutive equations have to be used:

$$\varepsilon_{xx} = (E_{xx}^{-1} - \nu_{xz}^2 E_{zz}^{-1})\sigma_{xx} - (\nu_{xy}E_{xx}^{-1} + \nu_{xz}^2 E_{zz}^{-1})\sigma_{yy} + (\alpha_x + \nu_{xz}\alpha_z) \cdot \Delta T \qquad (13)$$

$$\varepsilon_{yy} = -(\nu_{xy}E_{xx}^{-1} + \nu_{xz}^2 E_{zz}^{-1})\sigma_{xx} + (E_{xx}^{-1} - \nu_{xz}^2 E_{zz}^{-1})\sigma_{yy} + (\alpha_x + \nu_{xz}\alpha_z) \cdot \Delta T \qquad (14)$$

$$\varepsilon_{xy} = (1 + \nu_{xy})E_{xx}^{-1}\sigma_{xy} \qquad (15)$$

Therein the crack extension force G_I given in Fig. 11 shows the same characteristic behavior for each of the three material combinations mentioned above. Due to high values of the tension stresses in the vicinity of the fibre-matrix interface S_{fm} there already exists a local maximum value of the crack extension force for a small crack length a_o. Further, for an increasing crack length $a > a_o$ a decrease of the G_I-value takes place which can be explained by a deloading process of the self-stressed inhomogeneous material.

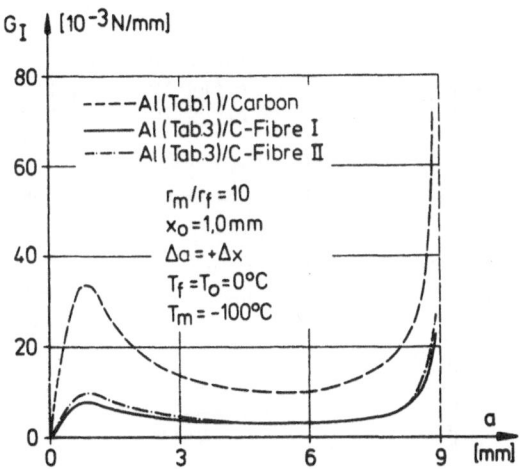

Fig. 11 - Crack extension force as function of crack length for a
crack starting at the interface S_{fm} and for three
different material combinations (circular unit cell)

However, if the leading crack tip reaches the neighborhood of the free
surface S_m a strong increase of the crack extension force can be observed.
This effect arises due to a local stress concentration in the vicinity
of the leading crack tip because the residual tension self-stresses
operate on a very small area between the crack tip and the free surface
S_m. Furthermore, the corresponding curves for the opening mode stress
intensity factor K_I given in Fig. 12 show a similar behavior as mentioned
before in case of the crack extension force G_I.

Moreover, Fig. 13 shows the crack edge displacement u_y^R of the upper
crack face S_R^+ for a material combination Carbon fibre/Aluminium matrix
(cf. table 1) as functions of crack length and of the distance x from the
interface S_{fm}. It can be seen that there exists a strong asymmetry of the
crack surface displacement in the vicinity of the crack tip adjoined to
the fibre. But the maximum value of the displacement u_y^R withdraws from
the fibre with increasing crack length. Finally, for a through crack the
crack edge displacement has a symmetric shape. In addition, Fig. 14 gives
the finite element mesh-work of the cracked composite structure before
and after thermal loading in the environment of the crack where the displacement values given in this graph were enlarged one hundred times.

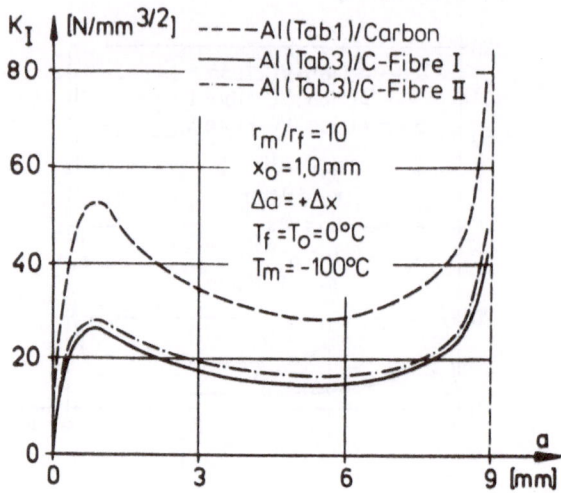

Fig. 12 - Opening mode stress intensity factor as function of crack
length for a crack starting at the interface S_{fm} and for
three different material combinations (circular unit cell)

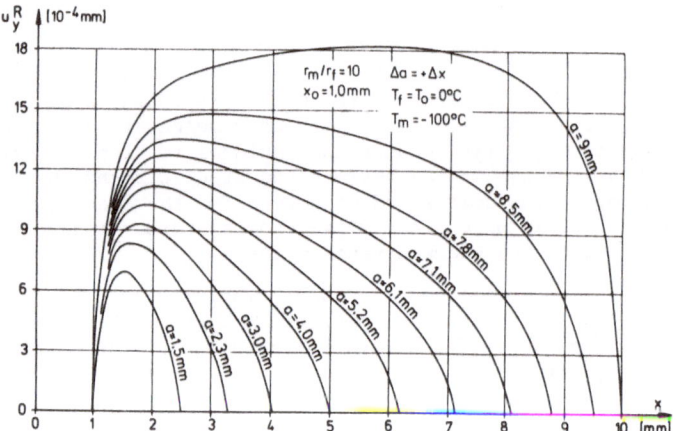

Fig. 13 - Crack edge displacement for the upper crack face S_R^+ as
functions of crack length and of the distance x from the
interface S_{fm} for the composite structure Carbon fibre/
Aluminium matrix (circular unit cell)

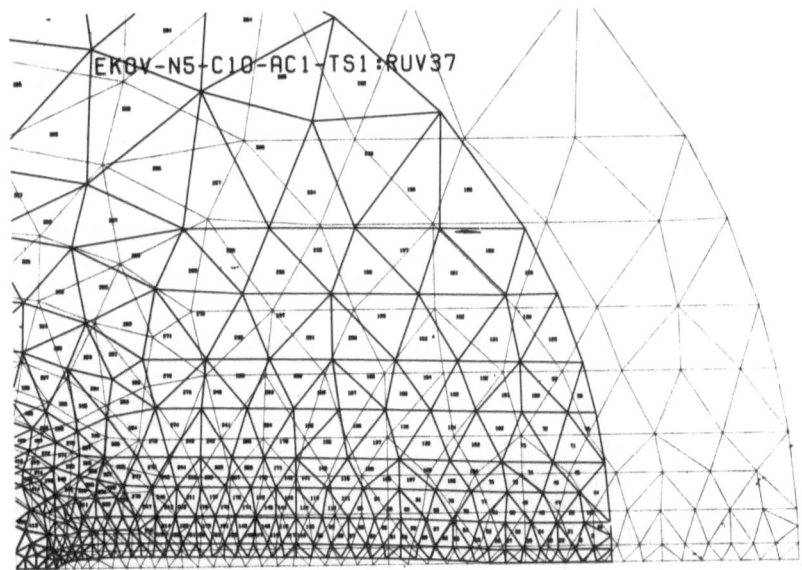

Fig. 14 - Finite element mesh-work for the cracked composite structure Carbon fibre/Aluminium matrix in the environment of the crack before and after thermal loading (circular unit cell)

Finally, the figures 15-16 show the crack extension force G_I and the opening mode stress intensity factor K_I for a cracked hexagonal unit cell (material combination Molybdenum fibre/Al_2O_3 matrix, cf. tables 5-6) and for an isothermal cooling of the specimen. By using the finite element mesh-work from Fig. 6 the numerical calculations were performed for se- veral starting positions x_o of a matrix crack. Therein Fig. 15 gives the crack extension force G_I as functions of the starting point x_o as well as of the quantity $(a-x_o)$. A comparison with the results obtained in Fig. 9 shows the same characteristic behavior of the crack extension force in de- pendence on crack length and for a fixed starting point x_o of a crack. How- ever, there exists a linear relationship between the crack extension force G_I and the variable x_o when other parameters are fixed. Therefore, the length of a crack traversing the fibre-matrix interface S_{fm} and arrested in the fibre is linearly dependent on the variable x_o. Fig. 16 gives the corresponding stress intensity factors in dependence on crack length and with the starting point x_o of a matrix crack as a parameter.

Molybdenum			
T [°C]	E [N/mm^2]	α [$10^{-6} \cdot K^{-1}$]	ν
0	324700	4.76	0.310
200	308000	5.05	0.313
400	288400	5.45	0.316
600	266800	5.92	0.319
800	240300	6.38	0.322
1000	212900	6.88	0.326
1200	168700	7.42	0.330

Al$_2$O$_3$			
T [°C]	E [N/mm^2]	α [$10^{-6} \cdot K^{-1}$]	ν
0	346700	7.8	0.260
200	340000	8.4	0.265
400	333400	9.0	0.270
600	324500	9.6	0.280
800	314900	10.2	0.310
1000	307700	10.8	0.365
1200	296500	11.4	0.450

Table 5-6 – Material properties of the composite structure
Molybdenum fibre/Al$_2$O$_3$ matrix in dependence
on temperature

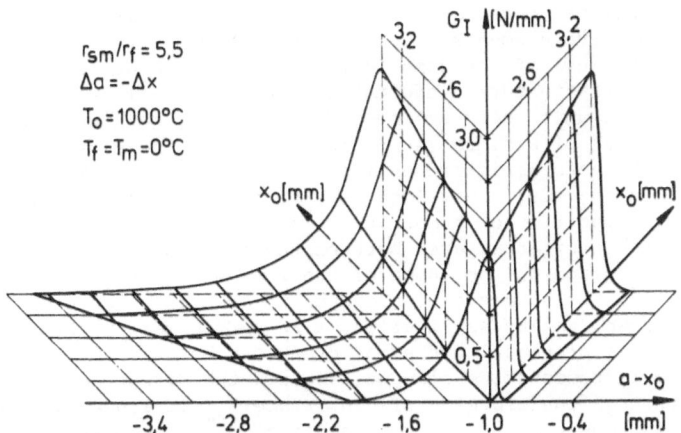

Fig. 15 - Crack extension force as functions of the starting position
x_o of a matrix crack and of the quantity $(a-x_o)$
(hexagonal unit cell)

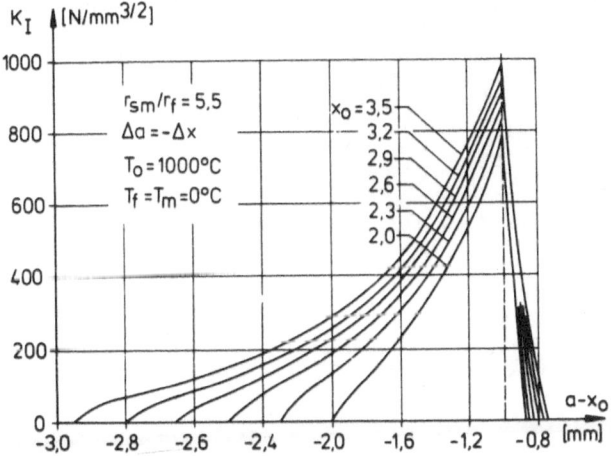

Fig. 16 - Opening mode stress intensity factor in dependence on crack
length and with the quantity x_o as a parameter
(hexagonal unit cell)

CONCLUSIONS

Investigations concerning the thermal shock resistance of two types of representative elements of a low-fibre concentration composite were performed. Because of the complicated shape of a cracked composite structure microstructural informations about thermal crack propagation in appropriate unit cells are needed in order to gain an understanding of the strength of a fibre reinforced material. Therefore, the determination of characteristic properties like as the crack extension force and the opening mode stress intensity factor in dependence on the length and starting position of a matrix crack as well as on the position and shape of the interfacial area between the individual phases of a compound material and on the shape of the free surface is useful for the assessment of the strength of a composite structure. Furthermore, it could be shown that by application of an appropriate macroscopic self-stress field a propagating thermal crack can be arrested inside of the fibre of a representative element. Therein the arrest length of a crack getting through the fibre-matrix interface S_{fm} is linearly dependent on the starting position x_o of the crack inside of the matrix material. These results represent a first step on the way of a fully understanding of the strength of fibre reinforced composites for which the interaction of an ensemble of cracked unit cells has to be investigated.

ACKNOWLEDGMENT

The research on the thermal fracture of composite materials reported herein was accomplished over a period of two years by the authors and was funded by the Deutsche Forschungsgemeinschaft of the Federal Republic of Germany.

REFERENCES

[1]   Corten, H. T., "Fracture mechanics of composites", Fracture,
      Vol. VII, H. Liebowitz, ed., Academic Press, New York, pp. 675-769,
      1972

[2]   Sih, G. C., Hilton, P. D., Badaliance, R. and Villarreal, G.,
      "Exploratory development of fracture mechanics of composite systems",
      Technical Report AFML-TR-70-112, Air Force Materials Laboratory,
      Wright-Patterson Air Force Base, Ohio, 1973

[3]   Broutman, L. J., ed., "Fracture and fatigue", Composite Materials,
      Vol. 5, Academic Press, New York, 1974

[4]   Erdogan, F., "Fracture of composite materials", Prospects of
      Fracture Mechanics, G. C. Sih, H. C. van Elst and D. Brock, eds.,
      Noordhoff International Publishing, Leyden, pp. 477-491, 1974

[5]   Cooper, G. A. and Piggott, M. R., "Cracking and fracture in
      composites", Fracture, Vol. 1, D. M. R. Taplin, ed., University of
      Waterloo Press, Waterloo, pp. 557-605, 1977

[6]   Hieke, M. and Loges, F., Z. Angew. Physik, Vol. 22, p. 14, 1966

[7]   Kassir, M. K. and Sih. G. C., "Three-dimensional thermoelastic
      problems of planes of discontinuities or cracks in solids",
      Developments in Theoretical and Applied Mechanics, Vol. 3,
      W. Shaw, ed., pp. 117-146, 1967

[8]   Brown, E. J. and Erdogan, F., Int. J. Engng. Science, Vol. 6,
      p. 517, 1968

[9]   Bruy, E., Ph. D. Dissertation, University of Stuttgart, 1973

[10] Bregman, A. M. and Kassir, M. K., Int. J. Fracture, Vol. 10, p. 87,
      1974

[11] Herrmann, K., Mechanics Research Communications, Vol. 2, p. 85,
      1975

[12] Herrmann, K. and Fleck, A., "Thermal fracture in compound materials",
      Fracture, Vol. 3, D. M. R. Taplin, ed., University of Waterloo
      Press, Waterloo, pp. 1047-1054, 1977

[13] Herrmann, K. and Fleck, A., Mechanics Research Communications,
      Vol. 4, p. 373, 1977

[14] Herrmann, K., "Interaction of cracks and self-stresses in a
      composite structure", Continuum Models of Discrete Systems, SM
      Study No. 12, J. W. Provan, ed., University of Waterloo Press,
      Waterloo, pp. 313-338, 1978

[15] Hasselman, D. P. H., J. of the American Cer. Soc., Vol. 52,
      p. 600, 1969

[16] Nakayama, J., "Thermal shock resistance of ceramic materials",
      Fracture Mechanics of Ceramics, Vol. 2, R. C. Bradt, D. P. H.
      Hasselman and F. F. Lange, eds., Plenum Press, New York,
      pp. 759-778, 1974

[17] Hegemier, G. A., "On a theory of interacting continua for wave
      propagation in composites", Dynamics of Composite Materials,
      E. H. Lee, ed., ASME, New York, pp. 70-121, 1972

[18] Herrmann, K., "On self-stresses in dissimilar solids I, II",
      Beiträge zur Spannungs- und Dehnungsanalyse, Vol. VI, K. Schröder,
      ed., Akademie-Verlag, Berlin, pp. 21-52, 1970 (in German)

[19] Herrmann, K. and Mattheck, C., "Thermal stresses in the unit cell
      of a fibre-reinforced material", to be submitted for publication

[20] Hayes, D. J., Int. J. Fracture, Vol. 8, p. 157, 1972

[21] Bueckner, H. F., ZAMM, Vol. 50, p. 529, 1970

[22] Braun, H., Fleck, A. and Herrmann, K., Int. J. Fracture, Vol. 14,
      R 3, 1978

MULTIPLE FRACTURE OF LAMINATES

A. Kelly

University of Surrey, Guildford, Surrey, England

Multiple fracture is of widespread occurrence and occurs in situations as diverse as the cracking of a brittle lacquer on a ductile metal [1] to the cracking of carbide whiskers in a high temperature eutectic [2]. It also occurs in high performance laminates [3,9]. A theory to account for many of the effects has been given by Aveston,Cooper and Kelly [5] and Aveston and Kelly [6,7].

The phenomenon occurs when in a multi-component system one component breaks at a smaller strain than the other and in addition there is sufficient evidence of the high elongation component to bear the total load at first cracking strain. This condition is

$$V_H \geq \frac{\sigma_{Lu}}{\sigma_{Hu}+\sigma_{Lu}-\sigma_H'} \tag{1}$$

where V represents volume fraction and σ the stress. The subscripts L and H refer to the lower elongation component (LE) and higher elongation HE respectively and u represents ultimate*. σ_H' is the stress on the HE component at failure of the LE component. In terms of strain, if the components are both elastic after the first cracking strain, ε_{Lu} then

$$\varepsilon_{Hu} \geq \varepsilon_{Lu}(1+\alpha) \tag{2}$$

where ε_u is the failure strain.

$$\alpha \equiv (\frac{E_2 V_L}{E_H V_H})$$

_____

*In what follows, the subscript c refers to the total composite (laminate).

where E is Young's modulus. This simple argument assumes there is no concentration of stress on the high elongation component when the component of lower elongation breaks, which is often a good assumption.

If equation (1) is obeyed and the composite is strained above ε_{Lu}, then a series of parallel cracks appears in the low elongation phase. The spacing of these and their opening depends on the absolute scale of size of the microstructure; that is, in a laminate on the dimensions of the individual lamellae. Expressions for the spacing of the cracks, for the separation of the faces of the crack and the cracking strain depend on whether the two components remain elastically bound to one another after the initial cracking. This depends on the maximum inter-facial shear stress at the crack and on the strength of the bond between the two components. In either case, however, the formation of a single crack results in relaxation of the broken component on either side of the crack, and hence an extension δl of the specimen. The additional load is borne by the high elongation component only over a short distance on either side of the crack, the transfer length. For a frictional bond, this transfer length depends on the limiting bond strength and for the elastic dase, where the bond remains intact, then it depends on the shear modulus. The important point is that as the lamellae of the low elongation phase are made smaller, the rate of load transfer from the component bridging the crack is increased because the area of interface between the two components per unit of volume of the composite is increased. It follows that the length additionally strained of this component and hence the displacement of the ends of the specimen δl is decreased. The product of the cracking stress $E_c \varepsilon_{Lu}$ and the displacement δl sets an upper limit to the work available from the loading system to form the crack. So, if the surface work of fracture $2\gamma_2$ remains constant and δl is decreased as a result of decreasing the size of the low elongation lamella, a point is reached where the strain on the composite must increase above the value of ε_{Lu} observed outside the composite before the required work of fracture can be extracted from the system.

Consider the 0°/90°/0° laminate in Figure 1 strained parallel to y. The 0° lamellae represent the HE component and if the composite is stretched, cracking of the 90° lamellae cannot occur unless a strain given by one of the two following formulae, Equations (3) and (4) are exceeded:

$$\varepsilon = \left\{\frac{6\gamma_L V_L \tau}{E_c E_H \alpha^2 b}\right\}^{1/3} \tag{3}$$

[8]. If at the interface between the 0° and 90° lamellae relative displacement occurs at a constant shear stress τ. If elastic continuity is maintained across the interface then

$$\varepsilon = \left\{\frac{2\gamma_L V_L \tau_{max}}{\alpha^2 E_c E_H b}\right\}^{1/3} \tag{4}$$

[7]. When cracking has occurred, the spacing of the cracks is given by

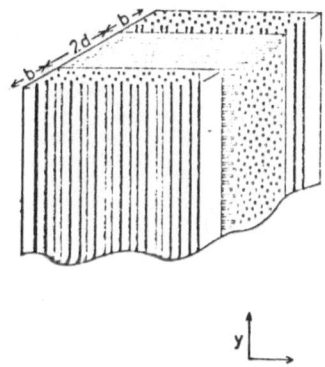

Fig. (1) - A 0°/90°/0° laminate - note dimensions b and d

$$x = \frac{\varepsilon_{Lu} E_H \alpha b}{\tau} \tag{5}$$

or by

$$x = \frac{\varepsilon_{Lu} E_H \alpha b}{\tau_{max}} \tag{6}$$

In formulae (4) and (6), the maximum interfacial shear stress is given by

$$\tau_{max} = \Delta\sigma_0 \left[\frac{G_L E_C}{E_H E_L V_L (d/b)} \right]^{1/2} \tag{7}$$

x is the minimum crack spacing, ($2x$ is the maximum), b is the thickness of the high elongation lamella, 2d the thickness of the low elongation lamella (Fig. 1). 2γ is the total fracture surface work of the low elongation lamella. The term $\Delta\sigma_0$ is the maximum additional stress spread thrown onto the high elongation lamellae as a result of the cracks in those of lower elongation; that is, it is the difference between the stress in the high elongation phase where it bridges the crack σ_a/V_H and the corresponding stress at a cross section where there is no crack, that is $(\sigma_a/E_C) \cdot E_H$ where σ_a is the applied stress on the laminate. Thus an expression for $\Delta\sigma_0$ is as follows:

$$\Delta\sigma_0 = \left\{ \frac{\sigma_a}{V_H} - \frac{\sigma_a E_H}{E_C} \right\} \tag{8}$$

If the two lamellae are <u>not</u> elastically bonded complete cracking of the LE lamella is predicted to occur at a specific strain ε_{Lu} and the final crack spacing is reached immediately. If the two phases are elastically bonded, the crack spacing lies between x and 2x where x is given by

$$x = - \left[\frac{E_H E_L bd^2}{G_L E_c (b+a)}\right]^{1/2} \ell n \left[1 - \frac{\varepsilon_{Lu} E_c}{\sigma_a}\right] \tag{9}$$

and so varies with the strain (applied stress) on the composite. Equation (9) reduces to equation (6) by using equation (8) for σ_a greater than a few times $\varepsilon_{Lu} E_c$. If the two phases are not debonded from one another, the value of τ_{max} can be deduced from formula (7). If debonding has occurred which is probable if τ_{max} is much greater than σ_{Lu}, then τ is due to sliding friction between the two components. This is the case considered by Stevens and Lupton [3]. Finer lamellae always mean more cracks per unit length of the specimen.

The maximum opening of each crack is

$$B \sim \varepsilon_{Lu}(1+\alpha)x \tag{10}$$

so it is clear that the crack opening decreases with x, i.e., with the crack spacing or transfer length, and hence inversely as τ or τ_{max}, equations (5) and (6). Obviously, thinner lamellae lead to smaller openings of the cracks.

If crack openings are to be minimized, then the two components and the interface between them must be able to withstand the value of τ_{max} without failure. That is, the shear strength of the interface and of the components must be large. τ_{max} is given by

$$\tau_{max} = \varepsilon_{Lu} E_L \left\{\frac{E_c G_L}{E_H E_L}\left(1 + \frac{V_H}{V_L}\right)\right\}^{1/2} \tag{11}$$

and hence increases with increase of the cracking strain the the LE lamella.

It may be desired to increase the cracking strain of a lamella by reducing its dimensions so that the first cracking is constrained and the value of the cracking strain increased above that found in an individual lamella outside the composite. Provided the value of τ_{max} does not lead to interfacial failure, so that the components must remain elastically bonded then the cracking strain varies inversely to the square root of the thickness being given by:

$$\varepsilon_{Lu} = \{\frac{2\gamma_L V_L}{E_c \alpha} \left(\frac{E_c G_L}{E_H E_L}\right)^{1/2} \frac{1}{V_H^{1/2} d}\}^{1/2} \tag{12}$$

which is equivalent to equation (4). However, if the ability to constrain the cracking is governed by the highest value attainable for the shear stress between the cracked and uncracked components, then the maximum value of the cracking strain which can be attained is given by equation (3) with τ a constant and so ε_{Lu} varies more slowly with the decrease in the thickness of the low elongation plate. When elastic coherence between the two components breaks down, τ_{max} is replaced by τ, the sliding friction at the interface, and the constant 6 then replaces 2 - compare equations (3) and (4) since additional frictional work has to be done by the applied stress when cracking occurs [6].

Cracking of the laminate can also arise due to any form of internal strain, such as due to a temperature change and to control cracking of the low elongation phase, one has to take into account the strains due to the various distortions such as that due to a temperature change plus that provided by the externally applied stress.

DIFFERENTIAL THERMAL STRESSES

A laminate is normally cured above its service temperature though considerable strains may exist in the lamellae as a result of differential thermal contraction. This problem is particularly acute with cross-ply laminates containing carbon fibre on account of its highly unisotropic thermal properties, for example, prepreg sheets of surface treated type I carbon fibres in shell D x 210 resin have a longitudinal thermal expansion coefficient of $-0.7 \times 10^{-6}/°C$ and a transverse expansion coefficient of $40 \times 10^{-6}/°C$. Each ply in a 0°/90°/0° laminate is then usually in tension normal to the fibres and since the failure strain in this direction is usually small - 0.47%, there is an obvious danger that advanced carbon fibre composites may be cracked before they enter service.

Aveston, Kelly and Cooper [5] have shown that the relation between cracking strain and the dimension of the low elongation phase is the same under external load and due to thermal loading, so the theory sketched above can be applied. The constrained thermal strain in each component of a 0°/90° lamella is given by:

$$\varepsilon_L^° = \Delta T \frac{E_H V_H}{E_c} (\alpha_L - \alpha_H) \tag{13}$$

and

$$\varepsilon_H^° = \Delta T \frac{E_L V_L}{E_c} (\alpha_H - \alpha_L) \tag{14}$$

where α is the thermal expansion coefficient and ΔT is the difference between the service temperature and that below which thermal stresses are frozen in. This temperature at which thermal stresses are frozen in is often difficult to determine. The stresses, as is seen from equations (13) and (14) are independent of fibre or laminate thickness. It is possible to control the cracking by specifying the maximum ply thickness that would yield an uncracked laminate for a given ΔT.

DIFFERENTIAL POISSON STRESSES

Consider the 0°/90°/0° laminate in Figure 1. It has not been pointed out in the literature before, as far as I am aware, that because of the small tensile cracking strain normal to the fibres, the presence of the 90° fibres parallel to x will tend to promote longitudinal splitting of the 0° lamellae. An example of the phenomenon is illustrated in Figure 2. As with thermal strains, the strains generated by this effect depend only on volume fraction but, whether these lead to cracking or not, can be made again to depend on the thickness of the lamellae by making the lamellae sufficiently thin.

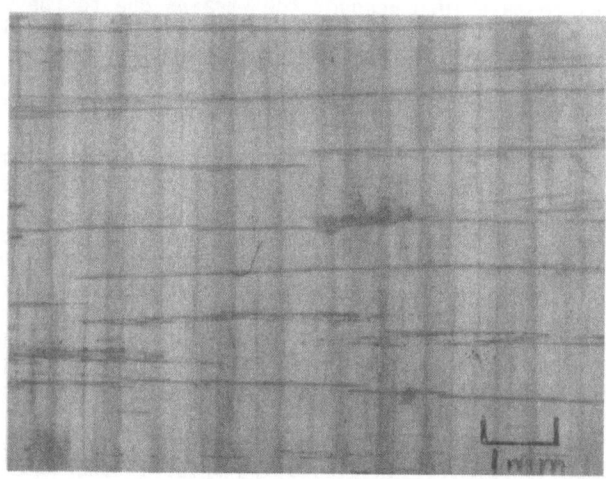

Fig. (2) - Transverse multiple cracking - vertical - and longitudinal splitting - horizontal - in an E-glass epoxy cross-ply laminate

To estimate the effect, Figure 3, let subscripts 1 and 2 refer to the directions in individual lamellae parallel and at right angles to the fibres respectively and let x and y be the principal stress directions for the total laminate. We call ν_{12} the contraction along 2 when the laminate is extended along 1, that is, it is the principal Poisson's ratio. For individual lamellae in plain stress

$$E_{11}\nu_{21} = E_{22}\nu_{12} \tag{15}$$

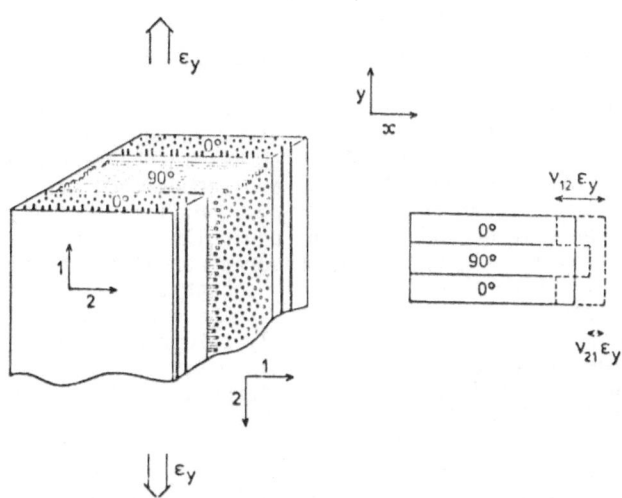

Fig. (3) - Differential Poisson strains in a cross-ply laminate

Referring to Figure 3, stress equilibrium in the direction yields

$$E_{22}\epsilon_x(0°)2b + E_{11}\epsilon_x(90°)2d = 0 \tag{16}$$

and strain compatibility in the x direction gives for a strain ϵ_y of the composite

$$\epsilon_x(0°) - \epsilon_x(90°) = (\nu_{12} - \nu_{21})\epsilon_y \tag{17}$$

and hence substituting into equation (16) we have

$$\epsilon_x(0°) = \frac{E_{11}\frac{d}{d+b}(\nu_{12}-\nu_{21})\epsilon_y}{E_{22}\frac{d}{d+b} + E_{11}\frac{b}{d+b}} = \frac{E_{11}\frac{d}{d+b}(\nu_{12}-\nu_{21})\epsilon_y}{E_{cL}} \tag{18}$$

where E_{cL} is the Young's modulus along the x direction of the laminate.

In general, E_{11} is much greater than E_{22} and therefore ν_{21} much less than ν_{12}. For example, in glass polyester of $\nu_j = 0.3$, $\nu_{12} = 0.3$, $E_{11} = 25$ GN/m²

$v_j = 0.3$, $v_{12} = 0.3$, $E_{11} = 25$ GN/m² and $E_{22} = 4$ GN/m² so $(v_{12}-v_{21}) = 0.25$. If
If say b=d, then $\varepsilon_x(0°)$ is approximately 0.22 ε_y so that in the absence of constraint effects, we expect to observe longitudinal splitting as a total laminate strain of some 4 to 5 times the first cracking strain. This is in fact observed in the experiments of Garrett and Bailey [4] who observed longitudinal cracks in the outer plies of a polyester-glass laminate made up as in Figure 1, the individual plies of which had failure strains normal to the fibres of about 0.2%. Its occurrence is now becoming well-known and recognised. It may obviously occur in all laminates whether these are composed of individual lamellae of the same or of different fibres and it may be controlled either by arranging for an increase in the tensile failure strain measured normal to the fibres or, if this cannot be done, by making the 0° layer thin enough to restrain the cracking according to the theory above. Again, it has to be borne in mind that the differential thermal expansion effects will usually take place, and put the 0° layer in a laminate under transverse tension. This makes it crack at a lower additional strain than expected, due to the differential Poisson effect. Bailey, Curtis and Parvizi [10] are investigating these effects.

The estimate of strain parallel to x induced in the 0° ply by stretching the composites along y given by equation (18) will be an underestimate because, unless the transverse failure strain of the 0° plies is much less than that of the 90° plies the latter will always crack first. The strain in the 90° layer will then be reduced becoming zero at the crack and the longitudinal strain in the 0° plies will be increased where they bridge the crack to a value $\varepsilon_y(1+\alpha)$ so that the maximum transverse strain in the 0° layers is

$$\varepsilon_x(0°) = v_{12}(1+\alpha)\varepsilon_y \qquad (19)$$

and the mean strain will lie between $v_{12}(1+\alpha/2)\varepsilon_y$ and $v_{12}(1+3\alpha/4)\varepsilon_y$ depending on whether the crack spacing is 2x or x. It follows that if α is greater than about 2, we expect longitudinal splitting to be initiated by the transverse cracking in the absence of constraint and with α greater than about 3 to be completed at the completion of transverse cracking. For given relative amounts of the two plies, reduction in the thickness of one of course automatically results in the reduction in thickness of the other so that longitudinal splitting may be reduced both on account of the increased transverse failure strain of the longitudinal plies and by the suppression of the initiating cracks in the 90° layers.

The glass epoxy specimen shown in Figure 2 consisted of ten 90° lamellae sandwiched between two outer 90° lamellae so that from equation (19), $\varepsilon_o(0°)$
= 0.9 ε_y. From equations (13) and (14) the maximum thermal tensile strains in the 0° and 90° layers are 0.28% and 0.11% respectively. From equation (12), the corresponding constrained cracking strains are 0.46% and 0.16%. This compares with the normal transverse cracking strain of 0.22%. We therefore expect transverse splitting to begin at a strain of $(0.46-0.28)/0.9 = 0.2$% and to be complete at the limit of multiple fracture. This is in fact observed. At the commencement of multiple cracking small cracks form at right angles to the transverse

cracks and by the completion of multiple cracking these join to form the pattern shown in Figure 2.

A very detailed application of the theory of transverse cracking of laminates has been carried out by Parvizi, Garrett and Bailey [11] who have studied cracking in 0°/90°/0° glass-epoxy laminates. Figure 4 shows the values of the

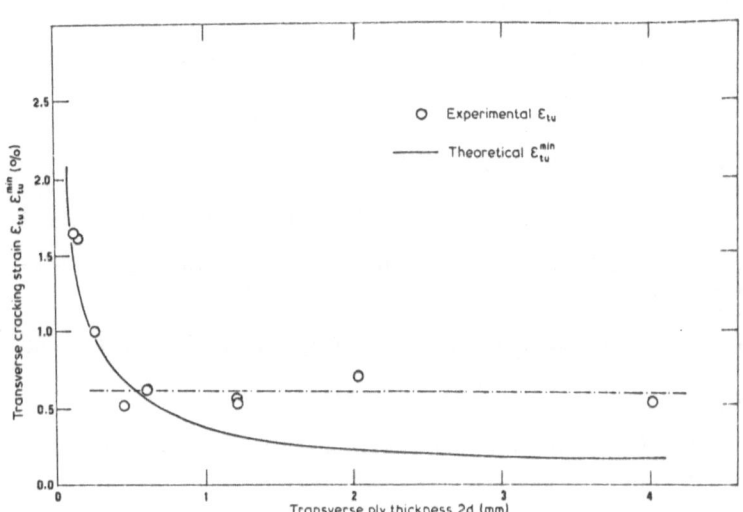

Fig. (4) - Plot of minimum cracking strain of 90° plies in a glass fibre epoxy cross-ply laminate versus ply thickness. The full line is equation (12) - Reference [11]

minimum cracking strain of the 90° plies as a function of ply thickness 2d. The horizontal line depicts the limiting values of ε_{tu} for large inner ply thicknesses, and the full line shows the theoretical curve. It is seen that very considerable increase in the cracking strain of the 90° plies obtained and the agreement with theory seems very satisfactory. The only real uncertainty is that these workers neglected thermal effects.

ACKNOWLEDGEMENTS

I am grateful to Mr. J. Aveston for permission to use our joint results in this paper and to Professor Bailey and his co-workers at the University of Surrey for allowing me to refer to their results.

REFERENCES

[1] Durelli, A. J. and Okubo, S., Proc. Soc. Exptl. Stress Analysis, 11, pp. 153-160, 1953.

[2] Bibring, H., Rabinovitch, M. and Khan, T., CR. Ac. Sc., 257c, pp. 1475-1478.

[3] Stevens, G. T. and Lupton, A. W., J. Mat. Sci., 12, pp. 1706-1708, 1977.

[4] Garrett, K. W. and Bailey, J. E., J. Mat. Sci., 12, pp. 2189-2194, 1977.

[5] Aveston, J., Cooper, G. A. and Kelly, A., Conference Proceedings, National Physical Laboratory. The Properties of Fibre Composites, IPC Science and Technology Press, pp. 15-16, 1971.

[6] Aveston, J. and Kelly, A., J. Mat. Sci., 8, pp. 352-362, 1973.

[7] Aveston, J. and Kelly, A., Proc. Royal Soc. London, in press, 1978.

[8] Argon, A. S. and Shack, W. J., Rilem Symposium Fibre Reinforced Cement and Concrete, Construction Press, Ltd., pp. 39-53, 1975.

[9] Garrett, K. W. and Bailey, J. E., J. Mat. Sci., 12, pp. 157-168, 1977.

[10] Bailey, J. E., Curtis, P. T. and Parvizi, A., in press, 1978.

[11] Parvizi, A., Garrett, K. W. and Bailey, J. E., J. Mat. Sci., 13, pp. 195-201, 1978.

POLYMER REINFORCEMENT FROM THE VIEWPOINT OF FRACTURE MECHANICS

W. G. Knauss
Graduate Aeronautical Laboratories, California Institute of Technology

H. K. Mueller
E. I. DuPont de Nemours & Co.

ABSTRACT

When cracks propagate in filled solids their advance is modified by the inclusion of the rigid filler phase. In this paper we estimate the reduction in average crack growth rate or, alternately, the elongation of the failure time scale due to the presence of aligned, rigid, well bonded cylindrical fillers. The filler volume fraction is assumed to be sufficiently small to prevent important filler-filler interactions and the crack propagates parallel to the filler particles. A simple, approximate equation is found relating the average rate of crack propagation v_f (time to failure) in the filled solid to the rate v_o for the unfilled solid as a function of volume fraction of filler by

$$v_o/v_f = 1 + 3.2\phi \quad (\phi < 0.2) .$$

INTRODUCTION

Fracture resistance is only one of many performance criteria by which materials in general, and reinforced polymers in particular, are evaluated for application in engineering design. Especially critical is such a property where primary load carrying ability is involved, whether the polymer is filled or unfilled.

The term polymer reinforcement carries several connotations. On the one hand one speaks of polymer reinforcement when the addition of a (solid) phase to a matrix results in a composite with increased rigidity due to the addition (modulus reinforcement). On the other hand one speaks of polymer reinforcement when failure or tear properties of the material are involved. Both types of reinforcement are, of course, not independent, as is well known, if not completely understood, from elastomer "reinforcement" with carbon-black.

In view of the work discussed later on it is important to distinguish "reinforcement" by size of the relatively rigid phase added. During the process of crack propagation a failure zone at the tip of the crack develops. The size of this zone varies from one material to another and it is necessary to distinguish between reinforcing filler material which is small compared to the dimension of this failure zone and filler material which is large compared to it. If the filler particles are small compared to the process or failure zone one expects that the major influence of the filler is imparted primarily through changes in the rheological (continuum) characteristics of the material. Specifically, the particle dimensions are then so small compared to the next larger and important dimension in the fracture process that the particle size does not affect the continuum aspects of the fracture process. While the appearance and extent of the fracture zone may well be changed one would expect that the time dependent deformation around the moving crack tip is influenced more strongly. The reinforcing particles affect fracture in this case primarily by changing the "rigidity" of the material and by changing the relaxation spectrum and thus the time-dependence of the material response characteristics.

An illustration of the latter effect is given by comparing the mechanical behavior of filled and unfilled elastomers. As one example we cite the relaxation behavior of H-C rubber (Thiokol Chemical Corp.) unfilled and highly filled (about 84% by weight) with ammonium perchlorate and aluminum powder (solid propellant). Figure 1 shows the relaxation modulus of the unfilled · rubber together with the response for the highly filled material. It is very obvious that the filler changed the relaxation times by many orders of magnitude but left the short and long time moduli practically unaffected. Therefore, if one were to compare the filled and the unfilled material at similar deformation rates one would find the filled material very much stiffer -- but merely by virtue of the changes in the rate of relaxation.

The carbon complexes in carbon-black reinforced elastomers are probably always small relative to any other microstructural feature of the reinforced material. Figure 2 shows crack growth data for pure and carbon-black filled (30% by weight) Butarex elastomer* and provides another example of the effect of filler induced changes in gross rheological properties on the crack propagation process. Clearly the crack propagates in the unfilled material considerably more rapidly than in the filled material. Alternately, in order to achieve the same rate of fracture the filled material needs to be stressed more than the unfilled one, thus giving the impression that under comparable

---

*The data were obtained some 15 years ago when the material was supplied by the then Rocketdyne Division of North American Aviation in McGregor, Texas.

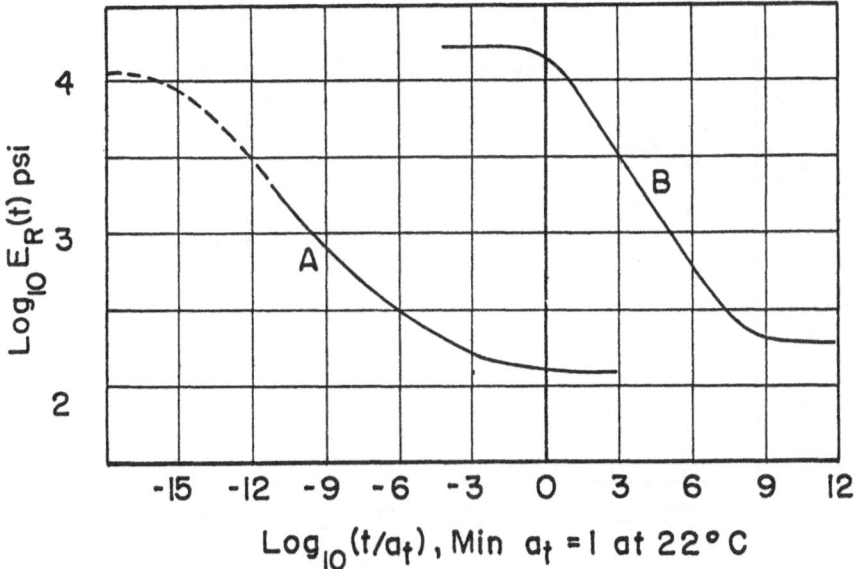

Fig. 1. Relaxation behavior of (A) unfilled PBAA elastomer and (B) filled PBAA (about 85% by weight of Ammonium Perchlorate; solid propellant). Dashed portion of A estimated; PBAA and propellant data provided by Thiokol Chemical Corp.

rates of failure ("times-to-failure") the filled material is "stronger." However, when very slow rates of crack growth are involved the (nearly) elastic equilibrium properties of the materials are involved and the influence of viscoelastic or time-dependent processes is greatly reduced or eliminated. Under these limit conditions one finds that the strength of the two materials is nearly the same. One is therefore led to conclude that the (apparent) increase in strength results from the changes in the time-dependence of the rheological properties.

The changes in the average properties of the highly strained material in that zone must be considered as a second phenomenon relating the changes in strength with filler particles which are small compared to the fracture zone at the crack tip. Recall the special example of crack growth in the filled

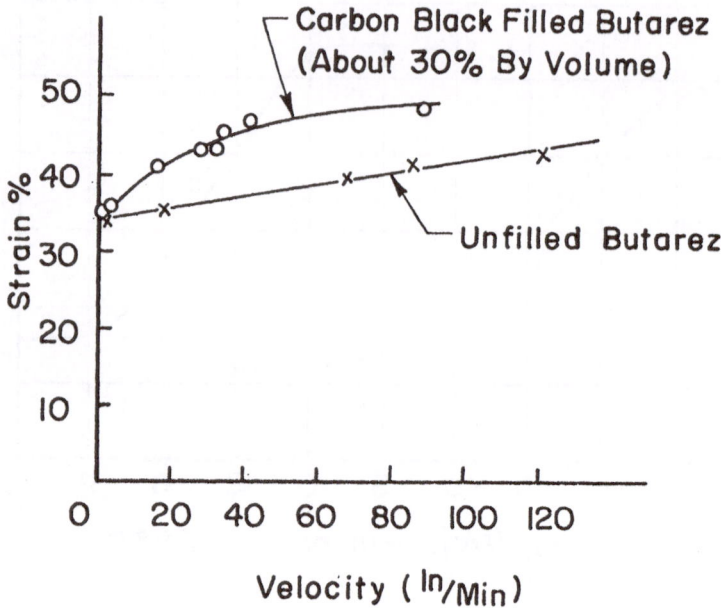

Fig. 2. Crack propagation data in constant-strain field in filled and
unfilled Butarez elastomer.

and unfilled Butarez material. This example seemed to indicate that the
"strengthening" derives primarily from the changes in the bulk rheological
properties. Nevertheless, in general one must allow that the addition of
finely dispersed particles will affect the ultimate cohesive strain inside
the fracture zone and, as a result the size of the fracture zone itself.
If this change is such that the cohesive zone increases then the likelihood
exists that more micro-fracturing within the cohesive zone occurs and this
may be relfected in an increase in the fracture energy. It appears that no
work has been reported in the open literature on the basis of which a more
factual evaluation could be based.

A third phenomenon affecting the time-dependent strength occurs when the
filler particles are large compared to the fracture zone at the tip of a
crack. In that event the stress field interaction between the tip of an

advancing crack and the filler particles becomes important. This paper deals with crack growth in this type of reinforced material.

As an initial attempt at examining the retardation of crack growth by relatively large and rigid particles we considered two-dimensional or planar geometries as an analogue to the three-dimensional case. This problem is of direct interest for fiber reinforced solids in which cracks propagate in dia- metrical planes of the cylindrical inclusions, such as shown in fig. 3. While we are not addressing specifically the problem of epoxy based fiber reinforced composites, the results are of some consequence even though we deal with solids loadings that are lower than commercial advanced composites.

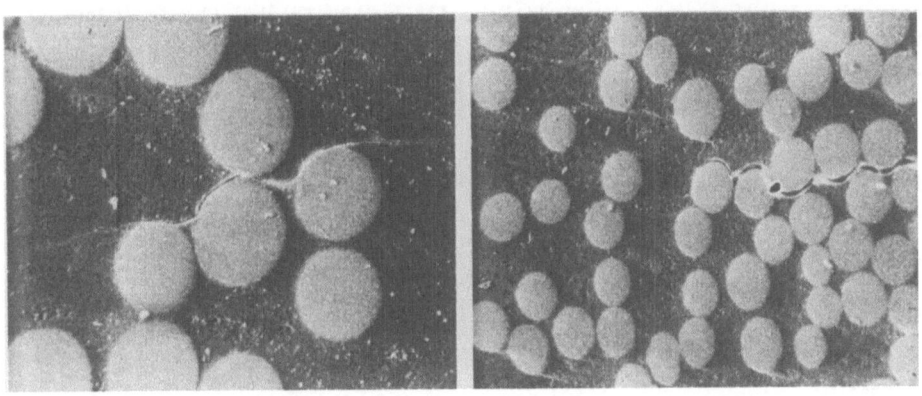

ε ≃ 0.2%   ε ≃ 0.4%

Fig. 3. Crack propagating parallel to glass fiber reinforcement
         (dia. = 10-14 μm) in polyester. Courtesy of H. Brintrup,
         IKV, Aachen, Germany.

If a crack approaches a hard and well bonded inclusion the stresses at the crack tip decrease by an amount depending on the alignment of the crack plane with the diametrical plane of the inclusion. We discuss first some experiments in sheet geometries containing circular rigid inclusions and holes. Next we consider analytically the growth of a crack approaching such single inclusions or holes along a diametrical plane; this is done by drawing on a model of crack propagation in viscoelastic solids developed in refs. 1 and 2 and substantiated in an exact way in ref. 3. Finally drawing on both these experimental and analytical results we derive a relation between the volume fraction of filler and the retardation of fracture (or life extension).

EXPERIMENTS

A convenient test geometry to determine the effect of the stress field around a rigid inclusion on the speed of crack growth is depicted and dimensioned in fig. 4. The specimen shown is sometimes referred to as a pure shear specimen (ref. 4). Although in the immediate crack tip vicinity the deformation field is essentially one of plane strain by virtue of the rigidity of the inclusion, some distance removed from it (~ 2.5 mm) one expects conditions more appropriately described by a state of generalized plane stress and we therefore assume that the latter state prevails everywhere. The test material is a polyurethane elastomer with a glass transition temperature of approximately $-18^{\circ}C$, commercially sold as a two component system (resin and catalyst) under the trade name Solithane 113 by the Thiokol Chemical Corporation. The composition used in this study is 50% resin and 50% catalyst (by volume).

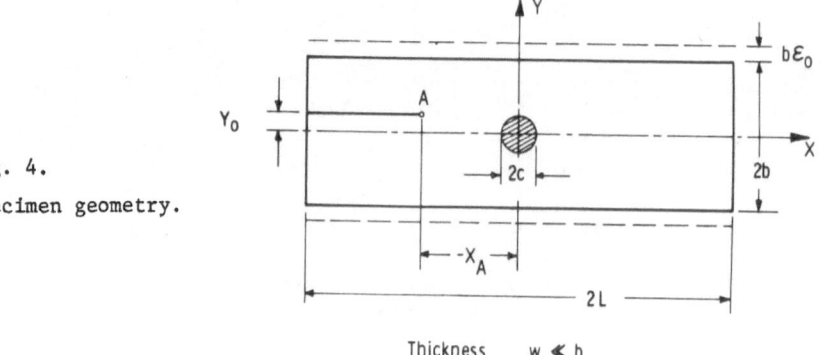

Fig. 4.

Specimen geometry.

The path of crack propagation in the vicinity of the inclusion depends on the initial approach path and on the strength of the interfacial adhesion. If the interfacial bond is "weak" the approaching crack tip will cause unbonding and crack propagation will be governed by the breakdown of adhesion instead of cohesion.

It was not an easy task to prepare satisfactory specimens. In fact, whether a specimen was satisfactory or not was often not evident until a test was performed; of course, only those tests which did not produce unbonding from the inclusion were useful in this study. As a direct consequence this type of difficulty in specimen preparation accounts for the relatively small number of test data reported here. We would have liked to record a more complete set of measurements but found them too time consuming and expensive.

Figure 5 shows typical crack path examples in the event that good interfacial bonding was achieved. Although crack propagation through imperfectly bonded specimens does not concern us in this study it is of interest to report here briefly on a phenomenon associated with imperfect bonding.

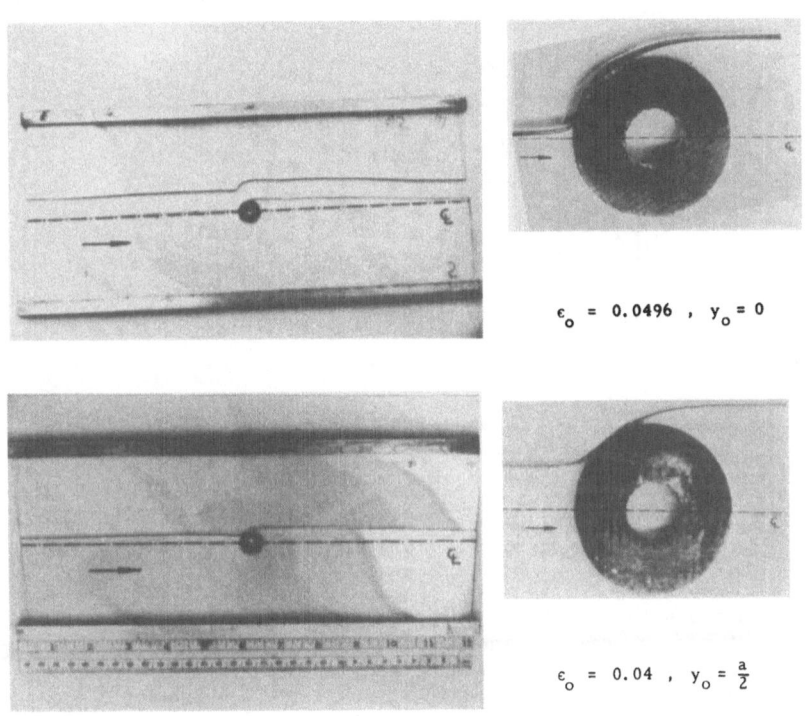

$\epsilon_o = 0.0496$, $y_o = 0$

$\epsilon_o = 0.04$, $y_o = \frac{a}{2}$

Fig. 5. Overall and detailed view of crack growth through inclusion-specimen and around inclusion. (a = inclusion radius; y_o = distance of initial crack path from diametral plane.)

In fig. 6 we record crack propagation data for a "perfectly" and "imperfectly" bonded inclusion with the initial crack path on the diametrical axis.
It is evident that the poor bond allowed the crack to follow along the interface for a longer distance than for the "perfect" bond, however, with a faster
rate of crack growth. Thus an imperfect bond may lead, on a time averaged
basis, to less retardation of crack growth than a "perfect" one. In some
instances not recorded here in detail, however, it was noticed that a very
poor bond would allow complete unbonding of the inclusion. The resulting
free circular boundary did not offer a sufficient stress riser for the crack
to propagate further. In such a particular problem average crack propagation
may be significantly reduced.

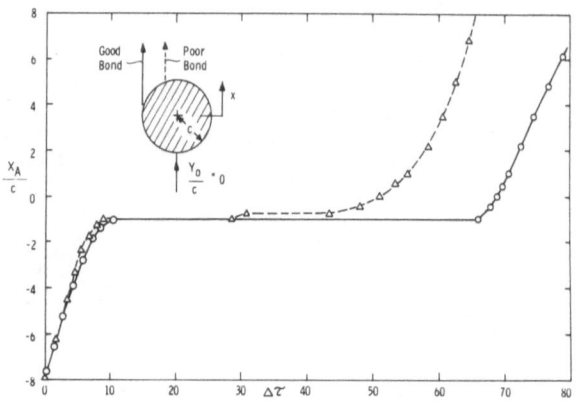

Fig. 6. Typical crack path and histories around a rigid circular
inclusion with good and poor bonding.

Figure 6 shows the crack tip position as a function of time. The steady
crack speed far away from the inclusion is denoted by v_0, it is governed by
the strain ε_0 applied to the test specimen, cf. fig. 4. The crack tip position
x_A is nondimensionalized by the inclusion radius c and the time interval $\Delta\tau$
between crack tip positions is nondimensionalized by the quotient c/v_0. In the
vicinity of the inclusion the slope does not represent the true crack tip
velocity, but only its projection on the x-axis. It is obvious that we may
view the effect of the inclusion as providing a time delay in the crack

propagation process. Most of this delay occurs during the transition from the crack approach into the interface. Once this transition is completed the crack proceeds at almost the same rate of propagation as before.

The crack propagation delay depends on the distance, y_o, of the initial crack path from the specimen centerline on which the inclusion center is located, cf. fig. 4. Growth histories for cracks propagating along paths with different distances y_o are shown in fig. 7. It is impractical to record here more than token examples of test results. Tests with double inclusions were performed to examine whether markedly different phenomena might occur; they did not occur. Many test records have been summarized in ref. 3.

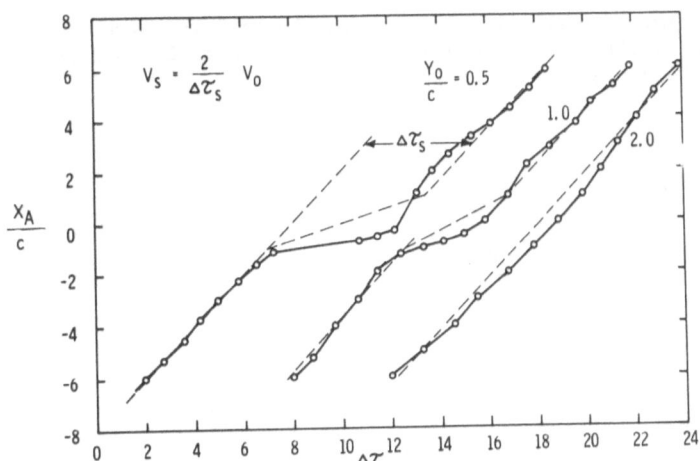

Fig. 7. Typical histories for crack tips approaching a rigid, circular inclusion at different distances from the centerline.

We may characterize the propagation around the inclusion by a reduced average speed v_s defined by the ratio of the inclusion diameter and the delay time (cf. solid lines in fig. 7). Then $v_s \equiv 2v_o/\Delta\tau_s$ is a function of the distance y_o and we may write $v_s/v_o = f(y_o)$. Figure 8 shows this velocity ratio as a function of y_o/c. It is clear from fig. 7 that crack propagation is practically unimpeded if the crack path is one diameter removed from the inclusion center.

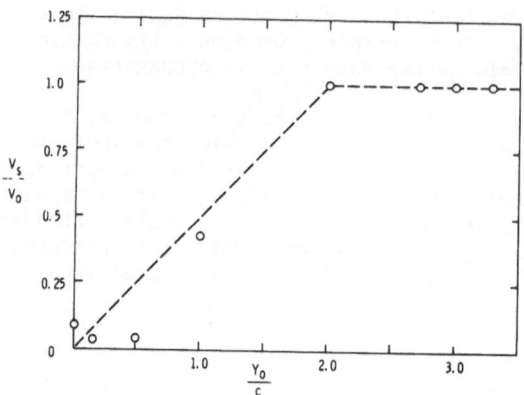

Fig. 8.   Relative crack velocity across the diameter of a rigid,
circular inclusion as a function of initial distance from
the centerline.

ANALYTICAL BACKGROUND

The rate of crack propagation v in a linearly viscoelastic solid is given
by the relation (1, 3)

$$I^2 \, \Theta \left(\frac{\alpha}{v} \right) \; = \; \text{const.} \tag{1}$$

where I is the stress intensity factor, $\Theta(\xi)$ a function very nearly equal to
the tensile creep compliance and α is the size of the cohesive zone in the
crack tip region in which the material undergoes fracture.   Derived originally
for constant rate of crack growth it is shown to be valid instantaneously as
a differential equation for crack length if

$$\frac{1}{I} \frac{dI}{dt} \; \ll \; \frac{v}{\alpha} \tag{2}$$

a condition that is easily met for the material and for the transient crack
motion under consideration.   For the Solithane material used in this study
and in the velocity range of interest one has as a good approximation

$$\Theta(t) \; = \; \text{const.} \cdot t^{1/2} \tag{3}$$

or, if t_o is some reference time

$$\frac{\Theta(t)}{\Theta(t_o)} = \sqrt{\frac{t}{t_o}} \qquad (4)$$

The crack propagation rate or at least its delay near the inclusion could be computed on the basis of (1) and a knowledge of the stress intensity factor I. A ready solution for the stress intensity factor for the general problem of arbitrary $y_o \neq 0$ was not available. However, from the experimental information we know the average reduction in speed of the crack past the inclusion. By virtue of the relation (2) we can relate the average stress intensity factor for $y_o \neq 0$ to the particular loading (ε_o) which gave rise to the experimental data. This allows us to estimate the average stress intensity factor and calculate the delay time under arbitrary loading.

We shall first calculate the crack propagation for a crack approaching a stiff inclusion and a hole (soft inclusion) on the axis of symmetry, i.e., for $y_o = 0$. Following that development we turn to an estimation of crack retardation for $y_o \neq 0$.

a) <u>Crack Growth Path on $y_o = 0$</u>: The inclusion diameter equals 1/5 of the specimen height. We therefore assume that the inclusion behaves as if it were embedded in an infinite sheet. With that notion in mind we consider the problem of a crack of length $|x_B - x_A|$ located on a ray of a circular inclusion as shown in the inset of fig. 9. The crack is embedded in the material having Young's modulus E_s and Poisson's ratio ν_s while the corresponding properties of the inclusion are E_i and ν_i. Let the stress intensity factor for the crack tip at x_A be I_A. The stress intensity factor I_o for a crack of length $|x_B - x_A|$ in a sheet without inclusion or hole is a well known quantity and the effect of the inclusion (hole) can be expressed by a correction factor defined as

$$K_A(x_A) = \frac{I_A(x_A)}{I_o} \qquad (5)$$

For the cases $E_i \gg E_s$ (rigid inclusion) and $E_i = 0$ (hole) the stress intensity correction factor K_A were calculated on the basis of ref. 5. The calculations were carried out for a very long crack with crack tip x_B fixed in position and far enough away from crack tip x_A or the inclusion to not affect the stress field around x_B. The stress intensity correction factor K_A was calculated as a function of position of x_A and was found to change noticeably as x_A/c decreases below 7. The results are presented graphically in fig. 9.

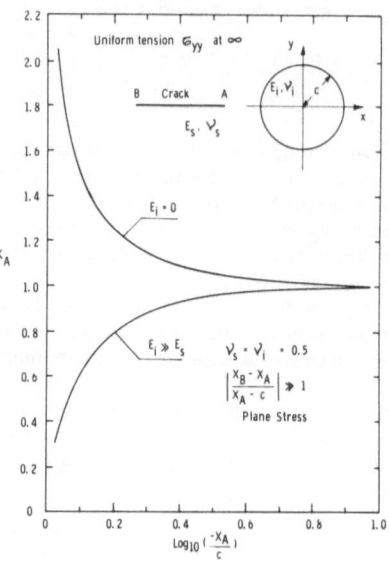

Fig. 9. Stress intensity correction factor for crack tip as a function of crack tip position.

The stress intensity factor for a very long crack in the plane stress strip geometry without inclusion is

$$I_{os} = \frac{E_s \epsilon_o \sqrt{b}}{\sqrt{1 - \nu^2}} \tag{6}$$

Assuming the specimen boundaries are sufficiently far removed from the inclusion to exclude first order modification of the crack-inclusion interaction we find the following approximate expression for the stress intensity factor for a crack approaching a circular inclusion on the centerline of a strip specimen

$$I_A = I_{os} \cdot K_A(x_A) = K_A(x_A) \frac{E_s \epsilon_o \sqrt{b}}{\sqrt{1 - \nu^2}} \tag{7a}$$

which, for Poisson's ratio $\nu = 0.5$, can be reduced to the simple expression

$$I_A = K_A(x_A) E_s \epsilon_o \sqrt{\frac{4}{3} b} \tag{7b}$$

From (1) there follows by virtue of (3)

$$K_A^2(x_A)E_s^2\varepsilon_o^2 \frac{4}{3} b \left(\frac{\alpha}{\frac{dx_A}{dt}}\right)^{1/2} = \text{const.} \qquad (8)$$

Also, for $x_A \gg c$, $K_A \to 1$ and $\frac{dx_A}{dt} = v_o$ $\qquad (9)$

and by dividing (9) into (8) one finds

$$\frac{dx_A}{dt} = v_o K_A^4(x_A) \qquad (10)$$

Integration of this equation leads to the time dependent crack tip position shown in fig. 10 together with experimental data for comparison purposes. A rigid circular inclusion which is bonded to the surrounding material is seen to slow the advancing crack until it comes, at least theoretically, to a

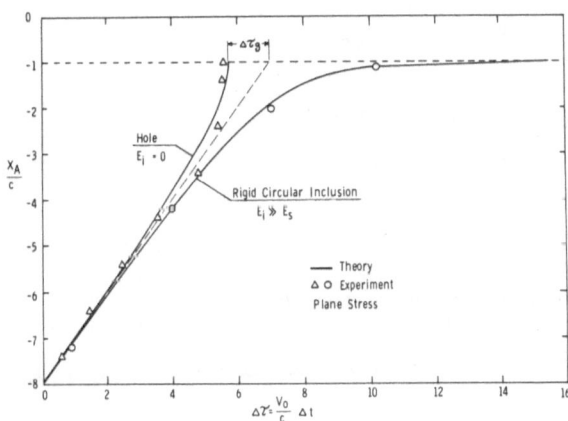

Fig. 10.   Histories for crack tips approaching a hole and a rigid circular inclusion.

complete stop. In practice, however, the stresses at the crack tip are never completely zero because the slightest deviation from the centerline introduces shear stresses which alter the crack direction and produce the gradual transition into the interface shown earlier in fig. 5. For an inclusion with finite modulus the stress at the crack tip never reduces to zero and the theory would predict a finite time for the arrival of the crack at the interface.

The rate of crack propagation increases if the crack moves towards a circular hole, this is also shown in fig. 10. The crack tip arrives earlier at the periphery of the hole than in the absence of the hole. This time differential is theoretically equal to $\Delta t_g = 1.27 \ c/v_o$.

While the assumptions which were made in the above derivation limit the validity of these estimates it is nevertheless clear that the crack propagation calculations are in acceptable agreement with experiments. We therefore feel encouraged to use the crack propagation relation (1) to generally compute crack retardation and proceed to the cases of

b) Crack Propagation Along a Line $y_o \neq 0$: We are interested in the value of the "average stress intensity factor" as the crack interacts with the inclusion. Denote this average value by $\overline{I}$ and the average velocity by v_s. The average velocity is assumed slow such that inertia effects do not come into play. We use equation (1) to define $\overline{I}$

$$\overline{I}^2 \ \Theta(\frac{\alpha}{v_s}) \ = \ \text{const.} \tag{11a}$$

Far away from the inclusion we have

$$I_o^2 \ \Theta(\frac{\alpha}{v_o}) \ = \ \text{const.} \tag{11b}$$

and in view of (3), it follows that

$$\overline{I} \ = \ (\frac{v_s}{v_o})^{1/4} \ I_o \tag{12}$$

We now recall that we consider the crack to propagate with velocity v_o under a constant stress intensity factor I_o before the crack tip interacts with the inclusion. Without changing the loading on the specimen (strain ε_o) the crack then propagates with average velocity v_s under an average stress intensity $\bar{I}$ around the inclusion or in its immediate vicinity. Thus, the same loading enters both $\bar{I}$ and I_0 and it follows that the ratio v_s/v_0 is independent of the loading. Therefore the experimentally determined function v_s/v_0 in fig. 8 applies to all loads on the strip geometry. We make use of this information in the next section.

c) Effect of Filler Volume Ratio on Crack Propagation: We now determine the average rate of crack propagation through a viscoelastic material with rigid, rod-like fillers bonded perfectly to the surrounding material. The fillers are aligned and their axes are parallel to the crack front, in other words, we are dealing again with an essentially two-dimensional problem. We assume a sufficiently small volume ratio to induce negligible interaction between the stress fields around each other. This means an average distance between fillers of about one filler diameter or more*. An initial crack is assumed to exist in the material. The filler volume fraction ϕ is defined as

$$\phi = \frac{S_f}{S_0} \tag{13}$$

where S_f denotes the total volume of filler particles in the specimen and S_0 is the total volume of the composite specimen. The average distance z between adjacent fillers is given by

$$\frac{c}{z} = \sqrt{\frac{\phi}{\pi}} \quad ; \quad \frac{c}{z} > 0 \tag{14}$$

An average of one filler particle will be found in a square (unit cell) with sides of length z. The crack is assumed to propagate unimpeded until it is within one diameter of the center of one of the filler rods at which time it propagates with the (lower) average velocity v_s. Based on these assumptions one finds that the crack requires the time t_z to travel across a unit cell

$$t_z = \frac{z - 2c}{v_0} + \frac{2c}{v_s} \tag{15}$$

Consequently the average velocity across a unit cell is given by

---

*That implies that the filler volume ratio must be about one-half, or less than that of the material shown in fig. 3.

$$v_{av} = \frac{z}{t_z} = \frac{v_o}{1 + 2\sqrt{\frac{\phi}{\pi}}\left(\frac{v_o}{v_s} - 1\right)} \tag{16}$$

Recall from fig. 6 that the ratio v_s/v_o is a function of y_o/c which, according to fig. 8, may for our present purposes by expressed as

$$\frac{v_s}{v_o} = \begin{cases} y_o/2c \; ; & y_o/2c < 1 \\ 1 & ; & y_o/2c > 1 \end{cases} \tag{17}$$

Assuming that the filler particles are perfectly dispersed and aligned, there will be an equal probability for any $0 < y_o < z/2$ within each square cell containing a filler particle in its center and having a side length z. The average rate of crack propagation $\bar{v}_{av}$ through the filled material can thus be determined by averaging equation (16) over y_o

$$\frac{\bar{v}_{av}}{v_o} = \frac{2}{z} \int_0^{z/2} \frac{v_{av}}{v_o} \, dy_o = \frac{2}{z} \int_0^{z/2} \frac{dy_o}{1 + 2\sqrt{\frac{\phi}{\pi}}\left(\frac{v_o}{v_s} - 1\right)} \tag{18}$$

Substitution of (17) and bearing in mind that a cell with side length z = 4c, i.e., $y_o < 2c$, corresponds to filler volume ratio of $\phi = \pi/16$ (cf. equation 14) one finds after integration of (18) that

$$\frac{\bar{v}_{av}}{v_o} = \begin{cases} 1 + \dfrac{8\frac{\phi}{\pi}}{1 - 2\sqrt{\frac{\phi}{\pi}}}\left[1 + \dfrac{\ln 2\sqrt{\frac{\phi}{\pi}}}{1 - 2\sqrt{\frac{\phi}{\pi}}}\right] & \text{for } \phi \leqslant \frac{\pi}{16} \\[4ex] \dfrac{1}{1 - 2\sqrt{\frac{\phi}{\pi}}} - \dfrac{8\frac{\phi}{\pi}}{1 - 4\sqrt{\frac{\phi}{\pi}} + 4\frac{\phi}{\pi}} \ln\left(1 + \dfrac{1 - 2\sqrt{\frac{\phi}{\pi}}}{8\frac{\phi}{\pi}}\right) & \text{for } \phi \geqslant \frac{\pi}{16} \end{cases} \tag{19}$$

where v_o is the crack propagation speed in the material without reinforcing fillers but under otherwise unchanged conditions. Conversely, if we denote by τ_f the time during which the crack propagates on the average through a unit distance in the reinforced solid and by τ_o the corresponding time for crack growth through the same unit distance in the unfilled solid, the ratio

$$\tau_f/\tau_o = v_o/\bar{v}_{av} \tag{20}$$

measures the expansion of failure times due to the addition of filler to the matrix. This "failure delay-ratio" is plotted in fig. 11 as a function of filler volume fraction ϕ. Because the preceding analysis has not included the interaction of stress fields between neighboring inclusions the above result is probably limited to filler volume ratios of less than thirty percent.

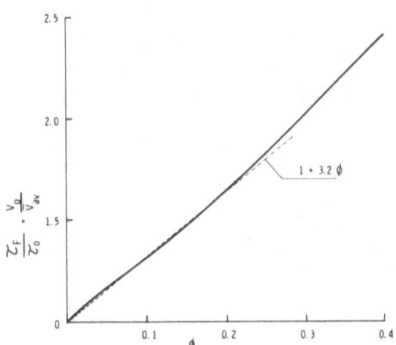

Fig. 11. Predicted failure delay ratio as a function of filler volume ratio (solid curve). The dashed line represents a close linear approximation for $\phi < 0.2$.

SUMMARY

We have provided a simple estimate of the change in failure time (inverse of the rate of crack propagation) which results from the addition of parallel, cylindrical fillers in a viscoelastic matrix. The filler particles are well bonded to the matrix and the crack propagates parallel to the filler axes. An almost linear increase in the time to failure is expected for small filler volume fractions, namely

$$\frac{\tau_f}{\tau_o} \cong 1 + 3.2 \, \phi \text{ for } \phi < 0.2$$

REFERENCES

(1) Mueller, H. K. and Knauss, W. G., "Crack Propagation in a Linearly Visco-elastic Strip," Journal of Applied Mechanics, Vol. 32, No. 2, pp. 48-488, 1971.

(2) Mueller, H. K. and Knauss, W. G., "The Fracture Energy and Some Mechanical Properties of a Polyurethane Elastomer," Trans. Society of Rheology, Vol. 15, No. 2, pp. 217-233, 1971.

(3) Knauss, W. G., "On the Steady Propagation of a Crack in a Viscoelastic Sheet: Experiments and Analysis," Deformation and Fracture of High Polymers, ed. by Kausch, Hassell and Jaffee, Plenum Press, 1974.

(4) Rivlin, R. S. and Thomas, A. G., "Rupture of Rubber. I. Characteristic Energy for Tearing," Journal of Polymer Science, Vol. 10, p. 291, 1953.

(5) Tamate, O., "The Effect of a Circular Inclusion on the Stresses Around a Line Crack in a Sheet under Tension," Int. Journal of Fracture Mechanics, Vol. 4, No. 3, pp. 257-265, 1968.

ACKNOWLEDGMENTS

This work was performed under Air Force sponsorship through research contracts F04611-67-C-0057 (Air Force Rocket Propulsion Laboratory) and F49620-77-C-0051 (Air Force Office of Scientific Research).

THE FRACTURE OF BORON FIBER REINFORCED 6061 ALUMINUM ALLOY

M. A. Wright, D. Welch and J. Jollay

University of Tennessee Space Institute
Tullahoma, Tennessee 37388

ABSTRACT

The fracture of 6061 aluminum alloy reinforced with unidirectional and crossplied 0/90°, 0/90/±45° boron fibers has been investigated. The results have been described in terms of a critical stress intensity, K_Q.

Critical stress intensity factors were obtained by substituting the failure stress and the initial crack length into the appropriate expression for K_Q. Values were obtained that depended on the dimensions of the specimens. It was therefore concluded that, for the size of specimen tested, the values of the critical stress intensity, K_Q, did not reflect any basic materials property.

INTRODUCTION

There have been a number of attempts to describe the failure of fiber-reinforced materials. The results of some early two-dimensional work of Hedgepeth [1] enabled calculation of the stress concentration factors associated with multiple adjacent broken fibers. Zender and Deaton [2] used Hedgepeth's results to calculate, and experimentally verify, the fracture strength of a reinforced polymer that contained cut Dacron fibers. The importance of matrix plasticity was noted by Hedgepeth and Van Dyke [3] who reported that the elastic stress concentration factors were appreciably reduced when the matrix began to flow.

Three-dimensional stress analysis is extremely difficult and time-consuming to carry out; thus, there have been a number of attempts to describe the failure of fiber-reinforced materials using the macroscopic concepts of linear elastic fracture mechanics. A number of authors have studied reinforced polymeric materials [4-7] and others have studied reinforced metals [8-11]. This type of approach appears to have been reasonably successful despite the fact that very few authors have commented on any size effect that might have influenced the results that they obtained.

In the following paper, experiments are presented in which the strengths of center notched and compact tension specimens of 6061 aluminum reinforced with unidirectional and crossplied 0/90°, 0/90/±45° boron fibers were obtained. The results indicated that the values of critical stress intensity (K_Q) depended on specimen size, and thus they did not reflect any basic materials property.

EXPERIMENTAL TECHNIQUE

A. Material Specification

The unidirectionally, reinforced material was fabricated by the AVCO Corporation, Lowell, Massachusetts, and the crossplied material was obtained from DWA Associates, Van Nuys, California. In general, both fabrication processes were similar and consisted of applying a pressure of several ksi to foil-filament arrays at a temperature of about 500°C. After fabrication, the material was cooled in air. Individual specimens were cut from the panels using a diamond-impregnated wheel. Notches having a root radius of 12.5×10^{-3} inches were cut using the electric discharge method (EDM).

B. Unidirectionally Reinforced Material

Three groups of 0.10 inch-thick specimens were tested. The width (w) of each group was 1.0, 2.0, or 4.0 inches, and the ratio of the gauge length to the specimen width was held constant at 3:1. The lengths of the center notches, 2a, were varied such that, in each group, a series of specimens were tested having values of 2a/w equal to 0.05, 0.10, 0.20, 0.40, and 0.60.

C. Crossplied Material

Compact tension specimens were cut from panels in which the fibers were oriented in the 0° and 90° directions. The outer ply was a 0° ply; thus, the thirteen-layer material contained six 90° layers. The dimensions of the compact tension specimens are shown in Figure 1. The width of the specimen was 2, 4, or 8 inches and the crack lengths were varied such that the effect of five crack length:specimen width ratios could be evaluated. The ratios used corresponded to a/b values of 0.2, 0.4, 0.6, 0.8, and 0.9.

A further series of specimens, similar to those described above, was prepared from material that contained fibers oriented in the 0/90/±45° directions. The layers were arranged in the following sequence: 0/90/+45/-45/0/90/0/-45/+45/90/0. Thus, a balanced laminate was obtained.

The geometry of the center notched specimens is shown in Figure 2. The gauge length of each test piece was approximately three inches and the width was 1, 2, or 4 inches, respectively. The crack lengths were varied such that the following crack length:specimen width ratios could be studied: 2a/w = 0.05, 0.10, 0.20, 0.40, and 0.60 for the one inch-wide specimens and 2a/w = 0.20 and 0.40 for the two- and four-inch-wide.

A further series of center notched specimens was fabricated from the material which contained boron fibers oriented using the 0/90/±45° sequence mentioned previously. These dimensions of these specimens were identical to those fabricated from the 0/90° material.

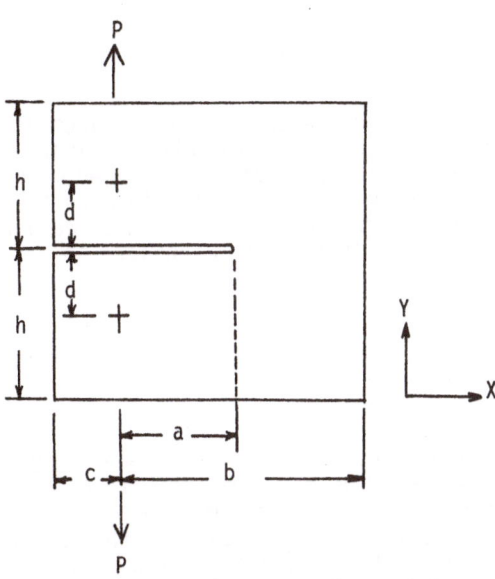

Fig. (1) - The geometry of the compact tension specimens used in this work.
P = load; E = elastic modulus; w = specimen width, 1, 2 or 4 in;
t = specimen thickness, approx. 0.10 in; b = 2, 4 or 8 in;
h = 1.2, 2.4 or 4.8 in; d = 0.55, 1.2 or 2.95 in.

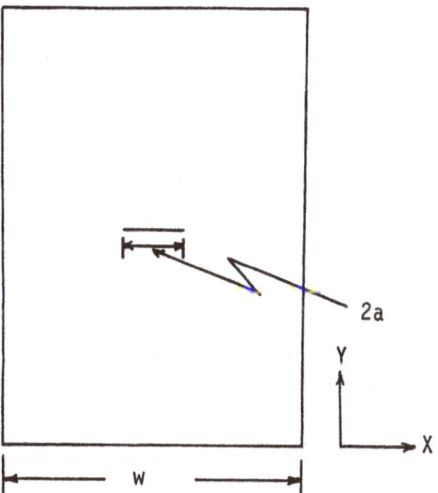

Fig. (2) - The geometry of the center notched specimens used in this work.
P = load; E = elastic modulus; w = specimen width, 1, 2 or 4 in;
t = specimen thickness, approx. 0.10 in; b = 2, 4 or 8 in;
h = 1.2, 2.4 or 4.8 in; d = 0.55, 1.2 or 2.95 in.

D. Fiber Testing

Individual boron fibers were extracted from selected specimens by dis-
solving the matrix material in a sodium hydroxide solution. The failure load
of each fiber was measured using an Instron screw driven machine. The pneumatic
grips of the machine were lined with aluminum to minimize crushing of the fibers
and to distribute the load evenly. The gauge length of each fiber was set by
using the gauge length adjusting dial of the machine.

E. Composite Testing

In order to protect the gripped portion and to minimize the tendency to
fail in that section, aluminum doubler plates were bonded to the ends of each
center notched specimen using Eccobond adhesive.

The load was applied to both types of specimen using an hydraulic materials testing system (MTS) operating at a loading rate of 500 lb/min. A conventional clip gauge was mounted on the specimen in such a way as to continuously
monitor the displacement of either the mid-point of the center notch or the edge
of the notch that was cut into the compact tension specimens.

RESULTS

The results obtained by Herring [12], Wright and Iannuzi [13] and Wright and
Wills [14] indicate that the strengths of brittle fibers tend to obey a Weibull
distribution characterized by

$$G(\sigma) - 1 - \exp \{-(L/D) \ [(\sigma-\sigma^*)/\sigma_0]\} \tag{1}$$

where $G(\sigma)$ = the probability of failure of a fiber subjected to a stress; σ^*
= lower limiting strength, assumed here equal to zero; σ_0 = distribution scale
factor that reflects the maximum strength that can be obtained from the fibers;
ω = distribution shape factor (describes scatter of data) and L/D = fiber length
to diameter ratio.

Accordingly, the tensile strength data obtained from the fibers tested in
this work is presented in Table 1, together with the Weibull constants. It is
noted that the strengths exhibited by the shorter fibers are larger than those
exhibited by the longer ones. In addition, the application of Student's 't'
test at the 95% probability level indicated that identical mean strengths were
exhibited by the fibers extracted from the materials prepared by either of the
composite suppliers.

A. Elastic Properties

Elastic compliance curves were obtained by dividing the applied load into the crack opening displacement as computed from the electrical output of the
clip gauge. The individual compliance values are shown plotted against the ratio of the crack length:specimen width in Figures 3 and 4. Included in these
graphs are those values of crack face compliance that would be expected from
isotropic center notched specimens. In this case, the displacement of the crack

TABLE 1 - THE STATISTICAL STRENGTH DATA OBTAINED FOR FIBERS EXTRACTED FROM COMPOSITE SPECIMENS OF DIFFERENT SIZE

Size of Original Composite		Fiber Gauge Length in	σ_0 ksi	ω	Mean Strength ksi	Bundle Strength ksi
Width in	Length in					
4	24	1	768.7	11.69	473.1	367.6
4	24	2	810.6	10.20	435.2	329.4
4	24	3	877.9	9.23	422.2	314.0
4	24	6	988.3	6.72	328.2	227.7
4	24	12	889.4	6.05	233.9	158.0
2	12	1	874.1	9.64	481.9	364.4

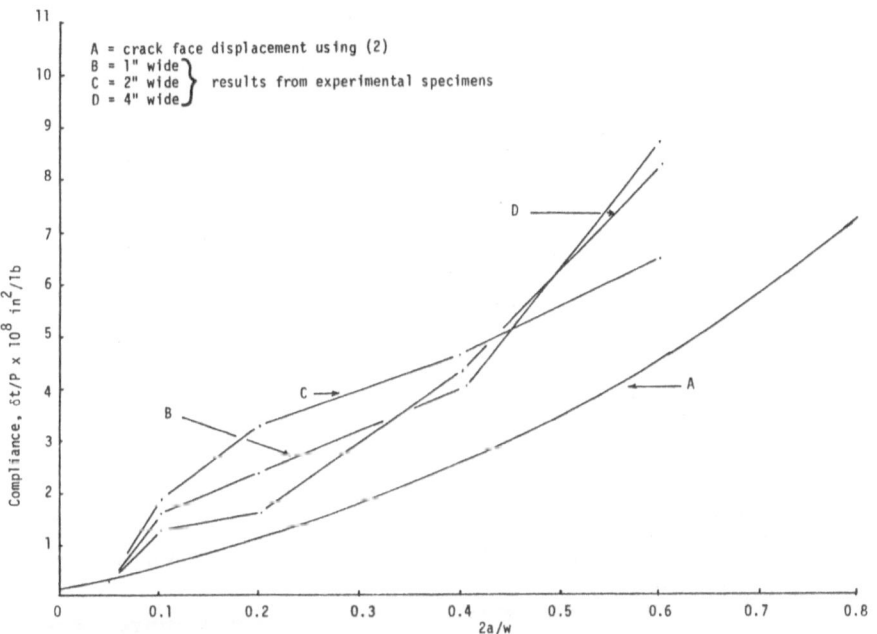

Fig. (3) - The compliance of various unidirectionally reinforced specimens expressed as a function of crack length to width ratio, 2a/w

faces, δ, was calculated using the expressions given by Tada et al [15], i.e.,

$$\delta = 4\sigma_{nom} \, a/E \, V(a/b) \tag{2}$$

226

where

$$V(a/b) = 0.071 - 0.535 \ (a/b) + 0.169 \ (a/b)^2 + 0.120 \ (a/b)^3$$
$$- 1.071 \ a/b \ \ln \ (1-a/b)$$

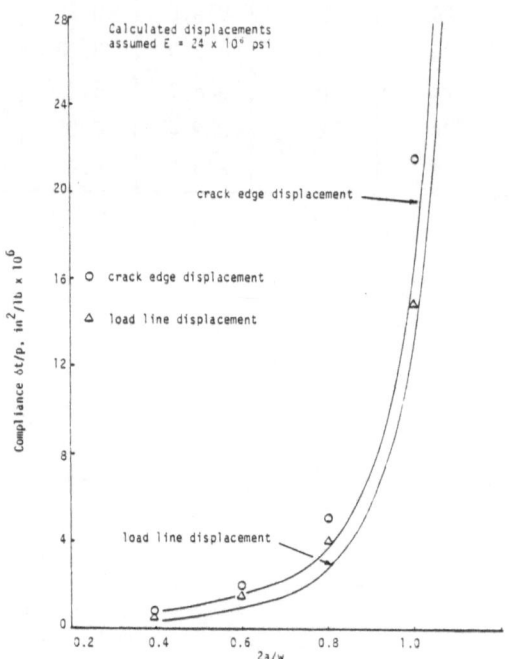

Fig. (4) - A comparison of the theoretical crack face compliance values
with those obtained from experiments carried out on crossplied 0/90° boron-aluminum

The corresponding expression that was used to describe the compliance of compact
tension specimens was:

$$\delta = P/E \ V_1 \ (a/b) \tag{3}$$

where the values of the function $V_1(a/b)$ are taken from the analysis of Roberts
[16].

The compliance values of the unidirectional specimens was obtained from the initial part of the load-extension curves. However, compliance values larger than those given by equation (2) were obtained using this method. Consequently, a small amount of load was reversed during each subsequent test carried out on the crossplied material and the compliance values were calculated from the slopes of the elastic recovery curves that were obtained. As can be observed, the elastic recovery technique enabled compliance values to be obtained from the pseudo-isotropic crossplied material that were very similar to those predicted using the isotropic formulations of Tada.

Experimental measurements of the strains generated at or close to the tip of the center notch present in the unidirectionally reinforced material were carried out using a linear array of traversely oriented strain gauges bonded adjacent to the crack tip. Each gauge, 0.031 inch long, was separated from the preceding one by a distance of 0.024 inch and the length of the total array was 0.25 inch.

According to Sih et al [17] the tensile stresses that exist at some reasonable distance from the tip of a long, thin crack in a homogeneous, orthotropic, elastic solid are given by:

$$\sigma_x = K_I/(2\pi r)^{1/2} \; (\beta_{22}/\beta_{11})^{1/2} \tag{4}$$

$$\sigma_y = K_I/(2\pi r)^{1/2} + \sigma_{nom} \cdots \tag{5}$$

where K_I is the Mode I stress intensity factor and σ_x and σ_y are the local stresses in the x and y directions as indicated in Figure 2. These directions also define the orthotropic axes of the material. Thus, $\beta_{11} = 1/E_x$ and $\beta_{22} = 1/E_y$, where E_x and E_y are the elastic modulus values in the respective directions.

Since for small applied loads the value of σ_y must approach σ_{nom} at some relatively large distance from the crack tip, it follows that a corresponding value for σ_x is given by:

$$\sigma_x = (\sigma_y - \sigma_{nom}) \; (E_x/E_y)^{1/2} \tag{6}$$

Use of the elastic constitutive equations enables the value of the transverse strain, ε_x, to be calculated for a plane stress specimen

$$\varepsilon_x = \sigma_x/E_x - \nu_{yx}\sigma_y/E_y \tag{7}$$

where ν_{yx} is the major Poisson's ratio.

Typical values of the elastic constants of the unidirectional material were found to be $E_{11} = 32 \times 10^6$ psi, $E_{22} = 18 \times 10^6$ psi, $\nu_{yx} = 0.23$. These were used to evaluate equations (5), (6) and (7). The results of calculations made after assuming that a load of 5000 lbs was applied to a four-inch-wide specimen containing a crack 0.2 inch long are included in Table 2. It is apparent from these results that the calculated strains are far greater than those measured experimentally. Thus, it can be concluded that since equation (6) does not predict the correct value of the strains at this loading, then the compliance will not be given by equation (2) either. This conclusion is not unexpected, for matrix plasticity occurs early in the loading history and this must result in appreciable stress relaxation taking place within the specimen.

TABLE 2 - THE EXPERIMENTAL STRAINS MEASURED ACROSS THE UNCRACKED LIGAMENT COMPARED TO THOSE CALCULATED USING EQUATION (7)

Gauge Number	σ Calculated using (4) ksi	σ_y Calculated using (5) ksi	ε_x Calculated using (7) $(\mu in)/in$	ε_x Measured $(\mu in)/in$
1	12.547	29.229	+487	+150
2	7.288	22.217	+245	+ 80
3	5.652	20.036	+170	0
4	4.779	18.872	+130	- 20
5	4.216	18.121	+104	- 50

Applied load = 5000 lbs; crack length = 0.2 in; specimen width = 4 in

B. Fracture Properties

In this work, an attempt was made to describe the failure of the composite specimens in terms of a single parameter function by substituting the maximum failure load, P, into one of the following expressions:

Center notched specimens

$$K_Q = \{P/wF\}\{\pi a \sec (\pi a/w)\}^{1/2} \qquad (8)$$

Compact tension

$$K_Q = \{Pa/bt\}\{29.6 - 185.5 \, a/b + 655.7 \, (a/b)^2 - 1017.0 \, (a/b)^3 + 638.9 \, (a/b)^4\} \qquad (9)$$

where K_Q is the critical stress intensity parameter.

Table 3 contains the failure loads and the associated critical K values that were obtained from unidirectionally reinforced boron-aluminum material. The results were obtained from specimens of three lengths, L, and three widths, w. However, the ratio L/w was maintained constant. It is to be noted that the nominal (gross) failure stresses exhibited by specimens containing identical crack sizes were smaller for the larger specimens. Thus, smaller K_Q values were obtained from larger specimens. This effect was most unexpected since, for metals, a specimen gauge length effect is not considered important.

The failure stresses obtained from crossplied boron-aluminum specimens are shown in Table 4. In this case, the lengths of the specimens were maintained constant and only the widths were varied. As can be seen, the nominal failure stresses of the specimens that contained identical crack sizes were larger for the wider specimens. Thus, K_Q values increased as the specimen width increased.

C. Effect of Specimen Length

The length effect has been discussed in a recent publication [18] in which it was concluded that, for unidirectionally reinforced composites, the stress concentration effect of the notch becomes constant and independent of notch size providing the notch is larger than some minimum value. It was also concluded that for identical large cracks, longer specimens were associated with larger stress concentration factors.

It is interesting to compare the conclusions that we have made using the experimentally determined strength values of center notched composites with the analytical work of Fichter [19]. Fichter pointed out that the stress concentrating effect of broken reinforcement fibers as calculated by Hedgepeth [1] was obtained by assuming that the fibers were contained in infinitely long and wide specimens. However, if the length of the specimen were finite, as measured by some distance from the notch plane, then Hedgepeth's results should be modified to indicate that the stress concentrating effect of the notch would not continue to increase but would become constant at some given crack size. Fichter's results are summarized in Figure 5. As can be observed, for a fiber-reinforced composite of finite length, the stress concentration values, K_r, will not continue to increase as the numbers of broken fibers increase. Rather, a limit to K_r is achieved, the value of which depends on the elastic properties of the composite and the length of the specimen.

A quantitative analysis of the experimental results reported herein is possible using Fichter's analysis. For instance, it can be noted that the absolute value of the stress concentration factor, K_r, is a function of the parameter k where

$$k = \ell(Gh/EAd)^{1/2}$$

TABLE 3 - THE MECHANICAL PROPERTIES OF UNNOTCHED AND CENTER NOTCHED SPECIMENS OF UNIDIRECTIONALLY REINFORCED BORON-ALUMINUM SPECIMENS

Specimen Gauge Length in	Specimen Width in	2a/w	2a	Failure Load lbs x 10^{-2}	Failure Gross lb	Stress Net lb	Failure Displacement in x 10^4	$\sigma_3{:}\sigma_6$	$\sigma_3{:}\sigma_{12}$	K_Q ksi$\sqrt{\text{in}}$
3	1	0.00	0.00	19.20	184.90	--	155.8	--	--	--
3	1	0.05	0.05	14.85	139.75	147.11	134.3	--	--	39.2
3	1	0.10	0.10	13.30	128.60	142.89	115.7[2]	--	--	51.3
3	1	0.20	0.20	11.60	110.75	138.44	99.8[2]	--	--	63.6
3	1	0.40	0.40	8.65	83.21	138.68	101.1[2]	--	--	73.3
3	1	0.60	0.60	5.60	53.40	133.50	82.3[2]	--	--	67.6
6	2	0.00	0.00	45.00	215.75	--	491.0[2]	--	--	--
6	2	0.05	0.10	28.20	113.22	140.21	338.2	1.02	--	52.9
6	2	0.10	0.20	21.67	103.04	114.49	248.7	1.21	--	58.1
6	2	0.20	0.40	18.32	87.92	109.90	295.3	1.26	--	71.5
6	2	0.40	0.80	14.17	68.18	113.63	286.0	--	--	85.0
6	2	0.60	1.20	7.29	34.90	81.25	163.0	--	--	62.5
12	4	0.00	0.00	38.80[1]	97.19[1]	--	--	--	--	39.1
12	4	0.05	0.20	28.47	69.68	73.35	341.0	--	1.89	47.7
12	4	0.10	0.40	24.05	59.81	66.46	291.4	--	2.09	55.4
12	4	0.20	0.80	19.60	48.24	60.30	220.0	--	--	57.8
12	4	0.40	1.60	13.37	32.77	54.62	190.1	--	--	60.0
12	4	0.60	2.40	9.55	23.70	59.25	157.8	--	--	--

Failure stresses are mean of two values except where noted by superscript 1. Failure displacements are single values; superscript 2 indicates mean of two values.

TABLE 4 - THE FAILURE CHARACTERISTICS OF CROSSPLIED BORON REINFORCED ALUMINUM

Specimen Number	Lay/up	Max Load lb	Max Stress psi	2a in	$\frac{2a}{w}$	w in	t in	K_Q ksi-$\sqrt{in}$	Net Failure Stress psi
1	0/90	11,000	114,592	0	0	.99753	.09623	--	--
2	0/90	11,750	120,954			.99737	.0974	--	--
3	0/90	8,350	98,003	.05	0.05	0.9925	0.0956	24,700	92,635
4	0/90	7,400	76,295	.1	.1	0.9992	0.09707	30,426	84,772
5	0/90	5,640	57,062	.2	.2	1.0006	0.09703	32,796	71,327
6	0/90	4,600	47,348	.4	.4	0.99727	0.0953	34,434	78,913
7	0/90	2,700	27,884	.6	.6	0.99957	0.09687	35,309	69,710
8	0/90	10,475	108,499			1.0039	0.09617		--
9	0/90	8,000	83,989	.058	.058	0.99707	0.09553	25,404	88,409
10	0/90	7,300	75,616	.096	.096	1.00083	0.09646	29,532	84,018
11	0/90	5,400	56,047	.210	.210	.99945	0.0964	33,095	70,059
12	0/90	3,675	37,959	.395	.395	1.00273	0.09655	33,149	63,265
13	0/90	2,640	27,422	.600	.600	1.00253	0.09603	34,724	68,555
14	0/90/±45	9,825	101,801	.059	.059	1.00507	0.09687	21,480	74,114
15	0/90/±45	6,850	70,408	.099	.099	.99813	0.0968	23,792	66,631
16	0/90/±45	5,800	59,968	.208	.208	1.0027	0.0969	28,985	61,686
17	0/90/±45	4,780	49,349	.398	.398	1.0003	0.0966	31,056	58,948
18	0/90/±45	3,400	35,369	0.60	.60	1.00020	0.0961	31,980	63,137
19	0/90/±45	2,425	25,255	0.000	0.000	0.9983	0.09583		--
40	0/90	9,000	94,106	0.000	0.000	0.9983	0.0958		--
41	0/90	8,100	84,515	0.055	0.055	1.0022	0.09563	24,888	88,963

TABLE 4 - CONTINUED

Specimen Number	Lay-up	Max Load lb	Max Stress psi	2a in	2a/w	w in	t in	K_Q ksi-$\sqrt{in}$	Net Failure Stress psi
42	0/90	6,900	72,304	0.10	0.10	0.9979	0.0965	28,834	80,338
43	0/90	5,650	58,590	0.205	0.20	0.9993	0.0965	34,136	73,237
44	0/90	3,800	39,566	0.40	0.40	1.0022	0.09583	34,869	65,943
45	0/90	2,450	25,456	0.60	0.60	1.0012	0.09613	32,234	63,640
46	0/90/±45	9,700	100,549	0.00	0.00	1.0025	0.09623	--	--
47	0/90/±45 (failed in grips)	6,950	72,384	0.056	0.056	0.9960	0.0964	21,510	76,194
48	0/90/±45	5,950	61,720	0.095	0.096	0.9928	0.0971	23,976	68,579
49	0/90/±	4,800	49,303	0.205	0.205	1.0013	0.09723	28,726	61,629
50	0/9C/±45	3,210	33,120	0.395	0.40	1.0054	0.0964	28,923	56,200
51	0/9C/±45	2,325	24,317	0.595	0.60	1.0022	0.0954	30,499	60,792
82	0/90	9,725	50,641	0.40	0.20	1.990	0.0965	41,058	63,301
83	0/90	9,300	48,346	0.40	0.20	1.975	0.0974	39,296	60,432
84	0/90	6,775	35,255	0.80	0.40	1.971	0.0975	43,938	58,758
85	0/90	6,900	35,812	0.80	0.40	1.966	0.0980	44,634	59,687
86	0/90	17,100	43,951	0.80	0.20	3.930	0.0990	50,521	54,939
87	0/90	17,090	44,499	0.80	0.20	3.939	0.0975	51,151	55,624
88	0/90	12,250	32,157	1.60	0.40	3.939	0.0975	56,678	53,595
89	0/90	31,085	31,085	1.60	0.40	3.939	0.0970	54,789	51,808

In this expression, ℓ is the specimen half length, G is the shear modulus of the matrix, E is the tensile modulus of the fibers of area A, h is the thickness of the specimen, and d is the distance between the fibers. If the value of G is assumed to be about 3×10^4 psi for an aluminum matrix that is deforming plastically, and if h is considered to be about equal to d for a material containing 50v/o reinforcement, then an approximate calculation indicates that K for a boron-aluminum material is given by

$$k \approx 5\ell$$

Further inspection of Figure 5 then indicates that maximum values of K_r will be about 2.7, 3.8 and 5.5 for 3, 6 and 12-inch-long specimens, respectively.

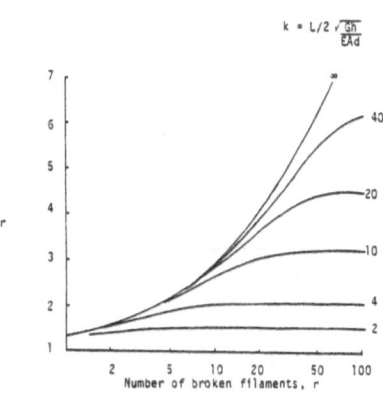

Fig. (5) - Stress concentration factor for fixed values of k expressed as a function of the number of broken filaments; uniform normal edge displacement

Kulkarni [20] has shown, using a finite element analysis technique, that the stress concentrating effect of cut fibers within a unidirectionally reinforced material is a maximum in the first adjacent fibers and then diminishes very rapidly as the distance from the cut fibers increases. Thus, it is reasonable to conclude that the product of the stress concentration factor K_r and the net stress at failure will represent the maximum stress in the fibers that are adjacent to the notch. The net failure stresses for unidirectionally reinforced boron-aluminum specimens σ_3, σ_6 and σ_{12} are shown in Table 3 for the 3, 6 and 12-inch-long specimens, respectively. As can be noted, the ratio of the strengths, i.e., $\sigma_3:\sigma_6:\sigma_{12}$, is about 1:1.2:2.0. These values are to be compared with 1:1.4:2.1 that would be predicted using the results of Fichter's analysis.

Additional observations can be made by noting that a mean net stress at failure of the unidirectional material is 140, 115 and 57 ksi for the 3, 6 and 12-inch-long specimens. Thus, the fibers in specimens containing 47% reinforcement carry respective net stresses of 298, 245 and 121 ksi and the first fibers adjacent to the notch will be subjected to maximum stresses of 804, 931 and 667 ksi.

234

These values closely approach the value σ_0, i.e., the maximum fiber strength values that can be described by a Weibull distribution. The same calculation can also be applied to the results obtained from the crossplied material and similar results are obtained. These are shown in Table 5.

TABLE 5 - CALCULATION OF THE MAXIMUM NET STRESS IN FIRST FIBERS ADJACENT TO A NUMBER OF FIBER BREAKS

Specimen Type	Length in	Width in	Thickness in	Mass Net Stress at Failure ksi	Maximum Net Stress in First Fibers Adjacent to Crack ksi
U.	3	1	0.1	140	804
U.	6	2	0.1	115	931
U.	12	4	0.1	57	667
0/90	3	1	0.1	74	794
0/90	3	2	0.1	61	650
0/90	3	4	0.1	54	576

D. Effect of Specimen Width

The variation in the fracture stress of thin metal specimens can be easily and simply discussed with reference to Figure 6. In this figure, the nom-

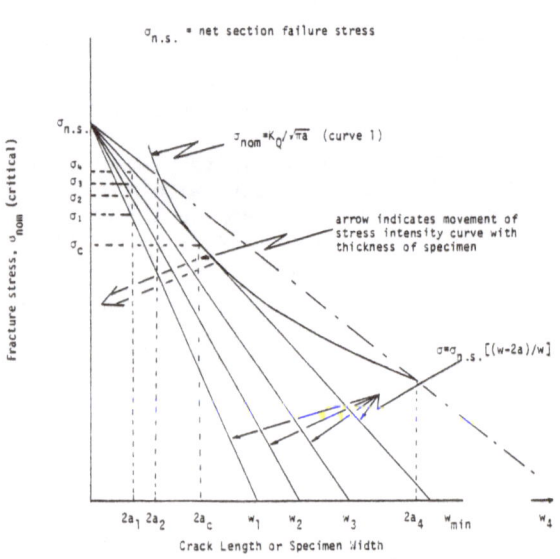

Fig. (6) - A curve illustrating the tendency of plane stress specimens to fail at a critical net section

inal failure stress is plotted against variations in the crack size and/or specimen width. The curve denoted 1 describes the failure stress of specimens that break when a critical stress intensity is generated at the crack tip. The effect of an increase in specimen thickness is to move the curve downwards and towards the ordinate, as shown in the figure. Essentially, this means that smaller K_Q values are exhibited by thicker specimens. Some thickness is eventually reached beyond which no further movement of the curve occurs; this value of K_Q is then termed K_{IC}. Specimen widths w_1, w_2, w_3, etc., can also be plotted on the horizontal axis of the graph. And, straight lines can be drawn joining the respective specimen widths to a stress which denotes net section failure. These lines then describe the failure stress of notch insensitive specimens. As can be noted, net section failure will occur in small specimens, i.e., w_1, w_2, w_3, even though a critical stress intensity may describe failure of larger specimens, i.e., w_4. However, even with large specimens, net section failure can occur if the crack sizes are smaller or greater than those denoted in the figure by $2a_2$ and $2a_4$. The values of K_Q that will be obtained from specimens failing in the net section will vary with specimen size, and/or crack size, as shown in Figure 7. Two effects can be observed: 1) for a given specimen width, the values of K_Q increase to a maximum before decreasing, and 2) larger values of K_Q are obtained from larger specimens.

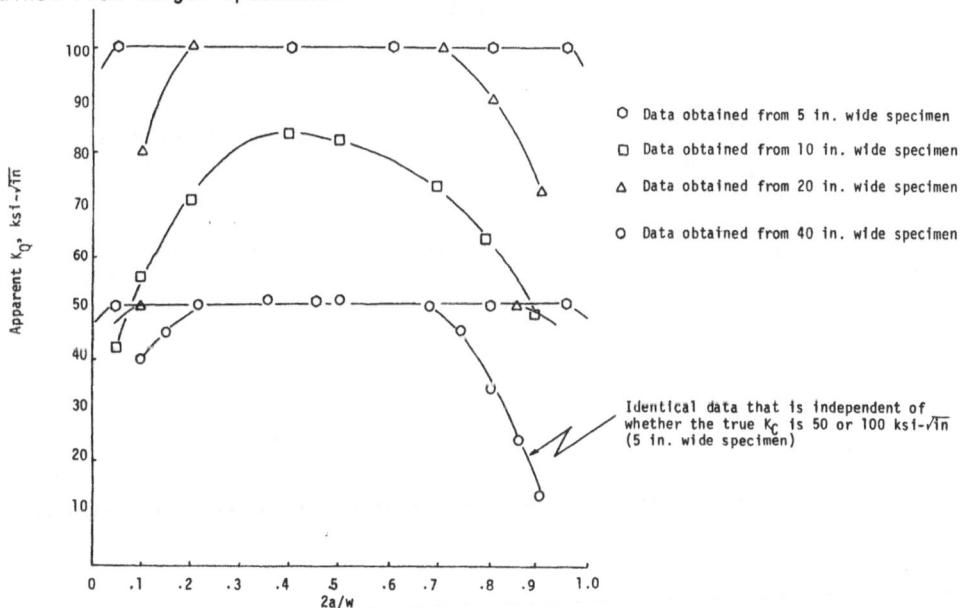

Fig. (7) - Effect of width on apparent K_C values (top curves calculated assuming K_C = 100 ksi-$\sqrt{in}$; bottom curve calculated assuming K_C = 50,000 ksi-$\sqrt{in}$)

The K_Q values that were obtained from the crossplied 0/90°, 0/90/±45° boron-aluminum composites are shown in Figures 8 and 9. In this case, since the

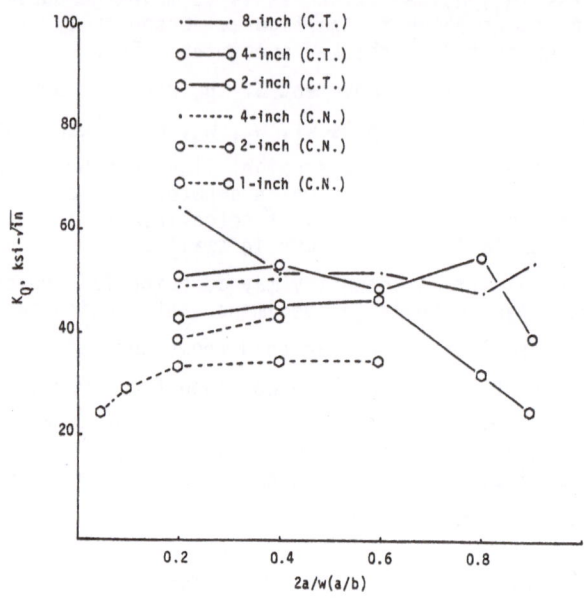

Fig. (8) - Critical values of the stress intensity factor calculated using the maximum load and the initial crack length; 0/90° reinforcement

stress concentrating effect of the notch is independent of notch length, it follows that failure of the composites will occur when the net section reaches some critical value. Thus, it is not surprising that the shape of these curves is identical to that expected from metal specimens failing by net section yield.

In summarizing the results of the above work, it can be stated that increasing the length of unidirectionally reinforced specimens reduced the K_Q values obtained. Conversely, increasing the width of crossplied specimens increased the K_Q values obtained. The effect of specimen width was not determined for the unidirectional material, neither was the effect of specimen length determined for the crossplied.

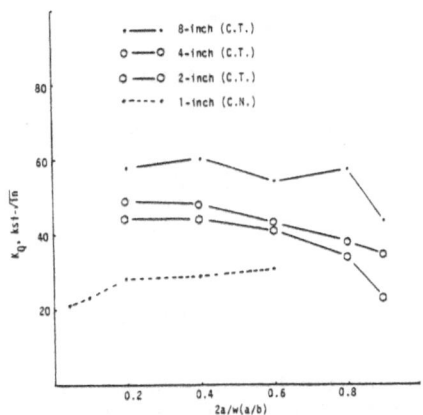

Fig. (9) - Critical values of the stress intensity factor calculated using the maximum load and the initial crack length; 0/90/±45° rein-forcement

ACKNOWLEDGEMENTS

The authors would like to acknowledge the support and encouragement of Dr. J. Newman who served as the technical monitor. The above work was carried out under the sponsorship of a NASA Research Grant NSG 1163, Supplement No. 2.

REFERENCES

[1] Hedgepeth, J. M., NASA Rep. No. TND-882, 1961.

[2] Zender, G. W. and Deaton, J. W., NASA Rep. No. TND-1609.

[3] Hedgepeth, J. M. and Van Dyke, P., J. Comp. Mat., Vol. 1, p. 294, 1967.

[4] Konish, H. T., Ph.D. Dissertation, The Carnegie Institute of Technology, Pittsburgh, 1973.

[5] Beaumont, P. W. R. and Phillips, D. C., J. Comp. Mat., Vol. 6, p. 32, 1972.

[6] Zimmer, J. E., J. Comp. Mat., Vol. 6, p. 312, 1972.

[7] Tirosh, J. and Berg, C. A., Comp. Mat. Testing and Design, ASTM STP 546, p. 663, 1974.

238

[8] Hancock, J. R. and Swanson, G. D., Comp. Mat. Testing and Design, ASTM STP 497, p. 299, 1971.

[9] Olster, E. F. and Jones, R. C., MIT, Res. Rep. No. R-70-75, 1970.

[10] Kreider, R. G. and Dardi, L. E., Failure Modes in Composites I, AIME, 1972.

[11] Sun, C. T. and Prewo, K. M., J. Comp. Mat., Vol. 2, p. 164, 1976.

[12] Henning, H. W., NASA Rep. No. TND-3202, 1966.

[13] Wright, M. A. and Iannuzi, F. A., Failure Modes in Composites II, AIME, 1974.

[14] Wright, M. A. and Wills, J. L., J. Mech. Phys. Solids, Vol. 22, p. 161, 1974.

[15] Tada, H. P., Paris, P. and Irwin, G. R., "The stress analysis of cracks handbook", Del Research Corporation, Hellertown, Pa., 1973.

[16] Roberts, E., Jr., Mat. Res. and Stds., Vol. 1, p. 27, 1969.

[17] Sih, G. C., Paris, P. C. and Irwin, G. R., Int. J. Fract. Mech., Vol. 1, p. 189, 1965.

[18] Wright, M. A. and Welch, D., Fiber Sc. and Tech., in press.

[19] Fichter, W. B., NASA TND-5947, 1970.

[20] Kulkarni, H. T., Ph.D. Dissertation, UTSI, 1976.

Section IV

FAILURE ANALYSIS

Section IV

FAILURE ANALYSIS

CRITERIA OF COMPOSITE MATERIAL STRENGTH

B. D. Annin and L. V. Baev

Institute of Hydrodynamics, Siberian Branch of the USSR Academy of Sciences
Novosibirsk

ABSTRACT

The fracture criterion of anisotropic body admitting transformation groups
is considered. Partial variants of the criteria applied to composites are derived. The experimental results on a complex loading of tabular specimens of
glass plastics subjected to tension and torsion are presented.

INTRODUCTION

One often simulates a composite material as anisotropic linear elastic body
until the limiting state for which we take the failure of material (lamination,
reinforcement fracture, etc.) is reached. The stressed state at a given point
in the body, at which the limiting state is reached, is considered independent
of loading history and can be described by some surface in space E_6: $\sigma = (\sigma_{11},
\sigma_{22}, \sigma_{33}, \sigma_{12}, \sigma_{13}, \sigma_{23})$

$$\Phi(\sigma_{ij}) = C \tag{1}$$

Here C is a constant and σ_{ij}, i,j = 1,2,3 are the components of the stress tensor in some fixed system of coordinates x_1, x_2, x_3 determined, for example, by
the material structure such as its properties of anisotropy. The condition of
limiting state in the case when the stress tensor is taken in other coordinate
system, is derived from equation (1) by substituting σ_{ij} for their expressions
through the components of stress tensor in the new system which is related to
the original system by the Euler angles.

Further, equation (1) will be referred to as the strength criterion. Many
papers are devoted to the construction of different strength criteria [1-6].
However, due to a great variety of the properties of composite materials, the
question of choice of the strength criterion can hardly be considered as completely solved.

A. The Analysis of Criteria

Let us consider the possible simplifications of equation (1) by making some additional assumptions on the properties of the limiting states and refer to the experimental results on the strength of glass-fibre-reinforced plastic under complex loading.

1. Let the limiting state be independent of hydrostatic pressure. This means that, if the point $\sigma \varepsilon E_6$ satisfies equation (1), the point σ' will also satisfy it, where

$$\sigma'_{ij} = \sigma_{ij} + p\delta_{ij} \tag{2}$$

Here, δ_{ij} is the Kronecker symbol and the parameter p is limited within $-\infty < p < +\infty$. The relations in equations (2) determine in space E_6 a one-parameter group of transformations with the operator

$$X = \frac{\partial}{\partial \sigma_{11}} + \frac{\partial}{\partial \sigma_{22}} + \frac{\partial}{\partial \sigma_{33}} \tag{3}$$

Under the action of this group the left part of equation (1) does not change. This means that it is the invariant with respect to the transformation, $X\Phi = 0$ [7]. In solving this equation we find that the condition in equation (1) must be of the form

$$\Phi_a(\sigma_{11}-\sigma_{22},\sigma_{11}-\sigma_{33},\sigma_{12},\sigma_{23}\sigma_{13}) = C \tag{4}$$

2. Let the state (σ'), determined by the transformation with the following operator, be also the limiting state together with each limiting state σ

$$Y = 2\sigma_{12}\frac{\partial}{\partial \sigma_{11}} - 2\sigma_{12}\frac{\partial}{\partial \sigma_{22}} + (\sigma_{22}-\sigma_{11})\frac{\partial}{\partial \sigma_{12}} + \sigma_{23}\frac{\partial}{\partial \sigma_{13}} - \sigma_{13}\frac{\partial}{\partial \sigma_{12}} \tag{5}$$

This transformation is induced by rotating the original system around the axis x_3 by an arbitrary angle [8], i.e., the plane x_1,x_2 is the plane of isotropy with respect to strength. By solving the equation $Y\Phi = 0$, we find that equation (1) must have the form

$$\Phi_\delta(J,J_2,J_3,T,\sigma_{33}) = C \tag{6}$$

Here

$$J = \sigma_{11} + \sigma_{22} + \sigma_{33}, \quad T = \sigma_{13}^2 + \sigma_{23}^2,$$

$$J_2 = (\sigma_{11}-\sigma_{22})^2 + (\sigma_{11}-\sigma_{33})^2 + (\sigma_{22}-\sigma_{33})^2 + 6(\sigma_{12}^2+\sigma_{13}^2+\sigma_{23}^2)$$

where J_3 is the cubic invariant of the stress tensor deviator.

3. Equation (1), which is simultaneously invariant with respect to transformations determined by the operators in equations (3) and (5), is constructed as the solution of the system $X\Phi = 0$ and $Y\Phi = 0$. It has the form

$$\Phi_{a\sigma}(\sigma_{33} - \frac{\sigma_{11}+\sigma_{22}}{2}, J_2, J_3, T) = 0 \tag{7}$$

Here J_2, J_3, and T have the same meaning as in equation (6).

The quadratic conditions of equations (5) and (6) were considered in [9] as plasticity conditions and in [10] as creep potential.

4. The condition in equation (1) when considered invariant with respect to a three-parameter group with the operators

$$Z_1 = \frac{\partial}{\partial\sigma_{13}}, \quad Z_2 = \frac{\partial}{\partial\sigma_{23}}, \quad Z_3 = \frac{\partial}{\partial\sigma_{33}}$$

has the form

$$\Phi_g(\sigma_{11},\sigma_{22},\sigma_{12}) = C \tag{8}$$

This is a common form of the limiting condition of plane stressed state.

5. Let the limiting state be invariant with respect to the following one-parameter group of transformations (the tensor components, which are actually transformed, are written out. α is parameter, where $-\infty<\alpha<\infty$)

$$\sigma_{11}' = \sigma_{11}+\alpha, \quad \sigma_{22}' = \sigma_{22}-\alpha$$

The transformation operator has the form

$$U = \frac{\partial}{\partial\sigma_{11}} - \frac{\partial}{\partial\sigma_{22}}$$

The strength condition which is invariant with respect to this transformation is such that

$$\Phi_e(\sigma_{11}+\sigma_{22},\sigma_{33},\sigma_{12},\sigma_{13},\sigma_{23}) = C \tag{9}$$

Let us show that a quadratic condition of the form in equation (8), taking into account equation (9)

$$A\sigma_{11}^2 + B\sigma_{22}^2 + C\sigma_{12}^2 + F\sigma_{11}\sigma_{22} + G\sigma_{11} + H\sigma_{22} + m|\sigma_{11} + \sigma_{22}| = 1 \qquad (10)$$

makes it possible to allow for the difference in strength under expansion and compression along the axes x_1 and x_2. The dependence of simple shear strength in the system S_{45} is obtained by a 45° around the axis x_3 with respect to the original system on the sign of shear stresses.

Let $\sigma_{11}^\pm$, $\sigma_{22}^\pm$ be the absolute values of tensile and compressive strength along the axes x_1, x_2; $\hat{\sigma}_{12}$ is the simple shear strength in the original system where τ_{45}^+, τ_{45}^- are the absolute values of strength limits under stress state of shear in the system S_{45}. By means of equation (10) in the assumption

$$\frac{\sigma_{11}^+}{\sigma_{11}^-} \neq \frac{\sigma_{22}^+}{\sigma_{22}^-}$$

we find

$$A = (\sigma_{11}^+\sigma_{11}^-)^{-1} - 2m(\sigma_{11}^+ + \sigma_{11}^-)^{-1}$$

$$B = (\sigma_{22}^+\sigma_{22}^-)^{-1} - 2m(\sigma_{22}^+ + \sigma_{22}^-)^{-1}$$

$$C = (\hat{\sigma}_{12})^{-2}$$

$$F = A + B - (\tau_{45}^+/\tau_{45}^-)^{-1}$$

$$G = (\sigma_{11}^+ + \sigma_{11}^-)/(\sigma_{11}^+\sigma_{11}^-) + m(\sigma_{11}^+ - \sigma_{11}^-)/(\sigma_{11}^+ + \sigma_{11}^-)$$

$$H = (\sigma_{22}^+ + \sigma_{22}^-)/(\sigma_{22}^+\sigma_{22}^-) + m(\sigma_{22}^+ - \sigma_{22}^-)/(\sigma_{22}^+ + \sigma_{22}^-)$$

in which m is single valued and determined from the relation

$$G = H = (\tau_{45}^- - \tau_{45}^+)/(\tau_{45}^+\tau_{45}^-)$$

The condition, which is similar to equation (10), has the following form in case of common stressed state

$$A_{ijkl}\sigma_{ij}\sigma_{kl} + B_{ij}\sigma_{ij} + m_1|\sigma_{22} + \sigma_{23}| + m_2|\sigma_{11} + \sigma_{33}| + m_3|\sigma_{11} + \sigma_{22}| = 1$$

B. Experiments

Let us give the results of experiments on the complex loading of tubular specimens made of naturally aged fabric glass laminate which had been subjected to the joint action of torque and axial force until they failed [11]. The specimens had an average diameter 22.5 mm, thickness 2.5 mm and the length between clamps is 100 mm. They were reinforced by a material with satin weave $ASTT(b)-C_2$. The ratio of the number of threads with respect to the warp and weft is $22\pm1\div13\pm1$ and E-40 epoxy was used as a binder. Two types of specimens were tested. The first type consisting of 25 specimens had the material warp along the specimen axis. The second type consisting of 20 specimens had the warp along the generatrix of a cylinder.

Figure 1 gives the values of tensile stress σ and shear stresses τ at the moment of failure corresponding to disintegration of specimens into pieces. In this figure, the crosses refer to the specimens of the first type, and the circles, to the specimens of the second type. The oblique crosses and dark circles

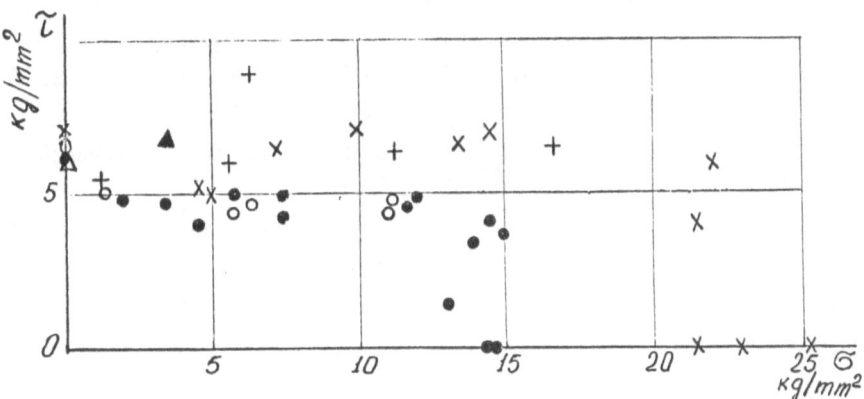

Fig. (1) - The values of stresses corresponding to the failure
of tubular samples subjected to tension and torsion

correspond to the specimens with proportional loading along a ray in the (σ,τ) plane. The straight crosses and light circles correspond to the specimens loaded according to the program in which initially the loading had been applied along the σ axis before it reached some value $\sigma=\sigma_0$ and then at constant σ_0 the shear stress had increased until the failure occurred.

In addition, the dark triangle gives the value of stresses for the first-type specimen tested according to the program in which at the beginning the loading had proceeded along the σ axis to $\sigma_1 = 17.4$ kg/mm^2, and then along the straight line in the $\sigma\tau$ plane with load with respect to τ and unload with respect to σ until the specimen failed. The light triangle refers to the second-type specimen, tested according to a similar program with $\sigma_1 = 11.1$ kg/mm^2.

The conducted experiments show that limiting stresses are independent of the history of loading. At $\sigma=0$, the shear strength of the first-type specimens is near that of the second-type specimens; the strength under complex loading in a sufficiently wide range of stresses is determined by the shear.

REFERENCES

[1]   Zaharov, K. V., "Kriterii prochnosti dlya sloistykh plasticov", Plastich-eskie massy, Vol. 8, p. 23, 1963.

[2]   Malmeister, A. K., Tamug, V. P. and Teters, G. A., "Soprotivlenie zhestkikh polimernyh materialov", Riga, 1967.

[3]   Goldenblat, I. I. and Copnov, V. A., "Kriterii prochnosti i plastichnosti construktsionnyh materialov", Moskva, 1968.

[4]   Ashcenazi, E. K. and Ganov, E. V., "Anizotropia construktsionnyh materialov. Spravochnik mashinostroenia", Leningrad, 1972.

[5]   Schneider, G. J., "Evaluation of lamina strength criteria by off axis ten-sile coupon tests", Fibre Sci. and Technol., Vol. 5, No. 1, pp. 29-35, 1972.

[6]   Sendeckyj, G. P., "A brief survey of empirical multiaxial strength cri-teria for composites", ASTM STP No. 497, 1972.

[7]   Ovsiannikov, L. V., "Gruppovye svoistva differentsialnyh uravnenie", Novosibirsk, 1962.

[8]   Annin, B. D., "Sovremennye modeli plasticheskih tel", Novosibirsk, 1975.

[9]   Hill, R., "Matematicheskaia theoria plastichnosti", Moskva, 1956.

[10] Rabotnov, J. N., "Polzuchest elementov construktsii", Moskva, 1968.

[11] Annin, B. D., Danilov, N. S. and Rabotnov, J. N., "Mashina na slozhnoe nagruzhenie s avtomaticheskim programirovaniem napryagennogo sostoiania", Ing. Zgurnal, Mehanika tverdogo tela, Vol. 6, pp. 161-162, 1966.

SOME PROBLEMS OF FAILURE OF THIN-WALLED FLEXURAL STRUCTURES FROM REINFORCED PLASTICS

Yu. V. Nemirovsky

Institute of Hydrodynamics, Siberian Branch of the USSR Academy of Sciences
Novosibirsk

INTRODUCTION

Recently the problems related to failure of reinforced plastics have received widespread attention. In a series of experimental investigations [1-5], different mechanisms of bending failure have been observed (e.g., classical fracture of relatively long elements, shear failure for medium length or short elements). These mechanisms cannot be explained on the basis of the classical equations for bending of thin-walled elements and the classical failure criteria for orthotropic materials. Certain modifications are necessary. In this paper, the modified equations of bending applied to reinforced curvilinear rods and beams will be derived, and the failure criteria describing the mechanisms of failure are proposed.

EQUATION OF BENDING OF CURVILINEAR RODS FROM REINFORCED PLASTICS

Let us introduce the coordinate system which is connected with the middle line of the curvilinear rod so that x is the arc coordinate and z the coordinate along the normal to the middle line. The displacements in these directions will be denoted by u and w, respectively. Neglecting compression and considering relatively shallow rods, we have

$$\varepsilon_x = \frac{1}{A_1}\frac{\partial u}{\partial x} + \frac{w}{R_1}, \quad \varepsilon_{xz} = \frac{\partial u}{\partial z} + \frac{1}{A_1}\frac{\partial w}{\partial x} \tag{1}$$

where A_1 is Lamé coefficient and R_1 the curvature radius. Assuming in the general case that (a) the reinforcement trajectories are not equidistant with respect to the middle line of the rod, (b) the fibers are one-dimensional or two-dimensional, (c) an ideal adhesion is observed, (d) all the elements of the substructures of the reinforced material are in an elastic state [6,7], we then have the following relations for stress in the reinforced rod:

$$\underline{\sigma}_x = a_{11}\underline{\varepsilon}_x + a_{13}\underline{\varepsilon}_{xz} - Ta_{11}^T$$

$$\underline{\sigma}_{xz} = a_{31}\underline{\varepsilon}_x + a_{33}\underline{\varepsilon}_{xz} - Ta_{12}^T \tag{2}$$

$$\underline{\sigma}_z = a_{21}\underline{\varepsilon}_x + a_{23}\underline{\varepsilon}_{xz} - Ta_{22}^T$$

Here, a_{ij} are the rigidity coefficients, a_{ij}^T the thermal rigidity coefficients, and T the temperature. The relation between a_{ij} and a_{ij}^T and the structural parameters of the reinforced rod can be obtained from the formulae in [6,7]. For the displacements of the rods, the following distribution law can be assumed:

$$u_z = w(x), \ u_x = u(x,z) = u_0 + u_1 z + u_2 z^2 + u_3 z^3 \tag{3}$$

Substituting these expressions into equations (1), (2) and satisfying the boundary conditions for the stresses $\underline{\sigma}_z$ and $\underline{\sigma}_{xz}$ at the edges $Z = \pm H$ (2H is the rod thickness, the relations are found:

$$a_{12}^T T_{\pm} = \frac{a_{31}}{A_1}\left(\frac{du_0}{dx} \pm H \frac{du_2}{dx} + H^2 \frac{du_2}{dx} \pm H^3 \frac{du_3}{dx} + \frac{A_1 w}{R_1}\right.$$

$$+ a_{33}\left(\frac{1}{A_1}\frac{dw}{dx} + u_1 \pm 2Hu_2 + 3H^2 u_3\right)$$

$$\tag{4}$$

$$p^{\pm} + a_{22}T_{\pm} = \frac{a_{21}}{A_1}\left(\frac{du_0}{dx} \pm H \frac{du_1}{dx} + H^2 \frac{du_2}{dx}\right.$$

$$\left. \pm H^3 \frac{du_3}{dx} + \frac{A_1 w}{R_1}\right) + a_{23}\left(\frac{1}{A_1}\frac{dw}{dx} + u_1 \pm 2Hu_2 + 3H^2 u_3\right)$$

where $T_{\pm} = T(\pm H)$, $p^{\pm} = \underline{\sigma}_z(\pm H)$. From equations (4), it follows that

$$\frac{du_0}{dx} = -\frac{A_1 w}{R_1} + \underline{\theta}_1, \ H^2 u_3 = -\frac{1}{2A_1}\frac{du}{dx} + \underline{\theta}_2$$

$$u_1 = \underline{\psi}_1 + C_1 + \frac{1}{2A_1}\frac{dw}{dx} - \underline{\theta}_2$$

$$4u_2H = \frac{a_{12}^T\Phi_1^-}{a_{33}} - \frac{2a_{13}H}{A_1a_{33}} \cdot \frac{d\Psi_1}{dx}$$

$$\underline{\theta}_1 = \frac{[\Phi_1^+(a_{23}a_{13}^T-a_{33}a_{22}^T)-\Phi_2^+a_{33}^T]A_1}{2(a_{31}a_{23}-a_{21}a_{33})} - \frac{1}{4}\frac{Ha_{12}^T}{a_{33}}\frac{d\Phi_1^-}{dx} + \frac{1}{2}\frac{a_{31}H}{a_{33}}\frac{d}{dx}(\frac{1}{A_1}\frac{d\Psi_1}{dx})$$

$$2\underline{\theta}_2 = \frac{\Phi_1^+(a_{12}^Ta_{21}-a_{22}^Ta_{31})-a_{31}\Phi_2^+}{2(a_{33}a_{21}-a_{23}a_{31})} - \underline{\Psi}_1 - C_1$$

$$2H\underline{\Psi}_1 = (a_{33}a_{21}-a_{23}a_{31})^{-1}\int_{x_0}^{x}[a_{33}\Phi_2^- + (a_{33}a_{22}^T - a_{23}a_{12}^T)\Phi_1^-]dx$$

$$\Phi_1^{\pm} = T_+ \pm T_-, \quad \Phi_2^{\pm} = p^+ \pm p^- \tag{5}$$

where C_1 is the integration constant. Taking into account equation (4), we shall obtain with the help of equations (1) to (3) the following equations for the stresses:

$$\underline{\sigma}_x = b_{11} + \frac{c_{11}}{A_1}\frac{dw}{dx} + z[b_{12} + c_{13}\frac{d}{dx}(\frac{1}{A_1}\frac{dw}{dx}) + z^2 [b_{13} - \frac{c_{12}}{A_1}\frac{dw}{dx}]$$

$$+ z^3 [b_{14} - c_{14}\frac{d}{dx}(\frac{1}{A_1}\frac{dw}{dx})] - Ta_{11}^T$$

$$\underline{\sigma}_{xz} = d_{11} + \frac{e_{11}}{A_1}\frac{dw}{dx} + z [d_{12} + e_{13}\frac{d}{dx}(\frac{1}{A_1}\frac{dw}{dx})] + z^2 [d_{13} - \frac{e_{12}}{A_1}\frac{dw}{dx}] \tag{6}$$

$$+ z^3 [d_{14} - e_{14}\frac{d}{dx}(\frac{1}{A_1}\frac{dw}{dx})] - Ta_{12}^T$$

$$\underline{\sigma}_z = f_{11} + \frac{g_{11}}{A_1}\frac{dw}{dx} + z [f_{12} + g_{13}\frac{d}{dx}(\frac{1}{A_1}\frac{dw}{dx})] + z_1^2 [f_{13} - \frac{g_{12}}{A_1}\frac{dw}{dx}]$$

$$+ z^3 [f_{14} - g_{14}\frac{d}{dx}(\frac{1}{A_1}\frac{dw}{dx})] - Ta_{22}^T$$

The following contractions have been made:

$$b_{11} = a_{11}\underline{\nu}_1 + a_{13}\underline{\nu}_2, \quad b_{12} = a_{11}\underline{\nu}_3 + a_{13}\underline{\nu}_4,$$

$$b_{13} = a_{11}\underline{\nu}_5 + a_{13}\underline{\nu}_6, \quad b_{14} = a_{11}\underline{\nu}_7,$$

$$c_{11} = a_{13}\underline{\nu}_8, \quad c_{12} = a_{13}\underline{\nu}_9, \quad c_{13} = a_{11}\underline{\nu}_{10},$$

$$c_{14} = a_{11}\underline{\nu}_{11}, \quad d_{11} = a_{31}\underline{\nu}_1 + a_{33}\underline{\nu}_2,$$

$$d_{12} = a_{31}\underline{\nu}_3 + a_{33}\underline{\nu}_4, \quad d_{13} = a_{31}\underline{\nu}_5 + a_{33}\underline{\nu}_6$$

$$d_{14} = a_{31}\underline{\nu}_7, \quad e_{11} = a_{33}\underline{\nu}_8, \quad e_{12} = a_{33}\underline{\nu}_9, \quad e_{13} = a_{31}\underline{\nu}_{10},$$

$$e_{14} = a_{31}\underline{\nu}_{11}, \quad f_{11} = a_{21}\underline{\nu}_1 + a_{23}\underline{\nu}_2, \quad f_{12} = a_{21}\underline{\nu}_3 + a_{23}\underline{\nu}_4,$$

$$f_{13} = a_{21}\underline{\nu}_5 + a_{23}\underline{\nu}_6, \quad f_{14} = a_{21}\underline{\nu}_7,$$

$$g_{11} = a_{23}\underline{\nu}_8, \quad g_{12} = a_{23}\underline{\nu}_9, \quad g_{13} = a_{21}\underline{\nu}_{10},$$

$$g_{14} = a_{21}\underline{\nu}_{14}, \quad \underline{\nu}_1 = \underline{\theta}_1 A_1^{-1}, \quad \underline{\nu}_2 = C_1 + \underline{\Psi}_1 - \underline{\theta}_2,$$

$$\underline{\nu}_3 = \frac{1}{A_1}\frac{d}{dx}(\underline{\Psi}_1 - \underline{\theta}_2), \quad \underline{\nu}_4 = \frac{1}{2H}(\frac{a_{12}\bar{\phi}_1}{a_{33}} - \frac{2Ha_{31}}{A_1 a_{33}}\frac{d\underline{\Psi}_1}{dx}),$$

$$\underline{\nu}_5 = \frac{1}{4HA_1}[\frac{a_{12}^T}{a_{33}}\frac{d\bar{\phi}_1}{dx} - \frac{2Ha_{31}}{a_{33}}\frac{d}{dx}(\frac{1}{A_1}\frac{d\underline{\Psi}_1}{dx}), \quad \underline{\nu}_6 = \frac{3\underline{\theta}_2}{H^2},$$

$$\underline{\nu}_7 = \frac{1}{H^2 A_1}\frac{d\underline{\theta}_2}{dx}, \quad \underline{\nu}_8 = \frac{3}{2}, \quad \underline{\nu}_9 = \frac{3}{2H^2}, \quad \underline{\nu}_{10} = \frac{1}{2A_1}, \quad \underline{\nu}_{11} = \frac{1}{2A_1 H^2}$$

The Lagrange variational principle will now be introduced to derive the appropriate boundary conditions.

$$\int_{L_1}^{L_2}\int_{-H}^{H} (\underline{\sigma}_x \delta\underline{\varepsilon}_x + \underline{\sigma}_{xz}\delta\underline{\varepsilon}_{xz})A_1 dx dz = \int_{L_1}^{L_2} \bar{\phi}_2 \delta wA_1 dx + \sum_{i=1}^{2} [P_x^o(L_i,z)\underline{\delta}u(L_i,z)$$

$$+ P_z^o(L_i,z)\underline{\delta}w(L_i)]dz$$

Using equations (1), (3) and (5), when calculating displacements and deformations, the usual procedure leads to the governing differential equation:

$$\frac{d}{d\underline{n}}\left(\frac{dT_1}{d\underline{n}} - Q_1\right) = \bar{\phi_2} \tag{7}$$

and the boundary conditions

$$\left\{\left[\int_{-H}^{H} P_x^o(L_i,z)dz\right]\underline{\delta u}_o\right\}\Bigg|_{L_1}^{L_2} = 0,$$

$$\left\{\left[\frac{dT_1}{d\underline{n}} - Q_1 + \int_{-H}^{H} P_z^o(L_i,z)dz\right]\underline{\delta w}\right\}\Bigg|_{L_1}^{L_2} = 0, \tag{8}$$

$$\left\{\left[T_1 - \frac{1}{2}\int_{-H}^{H} P_z^o(L_i,z)z(1 - \frac{z^2}{H^2})dz\right]\frac{d\underline{\delta w}}{d\underline{n}}\right\}\Bigg|_{L_1}^{L_2} = 0,$$

where

$$T_1 = B_1 + B_2\frac{d^2w}{d\underline{n}^2}, \quad Q_1 = B_3 + B_4\frac{dw}{d\underline{n}} \tag{9}$$

$$B_1 = \frac{2H^3a_{11}}{105A_1}\frac{d}{dx}(7\underline{\psi}_1 - 4\underline{\theta}_2) - \frac{5a_{11}^T}{4H}\int_{-H}^{H} T(1 - \frac{z^2}{H^2})zdz,$$

$$B_2 = \frac{4H^2a_{11}}{105}, \quad B_4 = \frac{6}{5}a_{33}, \quad d\underline{n} = A_1dx,$$

$$B_3 = \frac{2H}{5}(5d_{11} + d_{13}H^2) - \frac{3a_{12}^T}{4H}\int_{-H}^{H} T(1 - \frac{z^2}{H^2})dz$$

Taking into account equation (9) and after integration, we obtain the equation

$$A_2\frac{d^2w}{d\underline{n}^2} - A_4w = R_2 \tag{10}$$

$$R_2 = R_2^o + \int_{\underline{n}_o}^{\underline{n}} \left[R_1^o + B_3 - \frac{dB_1}{d\underline{n}} + \int_{\underline{n}_o}^{\underline{n}} \bar{\phi_2}d\underline{n}\right]d\underline{n}$$

where R_1°, R_2° are the integration constants.

Thus the problem of the determination of the rod deflection under an arbitrary loading is reduced to solving the equation for constrained oscillations and is not presented here. Taking into account the known solution of equation (10), the stresses and deformations of the reinforced rod is found from equations (6) with the help of equations (1), (3) and (5).

FAILURE CRITERION FOR REINFORCED RODS

In solving the problem of the reinforced structure failure, we shall follow the independence principle for the failure of the substructure elements of the reinforced material [6,8]. In this case, it is necessary to determine individually the stresses in the matrix and in the fibers. In the case of the simplest model, when the fibers and the matrix are assumed one-dimensional, the equation for the matrix can be written as

$$\sigma_x^c = E_c \varepsilon_2, \quad \sigma_z^c = -\nu_c E_c \varepsilon_x, \quad \sigma_{xz}^c = G_c \varepsilon_{xz} \tag{11}$$

and for the fibers as

$$\sigma_x^a = E_a \varepsilon_x \tag{12}$$

Then the simplest criteria for the failure of the matrix and the fibers can be written as

$$\max_{x_1 z} |\sigma_x^c| = \begin{cases} \sigma_c^+, & \text{if } \sigma_x^c(x_1^*, z_1^*) > 0, \\ \\ \sigma_c^-, & \text{if } \sigma_x^c(x_1^*, z_1^*) < 0, \end{cases} \tag{13}$$

$$\max_{x_1 z} |\sigma_z^c| = \begin{cases} \sigma_c^+, & \text{if } \sigma_z^c(x_2^*, z_2^*) > 0, \\ \\ \sigma_c^-, & \text{if } \sigma_z^c(x_2^*, z_2^*) < 0, \end{cases} \tag{14}$$

$$\max_{x_1 z} |\sigma_{xz}^c| = \tau_c^* \tag{15}$$

$$\max_{x_1 z} |\sigma_x^a| = \begin{cases} \sigma_a^+, & \text{if } \sigma_x^a(x_3^*, z_3^*) > 0, \\ \\ \sigma_a^-, & \text{if } \sigma_x^a(x_3^*, z_3^*) < 0, \end{cases} \tag{16}$$

where $\sigma_c^\pm$, $\sigma_a^\pm$ are the tensile strengths for the matrix and the fibers, respectively; x_k^*, z_k^* the coordinates of the corresponding maximum points. If the loading acting to the flexural reinforced rod changes proportionally to one parameter p as in equations (13) and (14), we then obtain a series of its limiting values p_i^* (i = 1,2,3,4). Each of them determines a possible initial failure mechanism, e.g., fracture of the matrix, lamination of the matrix caused by breakage, lamination caused by shear, fracture of fibers. The type of failure and the corresponding breaking loading p^* applied to the rod is determined by an equality

$$p^* = \min_i p_i^*$$

However, in order to obtain more refined theoretical results, it is expedient to use a model of the reinforced material where both the matrix and the fibers are considered two-dimensional. In this case, the failure condition obtained by P. P. Balandin or Mises can be used as a failure criterion for the substructure elements [9]. Meanwhile, it should be noted that the method becomes very complicated.

The above approach to solving the problem of failure of flexural reinforced structures, such as beams or curvilinear rods, was used in [10-12]. It was based on the equations which are very complicated when compared to those presented earlier. The calculated results were in good agreement with the experimental values. In [13] the analogous approach was used in the case of flexure of circular and ring plates, and in [14] the same approach was used for the case of rotational shells under an axisymmetrical loading.

REFERENCES

[1]   Tarnopolsky, Yu. and Rose, A., "On durability of oriented glass fibers being bent", Mekhanika polimerov, 4, pp. 535-542, 1966.

[2]   Tarnopolsky, Yu., Rose, A. and Shlitsa, R., "Testing of wound rings under concentrated forces", Mekhanika polimerov, 4, pp. 719-727, 1969.

[3]   Abramov, S., Mezentsev, N., Nikolaev, V. and Popov, V., "Prochnost pri izgibe stekloplastikov poluchennyh namotkoi", Problemy prochnosti, 10, pp. 62-64, 1975.

[4]   Menges, G. and Kleiholz, R., "Vergleich verschiedener Verfahren zum Bestimmen der interlaminaren Scherfestigkeit", Kunststoffe, 59, 12, pp. 959-966, 1969.

[5]   Sayers, K. and Harris, B., "Interlaminar shear strength of a carbon fibre reinforced composite material under impact conditions", Journal of Composite Materials", Vol. 7, pp. 129-136, 1973.

[6] Nemirovsky, Yu., "Ob uprugoplasticheskom povedenii armirovannogo sloya", Zhurnal prikladnoi mekhaniki i tekhnicheskoi fiziki", 6, pp. 81-89, 1969.

[7] Nemirovsky, Yu., "On the thermo-elastic bending theory of reinforced shells and plates", Mekhanika polimerov, 5, pp. 861-873, 1972.

[8] Nemirovsky, Yu., "Ob uslovii plastichnosti (prochnosti) dlya armirovannogo sloya", Zhurnal prikladnoi mekhaniki i tekhnicheskoi fiziki, 5, pp. 81-88, 1969.

[9] Andreev, A. and Nemirovsky, Yu., "K teorii uprugih mnogosloinyh anizotropnyh obolochek", Izv. Akad. Nauk SSSR, Mekhanika tverdogo tela, 5, pp. 87-96, 1977.

[10] Nemirovsky, Yu. and Reznikov, B., "On the fracture mechanism of reinforced beam under the action of bending load. 1. Shift Fracture", Mekhanika polimerov, 4, pp. 698-709, 1973.

[11] Nemirovsky, Yu. and Reznikov, B., "Izgib ploskih armirovannyh krivolineinuh sterzhnei i optimizatsiia ih struktury po nachalnomy razrusheniiu", In: Metody resheniia zadach uprugosti i plastichnosti", 7, g. Gorkii, pp. 106-124, 1973.

[12] Nemirovsky, Yu. and Reznikov, B., "Process of shear destruction of fibre reinforced rings in bending", Mekhanika polimerov, 3, pp. 435-444, 1976.

[13] Nemirovsky, Yu. and Reznikov, B., "On initial fracture of reinforced circular and ring-shaped plates", Izv. Akad. Nauk Armian. SSR Mekhanika, XXIX, 2, pp. 50-64, 1976.

[14] Nemirovsky, Yu. and Reznikov, B., "Problems of the fracture of bended reinforced structures", Mekhanika polimerov, 6, pp. 1029-1038, 1977.

OPTIMUM DESIGN AND STRENGTH OF LAYERED THIN COMPOSITES

I. F. Obraztsov and V. V. Vasil'ev

Moscow Institute of Aviation Technology
USSR

INTRODUCTION

High-performance composite materials are used today in different engineering applications. The strength and elastic constants of composites depend on the basic anisotropic properties of a single layer. The pattern of orientation of layers in a structural laminate may vary by a wide margin. The problem of design of composites with directional properties that satisfy certain specifications will be formulated. In particular, the design of thin composites with maximum strength is considered in this report.

PRELIMINARY EQUATIONS

Let us consider a composite material with an arbitrary number of layers κ having thicknesses h_i and angles of orientation ϕ_i as shown in Figure 1. The given stress resultants N_x, N_y, N_{xy} are referred to a cartesian coordinate system with axes x, y and stresses σ_1^i, σ_2^i, τ_{12}^i referred to a system of lamina principal coordinates 1,2. These quantities are related by the following equilibrium equations:

$$N_x = h \sum_{i=1}^{k} \bar{h}_i (\sigma_1^i \cos^2\phi_i + \sigma_2^i \sin^2\phi_i - \tau_{12}^i \sin2\phi_i)$$

$$N_y = h \sum_{i=1}^{k} \bar{h}_i (\sigma_1^i \sin^2\phi_i + \sigma_2^i \cos^2\phi_i + \tau_{12}^i \sin2\phi_i) \tag{1}$$

$$N_{xy} = h \sum_{i=1}^{k} \bar{h}_i [(\sigma_1^i - \sigma_2^i) \sin\phi_i \cos\phi_i + \tau_{12}^i \cos2\phi_i]$$

where

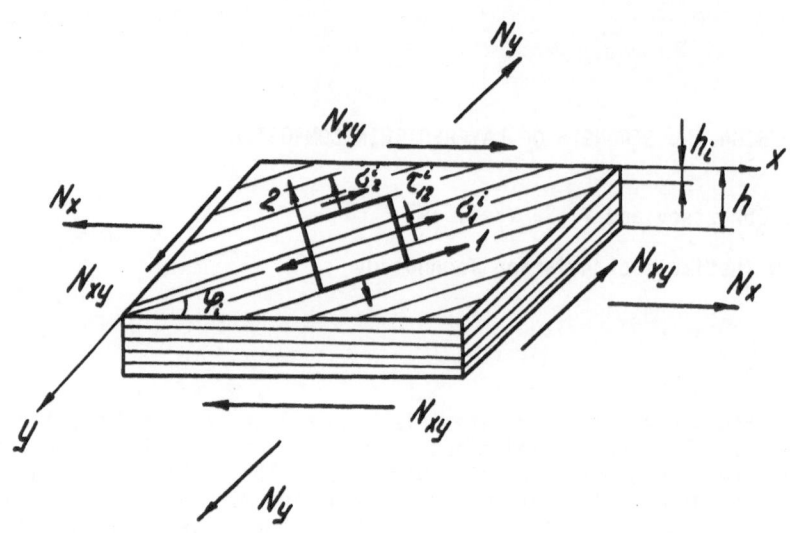

Fig. (1) - Notation of stress resultants and stresses
for a laminate composite

$$\bar{h}_i = h_i/h, \quad h = \sum_{i=1}^{k} h_i$$

Assuming that the average strains ε_x, ε_y, γ_{xy} are distributed uniformly through the thickness, we may obtain the following transformation equations:

$$\varepsilon_1^j = \varepsilon_x \cos^2\phi_j + \varepsilon_y \sin^2\phi_j + \gamma_{xy}\sin\phi_j\cos\phi_j$$

$$\varepsilon_2^j = \varepsilon_x \sin^2\phi_j + \varepsilon_y \cos^2\phi_j - \gamma_{xy}\sin\phi_j\cos\phi_j \qquad (2)$$

$$\gamma_{12}^j = (\varepsilon_y - \varepsilon_x)\sin 2\phi_j + \gamma_{xy}\cos 2\phi_j$$

The stresses and strains are related by the generalized Hooke's law

$$\sigma_1^j = \bar{E}_1(\varepsilon_1^j + \mu_{12}\varepsilon_2^j), \quad \sigma_2^j = \bar{E}_2(\varepsilon_2^j + \mu_{21}\varepsilon_1^j), \quad \tau_{12}^j = G_{12}\gamma_{12}^j \tag{3}$$

where

$$\bar{E}_{1,2} = E_{1,2}/(1-\mu_{12}\mu_{21})$$

For the given stress resultants N_x, N_y and N_{xy}, equations (1) to (3) define 3κ stresses σ_1^j, σ_2^j, τ_{12}^j, which are required to satisfy the following failure criterion in each layer:

$$F_j = F_{11}(\sigma_1^j)^2 + F_{12}\sigma_1^j\sigma_2^j + F_{22}(\sigma_2^j)^2 + F_{33}(\tau_{12}^j)^2 \leq 1 \tag{4}$$

where F_{mn} depend upon the ultimate stresses. For example, in the case of Hill's criterion, we have

$$F_{11} = -F_{12} = 1/\bar{\sigma}_1^2, \quad F_{22} = 1/\bar{\sigma}_2^2, \quad F_{33} = -1/\bar{\tau}_{12}^2$$

For fiber reinforced epoxy systems, the condition in equation (4) usually corresponds to resin crazing [1,2]. For properly designed laminate with fibers controlling the ultimate strength, resin crazing does not result in the failure of the composite material and its behavior may be described by equations (1) to (3) with $E_2 = G_{12} = \mu_{12} = \mu_{21} = 0$. The failure criterion in equation (4) then reduces to the maximum stress theory

$$F_{11}(\sigma_1^j)^2 \leq I$$

A PARALLEL-PLY LAMINATE OF MAXIMUM STRENGTH

Before considering the general problem, let us establish the optimum angle of orientation of a parallel-ply laminate. For $\kappa=1$, the stresses σ_1, σ_2, τ_{12} are statically determinate. For proportional loading with $N_x = \rho N_x^o$, $N_y = \rho N_y^o$, $N_{xy} = \rho N_{xy}^o$, equations (1) and (4) give

$$\left(\frac{h}{\rho}\right)^2 \geq \left(\frac{h}{\rho}\right)^2 = \frac{1}{4}[(F_{11}+F_{12}+F_{22})(N_x^o+N_y^o)^2 + F_{33}(N_x^o-N_y^o)^2 + 4F_{33}(N_{xy}^o)^2]$$

$$+ \left(\frac{N_x^o-N_y^o}{2}\cos2\phi + N_{xy}^o\sin2\phi\right)[(F_{11}+F_{22}-F_{12}-F_{33})\left(\frac{N_x^o-N_y^o}{2}\cos2\phi\right.$$

$$+ N_{xy}^o\sin2\phi) + (F_{11}-F_{22})(N_x^o+N_y^o)] \tag{5}$$

The conditions of maximum strength (or minimum thickness) may be written in the form

$$d(h/\bar{\rho})/d\phi = 0; \quad d^2(h/\bar{\rho})/d\phi^2 > 0$$

After differentiation and using equation (1), the following conditions are obtained:

$$\tau_{12}[(F_{11}+F_{22}-F_{12}-F_{33})(\sigma_1-\sigma_2) + (F_{11}-F_{22})(\sigma_1+\sigma_2)] = 0 \tag{6}$$

$$4(F_{11}+F_{22}-F_{33}-F_{12})\tau_{12}^{2\prime} - (\sigma_1-\sigma_2)[(F_{11}+F_{22}-F_{12}-F_{33})(\sigma_1-\sigma_2)$$
$$\times (\sigma_1-\sigma_2) + (F_{11}-F_{22})(\sigma_1+\sigma_2)] > 0 \tag{7}$$

The result for Hill's criterion was formerly obtained in [3]. For fiber reinforced epoxy systems with $\bar{\sigma}_1 \gg \bar{\sigma}_2$, $\bar{\sigma}_1 \gg \bar{\tau}_{12}$, $\sigma_1 > \sigma_2$, equations (6) and (7) have the following approximate forms:

$$\tau_{12}F_{33}\sigma_1 = 0; \quad 4\tau_{12}^2(F_{22}-F_{33}) + \sigma_1^2 F_{33} > 0$$

It follows that $\tau_{12} = 0$. Hence, a structure attains its maximum strength if the fibers are oriented along the directions of the maximum principal stresses. It is important to note that the conditions in equations (6) and (7) yield the optimum stress distribution in the material.

OPTIMUM DESIGN OF LAYERED THIN COMPOSITES

Consider a composite material consisting of κ layers. The first two expressions in equations (1) give

$$h = (N_x+N_y)/ \sum_{i=1}^{k} \bar{h}_i(\sigma_1^i+\sigma_2^i) \tag{8}$$

The stresses σ_1^i and σ_2^i must satisfy the strain compatibility equations, which following from equations (2) and (3), it is found that

$$L_j = \frac{1-\mu_{21}}{E_1}(\sigma_1^1-\sigma_1^j) + \frac{1-\mu_{12}}{E_2}(\sigma_2^1-\sigma_2^j) = 0 \tag{9}$$

$$M_j = \frac{1+\mu_{21}}{E_1}(\sigma_1^1\cos2\phi_1 - \sigma_1^j\cos2\phi_j) - \frac{1+\mu_{12}}{E_2}(\sigma_2^1\cos2\phi_1 - \sigma_2^j\cos2\phi_j)$$
$$- \frac{1}{G_{12}}(\tau_{12}^1\sin2\phi_1 - \tau_{12}^j\sin2\phi_j) = 0 \tag{10}$$

$$N_j = \frac{1+\mu_{21}}{E_1}(\sigma_1^1\sin2\phi_1 - \sigma_1^j\sin2\phi_j) - \frac{1+\mu_{12}}{E_2}(\sigma_2^1\sin2\phi_1 - \sigma_2^j\sin2\phi_j)$$
$$+ \frac{1}{G_{12}}(\tau_{12}^1\cos2\phi_1 - \tau_{12}^j\cos2\phi_j) = 0 \tag{11}$$

As discussed earlier, consider the optimum stress distribution that satisfy the strain compatibility equations (9) to (11) and the failure criterion in equation (4). The latter may be written in the following form [4]:

$$sinv_j - g_j = 0$$

where $g_i = 2F_j-1$. Consider the following function

$$F = h + \sum_{j=1}^{k} \lambda_1^j(sinv_j - g_j) + \sum_{j=2}^{k} (\lambda_2^j L_j + \lambda_3^j M_j + \lambda_4^j N_j)$$

where h is determined by equation (8) and the factors $(\lambda_1^j/\lambda_4^j)-\lambda$. Minimum h corresponds to minimum F and the variational equations are

$$\bar{h}_{1f} + 2\lambda_1^1(2F_{11}\sigma_1^1 + F_{12}\sigma_2^1) - \frac{1-\mu_{21}}{E_1}\sum_{j=2}^{k}\lambda_2^j - \frac{1+\mu_{21}}{E_1}\sum_{j=2}^{k}\xi_1^j = 0 \qquad (12)$$

$$\bar{h}_{1f} + 2\lambda_1^1(2F_{22}\sigma_2^1 + F_{12}\sigma_1^1) - \frac{1-\mu_{12}}{E_2}\sum_{j=2}^{k}\lambda_2^j + \frac{1+\mu_{12}}{E_2}\sum_{j=2}^{k}\xi_1^j = 0 \qquad (13)$$

$$4\lambda_1^1 F_{33}\tau_{12}^1 + \frac{1}{G_{12}}\sum_{j=2}^{k}\eta_1^j = 0 \qquad (14)$$

$$\bar{h}_j f + 2\lambda_j^1(2F_{11}\sigma_1^j + F_{12}\sigma_2^j) + \lambda_2^j \frac{1-\mu_{21}}{E_1} + \frac{1+\mu_{21}}{E_1}\xi_j^j = 0 \qquad (15)$$

$$\bar{h}_j f + 2\lambda_j^1(2F_{22}\sigma_2^j + F_{12}\sigma_1^j) + \lambda_2^j \frac{1-\mu_{12}}{E_2} - \frac{1+\mu_{12}}{E_2}\xi_j^j = 0 \qquad (16)$$

$$4\lambda_1^j F_{33}\tau_{12}^j + \frac{1}{G_{12}}\eta_j^j = 0 \qquad (17)$$

$$\lambda_j\sqrt{1-g_j^2} = 0 \qquad (18)$$

$$L_j = 0, \; M_j = 0, \; N_j = 0 \qquad (19)$$

where

$$f = \frac{N_x + N_y}{\left[\sum\limits_{i=1}^{k} \hbar_i (\sigma_1^j + \sigma_2^j)\right]^2}, \quad \xi_m^j = \lambda_3^j \cos 2\phi_m + \lambda_4^j \sin 2\phi_m, \quad \eta_m^j = \lambda_3^j \sin 2\phi_m - \lambda_4^j \cos 2\phi_m$$

This gives the $7\kappa-3$ equations (12) to (19) solving for the $7\kappa-3$ unknowns. They are the 3κ stresses σ_1^j, σ_2^j, τ_{12}^j; κ factors λ_1^j and $3(\kappa-1)$ factors λ_2^j, λ_3^j, λ_4^j. The solution of these equations specifies the optimum stress distribution in the material of maximum strength (or minimum thickness). The corresponding structural parameters may then be obtained from equations (1).

In view of the difficulties of obtaining a direct solution to the system of equations (12) to (19), a semi-inverse method will be considered. It was mentioned earlier that the fibers in a material with maximum strength should be aligned along the directions of the maximum principal stresses. With this in mind, τ_{12}^j in equations (12) to (19) are set to zero. Equations (14) and (17) then give $n_1^j = n_2^j = 0$. It is now possible to consider the following two cases:

(1) $\lambda_4^j/\lambda_3^j = tg2\phi_1 = tg2\phi_j$

(2) $\lambda_3^j = \lambda_4^j = 0$

Case (1) - Let $\lambda_4^j/\lambda_3^j = tg2\phi_1 = tg2\phi_j$. It follows that for a two-layered system, the angles of orientation are $\phi_1 = \phi$ and $\phi_2 = \phi + \pi/2$. Using the strain compatibility equations (19) or (9) to (11), the stresses σ_1^2 and σ_2^2 in the second layer may be expressed in terms of the stresses σ_1^1 and σ_2^1 in the first layer. Hence, the failure criterion in equation (4) for both layers may be written in terms of the stresses σ_1^1 and σ_2^1. The corresponding limit interaction curve for the graphite fibers in an epoxy matrix ($E_1 = 21.10^6$ psi, $E_2 = 1.7 \times 10^6$ psi, $\bar{\sigma}_1 = 15 \times 10^4$ psi, $\bar{\sigma}_2 = 0.76 \times 10^4$ psi and $\mu_{12} = 0.0227$) is shown in Figure 2. Curve 1 is for the first layer and curve 2 for the second layer. Equations (18) are satisfied by letting $\lambda_1^1 = 0$, $g_2 = 1$ (curve 2) or by $\lambda_1^2 = 0$, $g_1 = 1$ (curve 1). The case $\lambda_1^1 = \lambda_1^2 = 0$ is obviously impossible, and the case $g_1 = g_2 = 1$ corresponds to the point of intersection of curves 1 and 2. Thus, only four equations (12), (13), (15) and (16) remain unsatisfied. The corresponding solution determines the stresses σ_1^1, σ_2^1, σ_1^2 and σ_2^2. The angle of orientation and the thicknesses of the layers may be obtained by using equations (1):

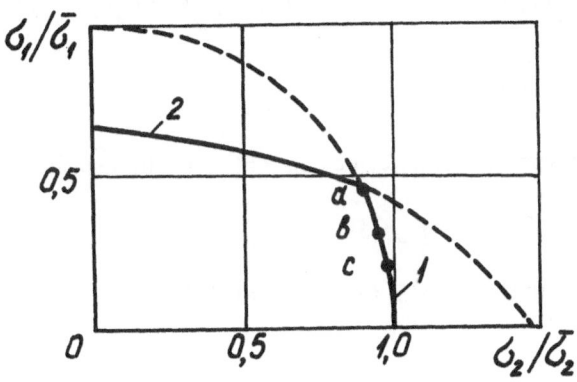

Fig. (2) - Limit interaction curve for the graphite fibers
embedded in an epoxy matrix

$$tg2\phi = \frac{2N_{xy}}{N_x-N_y} = \alpha$$

(20)

$$h_{1,2} = \frac{(N_x+N_y)(\sigma_1^{2,1}-\sigma_2^{2,1})+(N_x-N_y)\sqrt{1+\alpha^2}(\sigma_1^{2,1}+\sigma_2^{2,1})}{2(\sigma_1^1\sigma_1^2-\sigma_2^1\sigma_2^2)}$$

The results of calculations are shown by points, a, b, c in Figure 2. The two layered material is loaded by N_x=N, N_y=2N and N_{xy}=0. The point a corresponds to the optimum angle of orientation ϕ=0 with N_{xy} = 0.5N. The point b corresponds to ϕ = -13.5° and N_{xy}=N while point c to ϕ = -22°. The total thickness of the laminate is the same in all three cases.

The first of equations (20) specifies the angle of orientation ϕ for a system with statically determinate stress resultants N_x, N_y and N_{xy}. In the redundant structures, where N_x, N_y, N_{xy} depend upon ϕ, one simple particular case may be noted. Assuming that $h_1 = h_2 = h/2$, the following constitutive relations are obtained for a material with fibers directed along the trajectories of the principal stresses:

$$N_x = \frac{Eh}{1-\mu^2}(\varepsilon_x+\mu\varepsilon_y), \quad N_y = \frac{Eh}{1-\mu^2}(\varepsilon_y+\mu\varepsilon_x), \quad N_{xy} = \frac{Eh}{2(1+\mu)}\gamma_{xy}$$

where

$$E = \frac{1}{2(E_1+E_2)} [E_1(1+\mu_{12}) + E_2(1+\mu_{21})][E_1(1-\mu_{12}) + E_2(1-\mu_{21})],$$

$$\mu = (E_1\mu_{12}+E_2\mu_{21})/(E_1+E_2)$$

In this case, the stress resultants may be obtained by considering an equivalent isotropic material.

Case (2) - Consider $\lambda_2^j = \lambda_3^j = \lambda_4^j = 0$. The strain compatibility equations (19) are satisfied. With $\tau_{12}^j = 0$ in equations (9) to (11), there results the relations $\sigma_1^j = \sigma_1$, $\sigma_2^j = \sigma_2 = n\sigma_1$ where

$$n = \frac{E_2(1+\mu_{21})}{E_1(1+\mu_{12})}, \; j = 1,2,3,\ldots,\kappa) \tag{21}$$

With $\lambda_1^j = \lambda$ and $g_j = g$, equations (12) to (19) may be simplified to

$$\frac{N_x+N_y}{(1+n)(\sigma_1)^z} + 2\lambda(2F_{11}+nF_{12})\sigma_1 = 0 \tag{22}$$

$$\frac{N_x+N_y}{(1+n)(\sigma_1)^z} + 2\lambda(2nF_{22}+F_{12})\sigma_1 = 0 \tag{23}$$

such that

$$\lambda\sqrt{1-g^z} = 0$$

From equations (22) and (23), it follows that $\lambda \neq 0$ and $g=I$. The solution must satisfy the limit curve in equation (4). Substitution of the stresses in equation (21) into the failure criterion in equation (4) gives

$$\sigma_1 = \pm 1/\sqrt{F_{11}+nF_{12}+n^2F_{22}} \tag{24}$$

From equation (22), it follows that the mechanical properties of the material must satisfy the condition $n=n_o = (2F_{11}-F_{12})/(2F_{22}-F_{12})$, where n is defined earlier with reference to equation (21). For the fiber reinforced epoxy system $n>n_o$. According to the netting analysis $E_2 = \bar{\sigma}_2 = 0$ so that $n=n_o=0$. For a metal matrix composite $n\approx n_o$.

Substitution of the stresses in equation (21) into the equilibrium equations (1) gives the expressions that determine the parameters for the optimum structure:

$$\frac{N_x}{N_x+N_y} = \frac{1}{1+n} \left[n+(1-n) \sum_{i=1}^{k} \bar{h}_i \cos^2\phi_i\right]$$

(25)

$$\frac{N_{xy}}{N_x+N_y} = \frac{1-n}{1+n} \sum_{i=1}^{k} \bar{h}_i \sin\phi_i \cos\phi_i$$

$$h = \frac{N_x+N_y}{1+n} \sqrt{F_{11}+nF_{12}+n^2F_{22}}$$

(26)

Thus there exist a great number of structures with stresses given by equation (24) and the thickness in equation (26). It follows from equation (25) that the stress resultants must satisfy the conditions

$$\frac{n}{1+n} \leq \frac{N_x}{N_x+N_y} \leq \frac{1}{1+n} \cdot \left|\frac{N_{xy}}{N_x+N_y}\right| \leq \frac{1-n}{2(1+n)}$$

As an example, let us consider a pressure vessel in the form of a shell of revolution. For a uniform internal pressure q, the stress resultants are

$$N_x = \frac{qR_2}{2}, \quad N_y = N_x\left(2 - \frac{R_2}{R_1}\right), \quad N_{xy} = 0$$

(27)

For a two-layered cross-ply filament wound shell, $h_1 = h_2 = 0.5h$ and $\phi_1 = \phi$, $\phi_2 = -\phi$. The thickness of the shell is

$$h = \frac{ms}{2\pi r\cos\phi}$$

(28)

where m designates the number of filaments with areas s which intersect the arbitrary cross section r = const. The second of equations (25) is satisfied and the first in combination with equation (27) determine the shape of the meridian

$$2 - \frac{R_2}{R_1} = \frac{1-(1-n)\cos^2\phi}{n+(1-n)\sin^2\phi}$$

(29)

According to the netting analysis (n=0), the equation

$$2 - (R_2/R_1) = tg^2\phi$$

was formerly obtained in [6].

Let us now determine the angle of orientation $\phi(r)$ for isotensoid filament wound shell. Equations (24), (26) and (28) give

$$\frac{ms}{2\pi r cos\phi} = \frac{N_x + N_y}{(1+n)\sigma_1}, \quad N_x + N_y = \frac{N_x(1+n)}{n+(1-n)cos^2\phi}$$

Using equation (27) gives

$$\frac{ms\sigma_1}{\pi q} \frac{r}{R_2} = \frac{r^2 cos\phi}{n+(1-n)cos^2\phi} \tag{30}$$

This equation may be differentiated by assuming σ_1 = const. After some transformations with the help of equations (29), (30) and the Codazzi equation $d(r/R_2)/dr = I/R_1$, the following differential equation is found:

$$r \frac{d\phi}{dr} tg\phi \frac{n-(1-n)cos^2\phi}{1-(1-n)cos^2\phi} = 1$$

whose solution is

$$rcos^n\phi[1 - (1-n)cos^2\phi]^{\frac{1-n}{2}} = const$$

If n=0, the equation of the geodesic curve is then obtained for $rsin\phi$ = const.

REFERENCES

[1] Tsai, S. W. and Azzi V. D., ARS J., No. 2, 1966.

[2] Obraztsov, I. F., Vasil'ev, V. V. and Bunakov, V. A., "Optimum design of composite shells of revolution", Moskva, Mashinostroenie, 143 p., 1977.

[3] Brandmaier, H. E., J. Compos. Mater., V. 4, VII, 1970.

[4] Vorob'ev, L. M. and Vorob'eva, T. M., "Nonlinear transformations in applied variational problems", Moskva, Energia, 150 p., 1972.

[5] Halpin, J. C., J. Compos. Mater., V. 6, II, 1972.

[6] Zickel, J., ARS J., No. 6, 1962.

FRACTURE MODELS OF COMPOSITES WITH DIFFERENT FIBER ORIENTATION

R. B. Rikards, G. A. Teters and Z. T. Upitis

Institute of Polymer Mechanics
Academy of Sciences of the Latvian SSR, USSR

STRENGTH SURFACE

Optimum design of structures made of composite materials requires a knowledge of the strength of material as a function of parameters that characterize the composite in terms of its constituents. For a composite laminate, it is essential to know such variables as fiber orientation, intensity of reinforcement, volume content and properties of the fibers, etc. The present paper gives the experimental data on fracture of a composite laminate in the combined stress state. Fracture models are also constructed on the basis of micromechanical fracture analysis of the constituents.

Consider a composite laminate with fiber orientation $\pm\phi$ in the two-dimensional stressed state. The strength surface of the given material may be defined as follows [1]:

$$P_{\alpha\beta}(\phi_i,\theta_i,\mu)\sigma_{\alpha\beta} + P_{\alpha\beta\gamma\delta}(\phi_i,\theta_i,\mu)\sigma_{\alpha\beta}\sigma_{\gamma\delta} + P_{\alpha\beta\gamma\delta\epsilon\xi}(\phi_i,\theta_i,\mu)$$

$$\times \sigma_{\alpha\beta}\sigma_{\gamma\delta}\sigma_{\epsilon\xi} + \ldots = 1 \tag{1}$$

where $P_{\alpha\beta}$ and $P_{\alpha\beta\gamma\delta}$ are the strength tensors of the second and fourth order which depend on parameters characterizing the structure of the composite reinforcement. In equation (1), μ is the bulk reinforcement coefficient, θ_i the intensity of reinforcement in the ith direction and ϕ_i the ith angle of reinforcement.

First, the influence of fiber orientation on the strength tensors will be examined for $\phi_i \neq$ const, $\theta_i =$ const and $\mu =$ const. Experiments were carried out on glass-fiber-reinforced plastics with epoxy binder [2]. The specimens chosen were plates with fiber orientations $\phi_i = 0°$, $\pm10°$, $\pm20°$, $\pm30°$ and $\pm45°$

as shown in Figure 1. The bulk reinforcement coefficient is $\mu = 0.57/0.67$.

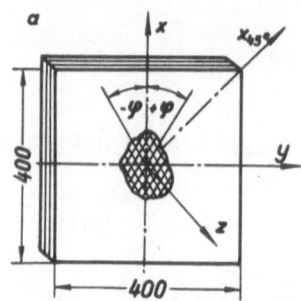

Fig. (1) - Structure of composite reinforcement

Determined from the experiments are the ultimate stress values for elongation, compression, shear and their combinations in the two-dimensional stressed state. The obtained results can be approximated by equation (1). Retaining first and second order terms, according to equation (1) for an orthotropic material, the following expression is obtained:

$$P_{11}\sigma_{11} + P_{22}\sigma_{22} + P_{1111}\sigma_{11}^2 + P_{2222}\sigma_{22}^2 + 2P_{1122}\sigma_{11}\sigma_{22} + 4P_{1212}\sigma_{12}^2 = 1 \quad (2)$$

The approximation of experimental results by the method of least squares [3] is represented in Figure 2, which shows good agreement between the experiments and theoretical predictions. It was found that the dependence of the strength tensor components on the fiber orientation angle can be approximated by the expression (Figure 3):

$$P(\phi) = \kappa_1 + \kappa_2 \, escp\{-\kappa_3(\phi-\phi_0)^{2\kappa_4}\} \quad (3)$$

where

$0 \leq \phi \leq \pi/2$ such that

$\phi_0 = \pi/2$ for P_{11}, P_{1111};

$\phi_0 = \pi/4$ for $2P_{1122}$, $4P_{1212}$; and

$\phi_0 = 0$ for P_{22}, P_{2222}.

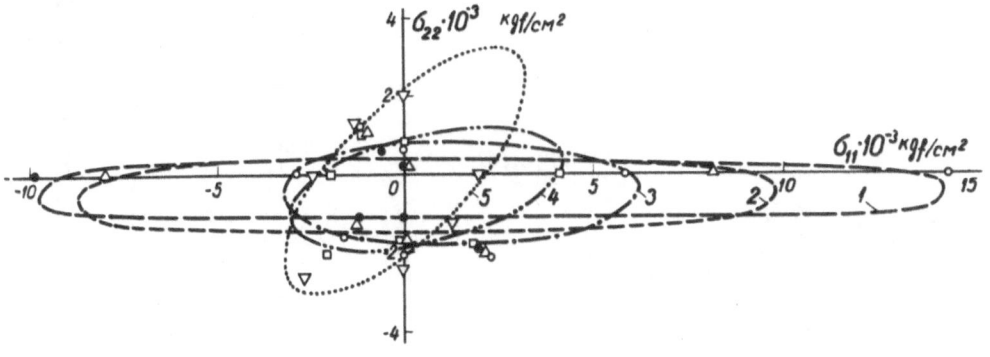

Fig. (2) - Strength surfaces of composites with different fiber orientation: Curve 1 - composite with fiber orientation $\phi = \pm 0°$ (o); Curve 2 - $\pm 10°$ ($\triangle$); Curve 3 - $\pm 20°$ (0); Curve 4 - $\pm 30°$ ($\square$); and Curve 5 - $\pm 45°$ (∇)

In addition, the influence of the intensity of reinforcement θ_i upon the strength tensor components is investigated [4]. In this case, μ = const and ϕ_i = const. The corresponding specimen is shown in Figure 4.

The structure of reinforcement for such composites can be represented in the plane $\theta_1 + \theta_2 + \theta_3 = 1$, Figure 5. Thirteen different composites with above mentioned structures were investigated. Each of the strength tensor components P_{ij} and $P_{ijk\ell}$ were found from equation (2). As an example, the dependence of P_{1111} upon the intensity of reinforcement θ_i is shown in Figure 6. Hence, the strength surface for different θ_i is

$$P_{ij}(\theta_i)\sigma_{ij} + P_{ijk\ell}(\theta_i)\sigma_{ij}\sigma_{k\ell} = 1 \tag{4}$$

The obtained components of the strength tensors which characterize fiber orientation and intensity may be employed for the optimum design of a composite laminate. For the given stressed state, the ultimate stress values can be characterized by a vector ρ:

$$\rho = \rho(\phi_i, \theta_i, \mu, P_{ij}, P_{ijk\ell}) \tag{5}$$

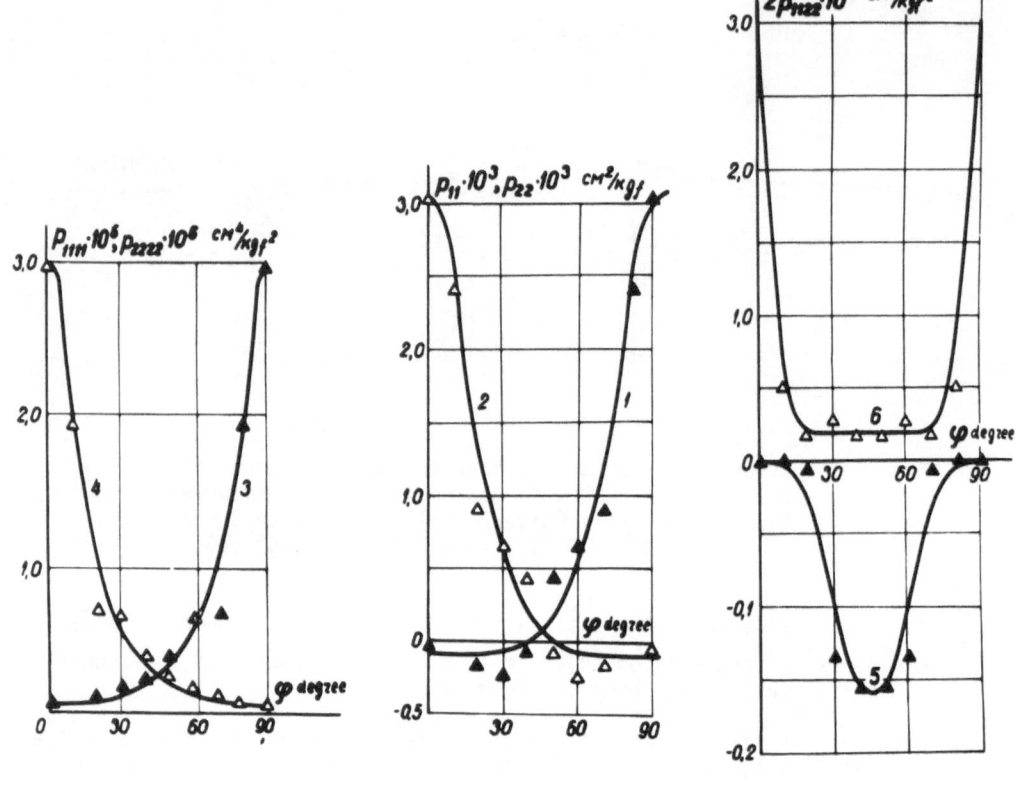

Fig. (3) - Components of strength tensors as functions of fiber orientation: Curve 1 - $P_{11}(\phi)$; Curve 2 - $P_{22}(\phi)$; Curve 3 - $P_{1111}(\phi)$; Curve 4 - $P_{2222}(\phi)$; Curve 5 - $2P_{1122}(\phi)$ and Curve 6 - $4P_{1212}(\phi)$

associated with the shapes

$$\sigma_{11} = \rho \sin\alpha_2 \cos\alpha_1,$$

$$\sigma_{22} = \rho \sin\alpha_2 \sin\alpha_1,$$

$$\sigma_{12} = \rho \cos\alpha_2,$$

which are different for composites with different reinforcement arrangements. The best structure corresponds to σ_{ij} with max ρ as indicated in Figures 7a and

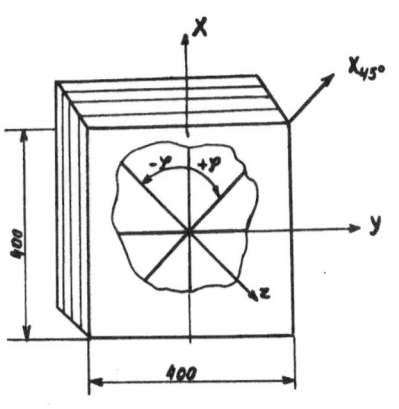

Fig. (4) - Structure of reinforcement of
the composite

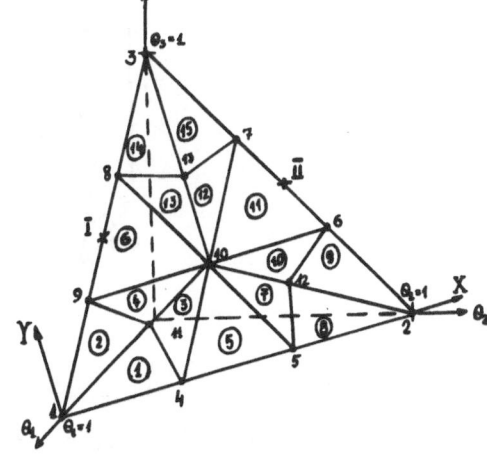

Fig. (5) - Description of different
reinforcement structures
of composites

7b.   The obtained results may also be used for prediction of the strength laminates with other structure properties.

FRACTURE SURFACE

The phenomenological approach is limited to defining the strength characteristics of certain specific composites with preassumed structures. A more general
approach involves a micromechanical analysis of the initial fracture of the composite. The region of microstresses in the composite may be determined by solving a three-dimensional problem employing the method of finite elements [5].
This approach determines the fracture surface of the unidirectionally reinforced
composite with linear, elastic and isotropic elements. Refer to Figure 8.

The composite may be divided into finite elements where the microstresses
are determined for a preassumed macrostress state. On the basis of the strength
criteria for each separate component of the composite, the element that first
fails is found. Figure 9 shows the grid pattern of the composite. Thus the ultimate macrostress determines the onset of fracture of the composite. Fracture
of the matrix is defined by a high order polynomial, which is derived from the
representation of a scalar function on the unit sphere by means of the tensors
[6]

$$\sum_{\kappa=0}^{\ell} W^{(\kappa)}(I_1,I_3)I_2^{-\frac{\kappa+1}{2}} = 1 \qquad (6)$$

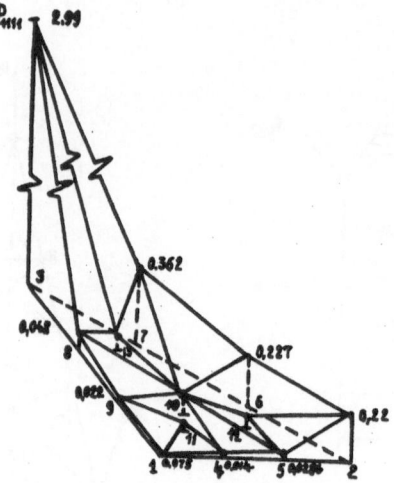

Fig. (6) - Dependence of strength tensor component
upon the intensity of reinforcement θ_i

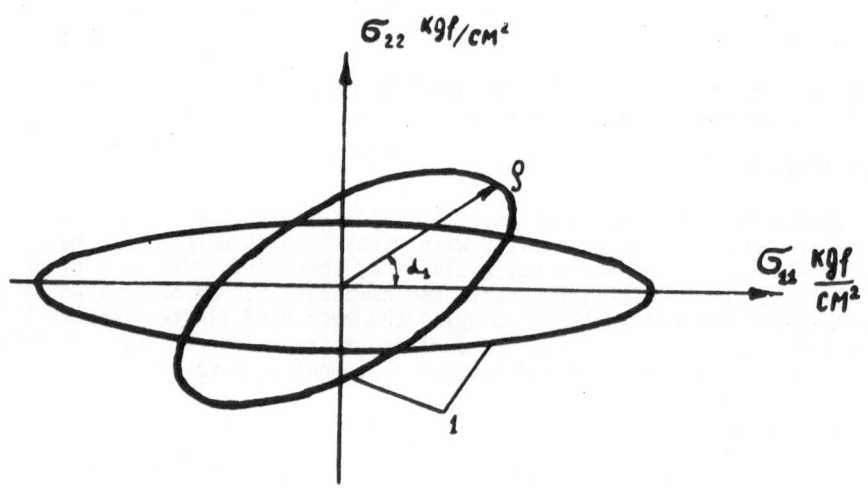

Fig. (7a) - Vector in stress tensor space

where

$W^{(K)}(I_1, I_3)$ is a polynomial of the invariants I_1, I_3 of order K. The fracture of the fibers is determined from the Mises criterion:

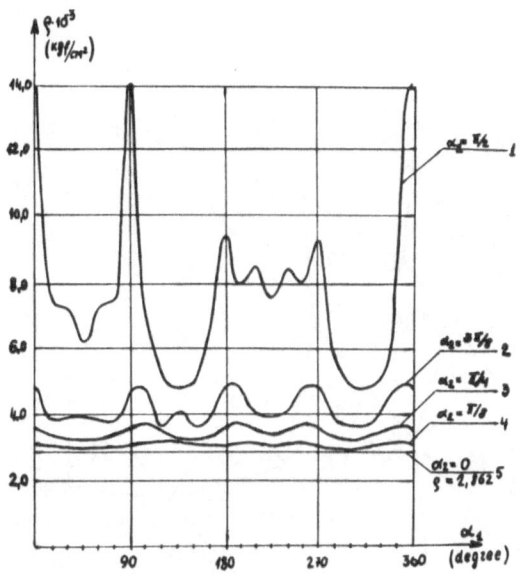

Fig. (7b) - Dependence of ρ upon α_1 and α_2

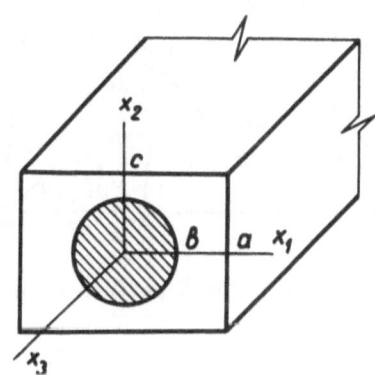

Fig. (8) - Element of a composite material

$$\frac{3}{2} I_{2D} = (\bar{r}_{100})^2 \tag{7}$$

where $\bar{r}_{100}$ is the strength of fibers during elongation and I_{2D} the second stress deviator invariant.

274

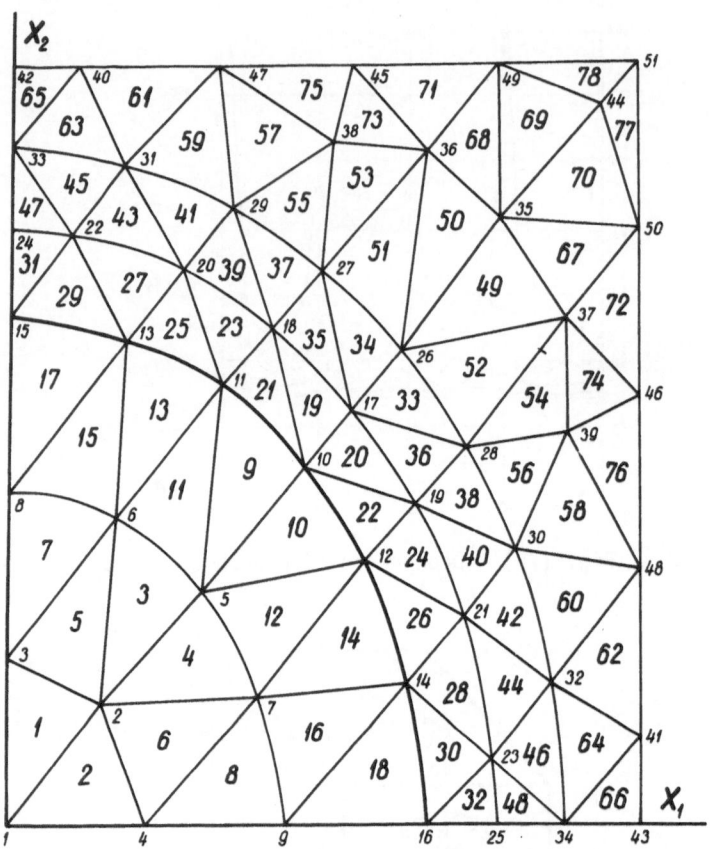

Fig. (9) - Finite element grid pattern of a composite

The theoretical calculations of the fracture surface have been compared with the experimental data where the initial fracture surface is determined by a mechanical luminescence method [7]. As shown in Figure 10, the theoretical and experimental results agree well.

MECHANICAL LUMINESCENCE STUDY

Now let us consider some experimental results of the mechanical luminescence study of GFRP in uniaxial tension [7]. "Dogbone" specimens were used in the experiments. Specimens were made with dimensions: 80 mm in length, 3 to 4 mm in the distance of the gage length and 0.35 to 1.5 mm in thickness. Light emission of the deformed specimen was measured by a photon counter. The multiplier Фзy-84

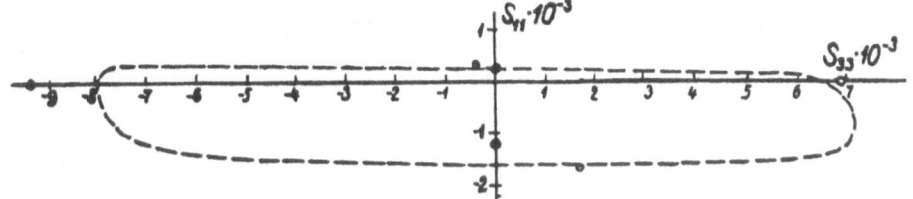

Fig. (10) - Fracture surface prediction with experimental points
determined by the mechanical luminescence method

served as a photoelectric receiver and, if required, was cooled by means of
liquid nitrogen steam. The rated running speed with linearity of 2% was equal
to 10^5 imp/sec for the noncorrelated photons. The impulse output did not exceed 10^4 imp/sec.

By comparing the strain curve and the mechanical luminescence diagrams, the
relation between luminescence and damage in GFRP can be established. Hence, the
curve $\sigma_{11}(\varepsilon_{11})$ for GFRP reinforced at angles $\phi = \pm45°$ (Figure 1) shows a distinct

nonlinear elastic intercept (Figure 11) followed by gradual failure of the material. On the boundary between the two parts of the curve, mechanical luminescenc

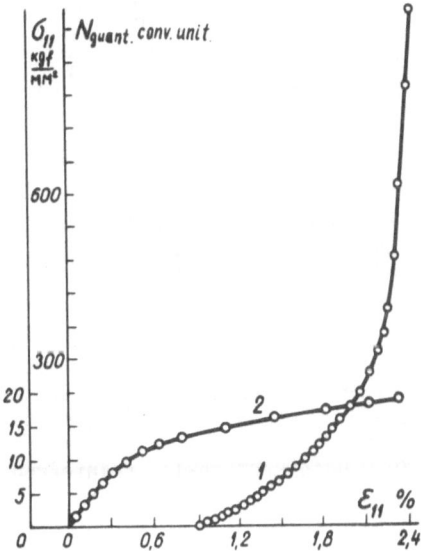

Fig. (11) - The dependence of mechanical luminescence for GFRP
with $\phi = \pm45°$. Curve 1 for stress and curve 2 for
strain

was detected and light emission reached 920 reference units. If the above experiment predicts the onset of damage and its development, say in an approximate manner, then the curve $\sigma_{11}(\varepsilon_{11})$ for the unidirectional GFRP is a straight line (Figure 12) and gives no information on failure process during loading.

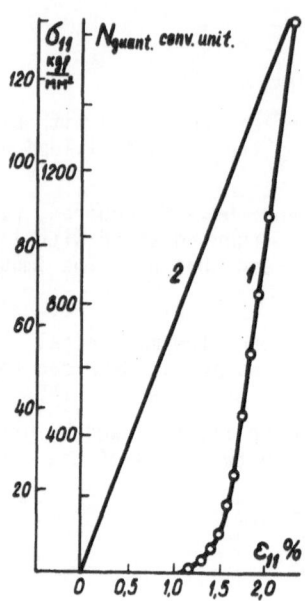

Fig. (12) - The dependence of mechanical luminescence for the unidirectional GFRP: Curve 1 for stress and curve 2 for strain

However, the diagram of mechanical luminescence shows that the first failure starts at 58.1 ± 2.4% of ultimate strain, but the accumulation of light quanta proceeds with velocity of 38 reference units per second in the region of intensive failure. The obtained data show a fairly good correlation with the theoretical calculations obtained from the finite element technique for predicting the onset of cracking in a unidirectionally reinforced material. This is shown in Figure 10.

REFERENCES

[1] Malmeister, A., "Geometry of strength theories", (in Russian). Mechanika Polimerov, No. 4, pp. 519-534, 1966.

[2] Upitis, Z. T. and Rikards, R. B., "Dependence of the composite materials strength on the fiber orientation in the plate stressed state", (in Russian). Mekhanika Polimerov, No. 6, pp. 1010-1017, 1976.

[3] Upitis, Z., Brauns, Y. and Rikards, R. B., "Determination of the components of yield surface tensors by the least square method", (in Russian). Mekhanika Polimerov, No. 3, pp. 552-554, 1974.

[4] Upitis, Z. T. and Rikards, R. B., "Strength and deformation properties of GFRP as functions of a reinforcement layup in the plane stressed state", (in Russian). Mekhanika Polimerov, No. 5, pp. 714-723, 1978.

[5] Rikards, R. B. and Chate, A. K., "Initial strength surface of unidirectionally reinforced composite in the plane stressed state", (in Russian). Mekhanika Polimerov, No. 4, pp. 633-639, 1976.

[6] Tamuzh, V. P. and Lagzdinsh, A. A., "Variant of phenomenological theory of rupture", (in Russian). Mekhanika Polimerov, No. 4, pp. 638-647, 1968.

[7] Krauja, U. E., Laizan, J. B., Upitis, Z. T. and Tutan, M. J., "Mechanical luminescence in glass-fiber reinforced plastics during tension", (in Russian). Mekhanika Polimerov, No. 2, pp. 316-320, 1977.

PECULIARITIES OF ORGANIC FIBERS REINFORCED PLASTIC FAILURE AND THEIR INFLUENCE
ON THE STRENGTH

B. V. Perov, A. M. Skudra, G. P. Mashinskaja and F. J. Bulavs

Institute of Aviation Materials, Moscow
Riga Polytechnic Institute, Riga

ABSTRACT

Experiments show that the failure of organic fiber reinforced plastics is,
almost in every case, connected with fiber breaking. This peculiarity of ORFP
failure has not received sufficient attention and should be accounted for in
the development of strength criteria.

INTRODUCTION

Plastics reinforced with organic fibers are one of the more effective new
composite materials. One of the advantages of OFRP in comparison with the other
reinforced plastics is their lower weight to volume ratio. They differ also because they fail in a peculiar manner. This peculiarity in comparison with glass,
boron and carbon fibers reinforced plastics lies in the fact that OFRP failure
in shear, tension and compression is intimately connected with the breaking of
fibers.

The purpose of the present paper is to investigate the failure mechanism and
its influence on the strength of OFRP in tension, compression and shear.

AXIAL AND TRANSVERSE TENSION OF PLASTICS REINFORCED BY ORGANIC FIBERS

Organic fibers are anisotropic in strength. It is difficult to define the
strength and elasticity properties of organic fibers in a simple way. Experiments show that the organic fibers possess very low stiffness and strength in
the transverse direction. Refer to the microphotos in Figures 1 referring to
the failure of the fibers (type SVM). Figure 1a shows the transverse cracking
of fibers in the case of axial tension. Figure 1b shows fibers undergo plastic
deformation in the transverse direction under compression. Note that the round
sections are deformed into rectangular ones. This figure shows the fibrous
structure as a result of failure of the fibers in the transverse direction into
fibrillae.

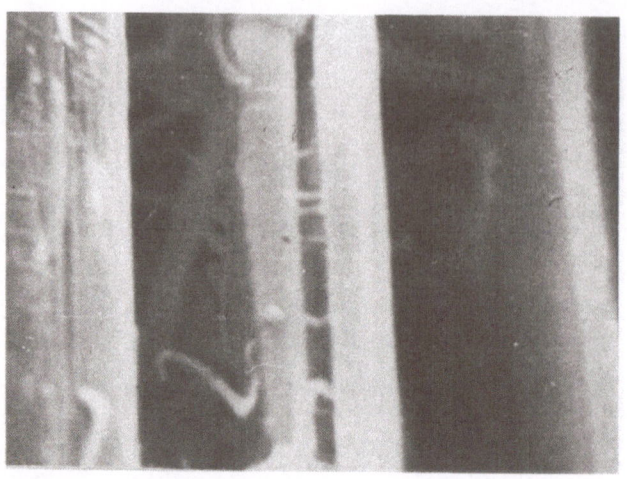

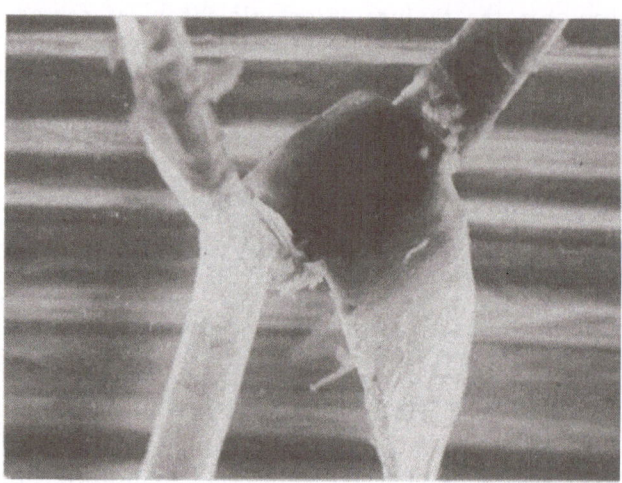

Fig. (1) - Organic fibers failure: (a) axial tension magnified
10,000 times and (b) deformation under compression
in the transverse direction magnified 2,000 times

As a structural unit, the organic fibers are known as stiff macromolecules.
The high strength and stiffness of these fibers in tension depends on

(1) high degree of orientation of macromolecules along the axis of fiber
and,

(2) high energy of dissociation of chemical bonds in the chain of the initial polymer.

The failure character of the orientated fibers in axial tension depends on the macroheterogeneity of the initial polymer. As a rule, the fibers do not transmit the same amount of load. This results in stress gradients among the neighboring fibers and shear stresses along the interfaces. Microfibrillization then prevails which leads to fiber breaking. Under uniaxial tension, cracks between fibrillae are formed in the direction of loading. Splitting of the fibers is accompanied by fiber breaking. The remaining bundle of fibers then carries the load until complete failure of the specimen occurs. During the stage of final failure, intensive splitting and swelling of the specimen are observed.

Figure 2 shows a typical form of failure for unidirectionally reinforced OFRP stretched in the direction of reinforcement. As the polymer matrix trans-

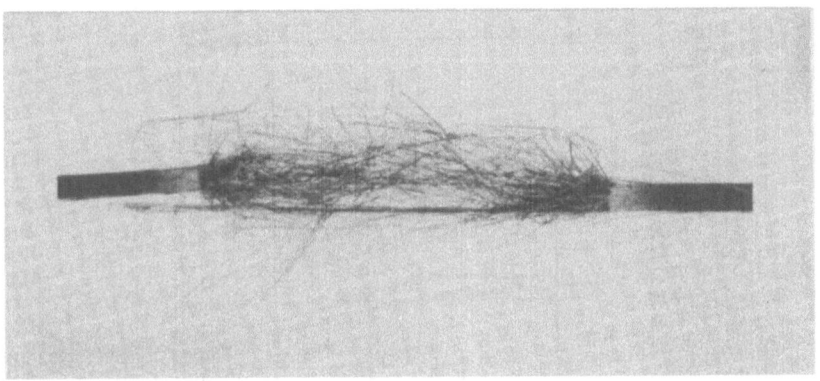

Fig. (2) - The mode of OFRP failure caused by tension in the
direction of reinforcement

mits the load from one fiber to the next, a redistribution of the stresses takes place in OFRP under tension. Stresses also change owing to the interaction of separate fibrillae bundles stressed to various degrees within one of the heterogeneous fiber. The strength of the unidirectional OFRP stretched in the direction of reinforcement can be determined from the law of summation:

$$R_{11}^* = [\psi E_{BZ} + (1-\psi)E_A]\varepsilon_{BR}^+ \tag{1}$$

where ψ is the volume content of fibers, E_A the modulus of elasticity of resin, and ε_{BR}^+ the critical tensile strain of fibers.

It is important to note that the law of summation applies conditionally in the case of OFRP. The modulus of elasticity of the fibers and the critical strain depend considerably on the applied tensile load and the rate of loading.

282

In the OFRP, the fiber strength will also be affected by the physical-chemical interaction of the fiber with the resin. It follows that equation (1) for OFRP applies only in the case of uniform tensile load and fixed rate of loading.

The critical strain ε_{BR}^{+} is defined experimentally from the condition that the fibers and the resin are strained together as a unit. There is no slipping. The strain of the fibers at the moment of failure is the same as that of the reinforced plastic. It should be noted that ε_{BR}^{*} stands for the strain that causes avalanche-like failure of the fibers.

The large surface of failure created by the organic fibers indicates high energy of dissipation of the composite material. This is unlike the failure of plastics reinforced by the common types of fibers. This peculiarity of failure of OFRP explains their relatively high strength at low cyclic fatigue. This is shown by the experimental data in Figure 3.

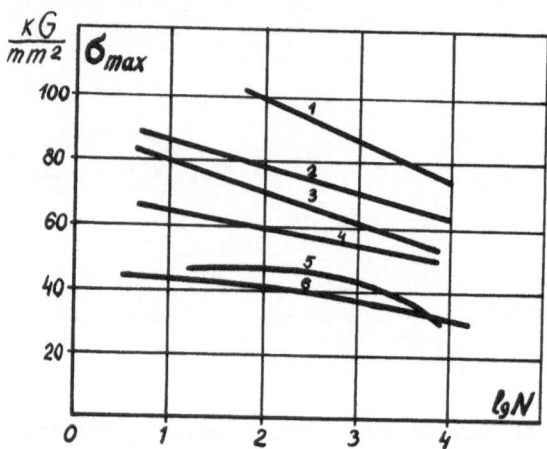

Fig. (3) - Fatigue strength of different types of materials in tension:
Curve 1 - unidirectionally reinforced organic plastics (ψ = 0.40); Curve 2 - unidirectionally reinforced carbon plastics (ψ = 0.65); Curve 3 - unidirectionally reinforced boron plastics (ψ = 0.60); Curve 4 - fabric reinforced organic plastics (ψ = 0.45); Curve 5 - aluminum alloy; and Curve 6 - plastics reinforced with glass fiber fabric. All reinforced plastics are made on the basis of epoxy resin. The characteristics of the cycle r = 0.1

The strength of unidirectionally reinforced glass, boron and carbon fibers under transverse tension is defined by the following relationship:

$$R_{\perp}^{+} = \frac{R_{A}^{+}}{\bar{\sigma}_r \sqrt{1-\nu_A^2}}$$

(2)

where $\bar{\sigma}_r$ is the stress concentration coefficient being dependent on elastic properties of the constituent, the fiber volume fraction and the fiber distribution. This formula is based on the assumption that under transverse tension, failure takes place in the polymer matrix. Figure 4 shows the form of OFRP failure under transverse tension. It is seen from the photograph that the organic fibers fail under transverse tension.

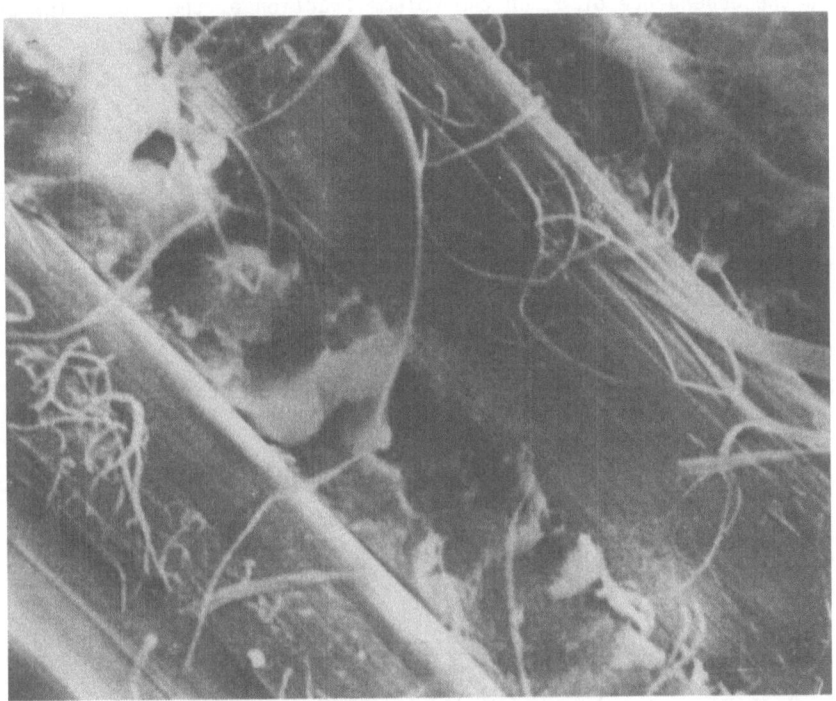

Fig. (4) - The mode of OFRP failure under transverse tension
magnified 3,000 times

Depending on the nature and technology of making the initial polymer, it is possible to observe on the interface of the newly-formed fibers containing different surface defects such as longitudinal folds, cracks, pores, etc. The making of high modulus polymer fibers, e.g., by weaving or winding introduces friction and electrization of fibers and can cause partial surface failure such as the splitting of fibers resulting in fibrillae. The above mentioned defects in the fibers increase the interface of contact with resin and play a definite role in improving the mechanical bond of fibers with the matrix. The sheared off fibrillae play the role of reinforcing element.

284

The transverse strength of the organic fibers is usually lower than that of the polymer matrix. Hence, fiber failure will first, under transverse tension, take place and in such a case, equation (2) becomes

$$R^+ = \frac{R^+_{Br}}{\bar{\sigma}_r\sqrt{1-\nu_A^2}} \tag{3}$$

in which the dependence of $\bar{\sigma}_r$ on the volume fraction of the fibers for OFRP is shown in Figure 5. The dependence of the strength R^+ on the volume fraction of the fibers is shown in Figure 6.

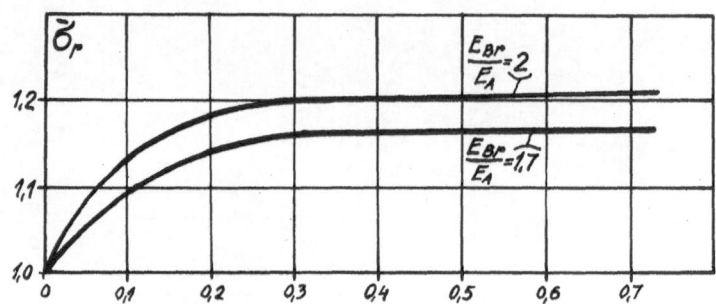

Fig. (5) - The dependence $\bar{\sigma}_r$ on volume content of organic fibers

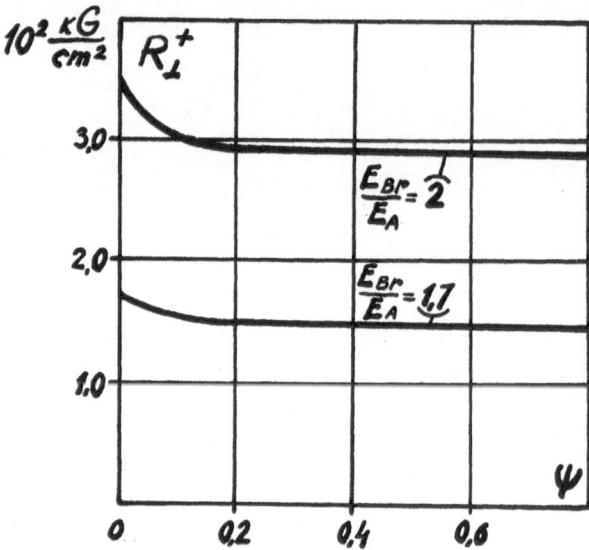

Fig. (6) - The influence of volume fraction of fibers on the strength of OFRP under transverse tension

ORTHOTROPIC REINFORCEMENT

When the plastics are reinforced in two orthogonal directions and loaded in one of the reinforced directions, the failure takes place in two steps. Failure is first observed in the layer with reinforcement normal to the tensile load. In the case of glass, boron or carbon reinforced plastics, this layer corresponds to the polymer matrix or the interface. In the OFRP, however, fiber failure is observed due to the effect of transverse tension. The material then loses its continuity. The critical strain that accounts for the unusual failure of OFRP is given by

$$\varepsilon_R = \frac{1-m}{E_1 - mE_{11}} \cdot \frac{R_{Br}^+}{\bar{\sigma}_r \sqrt{1-\nu_A^2}} \tag{4}$$

where E_1 is the modulus of elasticity of the orthogonally reinforced OFRP in the direction of tension and E_{11} is the modulus of elasticity of the unidirectionally reinforced layer.

COMPOSITE UNDER LONGITUDINAL SHEAR

The failure of glass, boron and carbon fibers reinforced plastics under longitudinal shear is caused by breaking of the polymer matrix and/or interface. It is found that when the bond strength is greater than that of the polymer matrix (e.g., glass fiber reinforced plastics), the failure of the polymer matrix is associated with plastic yielding. The strength of the unidirectionally reinforced plastics subjected to longitudinal shear is almost the same as the strength of the polymer matrix in shear; that is

$$T_{11} = T_A \tag{5}$$

In the case of plastics reinforced by organic fibers, the bond strength is higher than that of the polymer matrix. What is unusual is that the organic fibers break first in longitudinal shear. Typical form of this phenomenon is shown in Figure 7. The strength of the unidirectionally reinforced OFRP undergoing longitudinal shear can be estimated from

$$T_{11} = \frac{T_{Brz}}{\bar{\tau}_{rz}} \tag{6}$$

where $\bar{\tau}_{rz}$ is the coefficient of concentration of the longitudinal shear stress. Dependence of $\bar{\tau}_{rz}$ on the fiber volume fraction is shown in Figure 8. Unlike the glass fiber reinforced plastics, the longitudinal shear strength of OFRP decreases as the fiber volume fraction is increased.

COMPRESSION OF FIBER REINFORCED PLASTICS

One of the basic shortcomings of OFRP is the low compressive strength in the direction of the fibers. A common feature of all reinforced plastics is that a

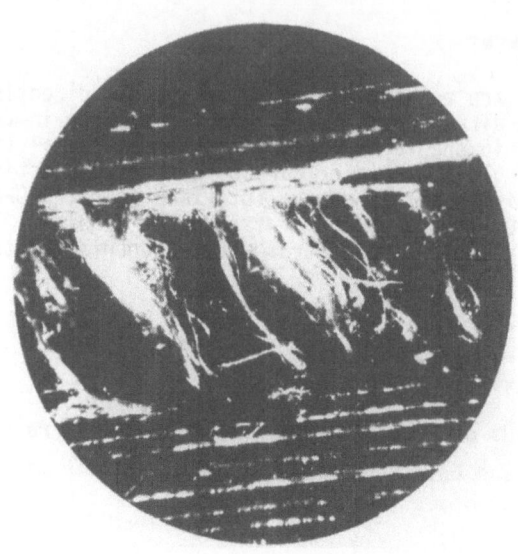

Fig. (7) - Typical failure mode of plastics reinforced by fibers under longitudinal shear

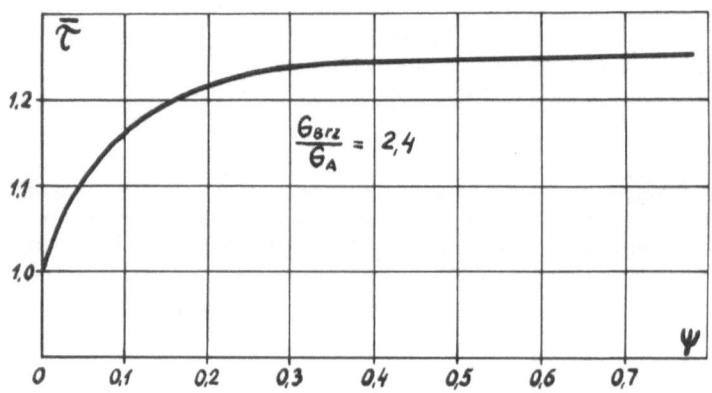

Fig. (8) - The dependence of $\bar{\tau}_{rz}$ on the fiber volume fraction

shear type of failure is observed under compression. The typical form of OFRP failure is shown in Figure 9. The mechanism causing the fibers to loose their load carrying capacity under compression is complex and has not received sufficient attention in the past. Figure 9a shows the compressive failure mode of OFRP. It is seen from the photograph that the failure is of the shear type. The compressive failure of glass, carbon, boron fibers reinforced plastics differs from that of OFRP in that the fibers in the plane of failure do not shear

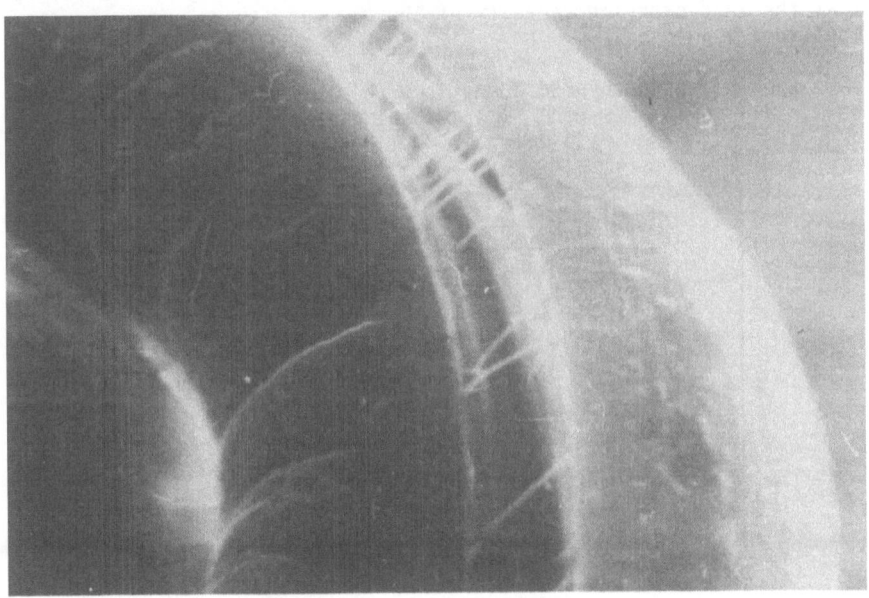

Fig. (9) - A typical failure mode of plastics reinforced by organic
fibers under compression: (a) macroscale magnified 120
times and (b) fibers bending at the microscale level
magnified 10,000 times

off but only bend. Refer to Figure 9b. As a result of bending in the compressed
region, the fibers form ring-shaped cross folds. Fiber splitting is also observed. This is accompanied by formation of cross links between the cracks.
The bending of the fibers is assumed to occur when the axial strain reaches a
critical value. Using the law of summation, the strength is

$$R_{11}^{-} = [\psi E_{Br} + (1-\psi)E_A]\varepsilon_{BR} \tag{7}$$

where ε_{BR}^{-} is the fiber strain corresponding to the loss of load carrying capacity
of the fibers. As an approximation, this strain is assumed to be equal to the
strain of the whole reinforced plastic at the moment of failure.

It is important to remark that the conditional usage of the summation law
for OFRP under compression is more suitable.

It should be remarked in passing that the summation law for the strength
prediction of OFRP is even more justified in the case of compressive loading as
compared with the case of tensile loading. The strain ε_{BR}^{-} does not correspond
to the breaking of fibers under compression but to the bending of fibers. It
follows that the load carrying capacity of the organic fibers can be increased
by delaying fiber bending.

The same applies to the hybrid composites reinforced with both glass and organic fibers. Figure 10 shows the experimental data. In these experiments, both

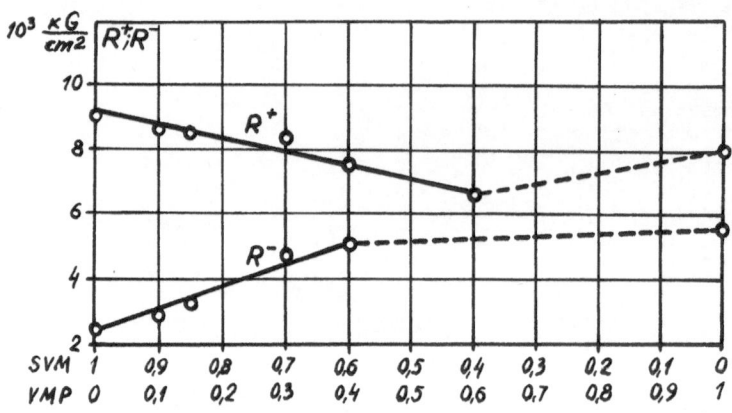

Fig. (10) - The dependence of the strength of plastic reinforced
with hybrid fabric on the volume fraction of glass
fibers (VMP) and organic fibers (SVM) embedded in the
DEN-6 resin.

organic and glass fibers were used to reinforce the resin. The results show that the strength of the hybrid material decreases to a minimum as the volume fraction of the glass fiber is increased. This is because the critical strain for the organic and glass fibers in tension are not the same.

Increase of the volume fraction of the organic fibers, however, tends to increase the strength of the hybrid composite. This implies that the critical strain ε_{BR}^- corresponding to the bending of organic fibers is increased. In this respect, it would be desirable to reinforce the composite entirely by the organic fibers.

CONCLUSION

In conclusion, we may remark that for the effective use of OFRP it is necessary to recognize its unusual feature of failure. It would be desirable to develop a theory for predicting the strength of composites reinforced by organic fibers.

ANALYSIS OF FREE EDGE INDUCED FAILURE OF COMPOSITE LAMINATES

Frank W. Crossman

Lockheed Palo Alto Research Laboratory, Palo Alto, California, USA

ABSTRACT

Eight different stacking sequences of T300 Graphite Fiber/934 Epoxy $(0/\pm45/90)_{2c}$ quasi-isotropic laminates were constructed from prepreg tape and fabric (HMF 330C) and tensile tested. Tensile strength was found to be stacking sequence-dependent and ranged from 60 to 90 ksi. The average through thickness Poisson ratio ν_{xz} ranged from .85 to -.30. Delamination was a prevalent failure mode in those stacking sequences where a near-zero or negative ν_{xz} was measured. Metallography was employed to determine the fracture sequence as a function of tensile strain and laminate stacking. The propensity for delamination in a particular stacking sequence correlated with the magnitude of the free edge normal stress σ_z in laminated composites as examined by finite element analysis, which determines the independent contributions of uniaxial mechanical loading, temperature change and moisture absorption to the free edge stress state. These contributions when superposed demonstrate the importance of prior hygrothermal history on the fracture behavior.

INTRODUCTION

The stiffness and strength of multidirectionally reinforced fiber composite laminates is dependent not only on the relative fraction of fibers oriented in each direction but also on the particular sequence with which the unidirectional fiber reinforced layers are stacked in the laminates. Before the engineer can make use of the macroscopic mechanical response of a given laminate construction in his analysis of full scale structures, it is necessary to perform, first, what is essentially a microstructural analysis. Approaches based on classical laminated plate theory (CLPT), which assumes a state of plane stress within each composite lamina, have successfully predicted the stiffness characteristics of arbitrary laminates; but predictions of fracture strength based on CLPT have been relatively unsuccessful due in large part to the multiplicity of available fracture modes and the existence of significant out-of-plane stress components [1] near the traction-free surfaces of the laminate. These stresses are not accounted for in the plane stress laminated plate theory.

Scanning electron microscopy of fracture surfaces, TBE enhanced x-radiography, surface photomicrography, subsurface photomicrography, ultrasonic scanning, acoustic imaging, video thermography, and surface replication fractography have been employed to determine the fracture modes and sequence of damage accumulation in composite laminates [4]. With the exception of the work by Reifsnider and colleagues [2], most of these studies have provided a phenomenological description of the fracture process in particular laminates, but have not developed a concurrent analysis of the mechanics of damage accumulation at the laminate level which is necessary for the ultimate prediction of fracture in laminates of different construction and under more complex loading conditions.

Recently, a generalized plane strain finite element analysis procedure for calculation of free-edge stresses due to mechanical and thermal loads has been coupled to a two-dimensional finite element analysis of moisture diffusion into the laminate to examine the effect of hygroscopic swelling on free-edge stresses [3]. In the research described in this paper, free-edge effects in quasi-isotropic T300/934 laminates were examined experimentally [4] and analytically [5]. The dependence of laminate static tensile strength and failure modes on the ply stacking sequence is documented. Finite element analysis is then employed to show that the stacking sequence effects on the fracture modes, strength, and through thickness deformation are related to the out-of-plane free-edge stress components in the tensile coupon.

EXPERIMENTAL STUDIES OF STACKING SEQUENCE EFFECTS

A series of tensile tests were performed at room temperature to determine the effect of stacking sequence on the tensile strength of quasi-isotropic laminates fabricated from either unidirectional Fiberite T300/934 prepreg tape or HMF330C/34, a crossplied eight-harness satin woven graphite prepreg [4]. Figure 1 shows the tensile strength of a number of tape and fabric reinforced quasi-isotropic laminates. A single layer of crossplied HMF fabric is designated by (0/90) or ($\pm$45) in the figure. With the exception of panel 4 the average tensile strength and standard deviation are based on three specimens. The highest tensile strength of 87 KSI (600 MPa) was obtained in panel 15 with a $[90/45/-45/0]_s$ stacking sequence while panel 16 containing a $[+45/-45/0/90]_s$ sequence averaged only 63 KSI (434 MPa), a value of 28 percent less than the strongest configuration. The fabric reinforced laminates exhibited generally lower strengths than the tape-reinforced laminates and relatively little dependence on stacking sequence.

On several specimens, a 0.10 cm long strain gage was mounted on the edge of the tensile specimen to measure the through thickness normal strain ϵ_z during tensile loading. Figure 2 shows the variation of through thickness strain ϵ_z versus tensile strain ϵ_x. Panels 14, 15, 17, 2 and 4 showed a greater Poisson contraction during tensile loading than was found for a unidirectional composite. Panel 13 showed nearly zero Poisson contraction prior to delamination, while panels 1, 3, 16 and 18 expanded in the through thickness direction during tensile loading. Note that the response of the HMF panel 2 lies between that of the tape configurations - panel 14 and panel 17 -

MATERIAL	THICKNESS (IN.)	PANEL NO.	LAYUP SEQUENCE	AVERAGE TENSILE STRENGTH (ksi) ± STANDARD DEVIATION
T300/934	0.042	13	$[0/+45/-45/90]_s$	
T300/934	0.042	15	$[90/45/-45/0]_s$	
T300/934	0.042	14	$[0/90/45/-45]_s$	
HMF330C/34	0.054	2	$[(0/90)/(\pm45)]_N$	
HMF330C/34	0.108	4	$[(0/90)/(90/0)/(\pm45)/(\mp45)]_s$	
T300/934	0.042	18	$[45/-45/90/0]_s$	
T300/934	0.042	16	$[45/-45/0/90]_s$	
HMF330C/34	0.054	1	$[(\pm45)/(0/90)]_s$	
HMF330C/34	0.108	3	$[(\pm45)/(\mp45)/(0/90)/(90/0)]_s$	

Figure 1. Tensile strength of quasi-isotropic T300 graphite fiber -
934 epoxy matrix laminates as a function of ply stacking
sequence. Laminates designated HMF were fabricated from
woven graphite fabric.

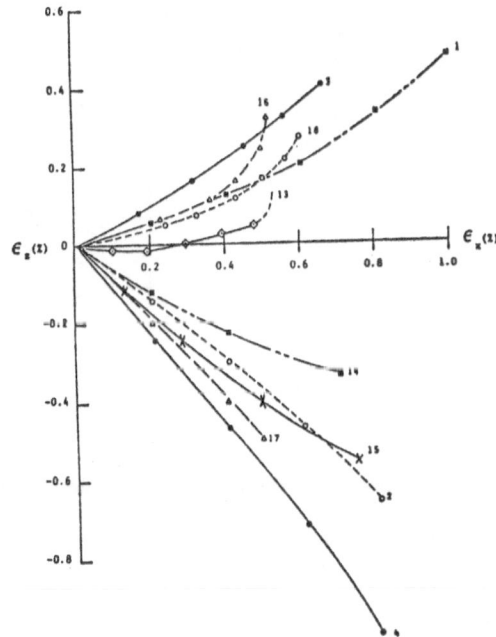

Figure 2. The through thickness
normal strain ϵ_z vs.
applied tensile
strain ϵ_x as a function of T300/934
panel type described
in Figure 1.

which most closely approximate the fabric stacking sequence. A similar relationship is found between the ϵ_z vs ϵ_x response of HMF panel 1 and the tape panels 16 and 18.

EXAMINATION OF TENSILE FAILURE MODES

In order to understand the causes of so large a variation in laminate strength and through thickness Poisson strain behavior, three tape laid laminates (13, 15 and 16) were examined in greater depth. Four specimens of each panel were tensile loaded to a stress below ultimate strength, unloaded and then sectioned for microscopic examination. The three plane sections examined were: (1) a longitudinal x-z plane section at the specimen free edge, (2) a longitudinal x-z plane section near the center line of the specimen, and (3) a transverse section in the y-z plane. Figure 3 tabulates the location of damage in each laminate as a function of strain. Prior to ultimate failure, two distinct fracture modes were observed: (1) transverse cracking of 90° and 45° oriented plies, with cracks running in the matrix and along fiber-matrix interface, and (2) delamination cracks between plies. Transverse cracking of 90° plies was observed to occur in all laminates examined, but delamination modes were strongly dependent on position and stacking sequence. Some delamination modes such as those at the (+45) interface in panel 13 and at the (90/45) interface in panel 15 occurred early in the strain history but were clearly not catastrophic in terms of laminate ultimate strength.

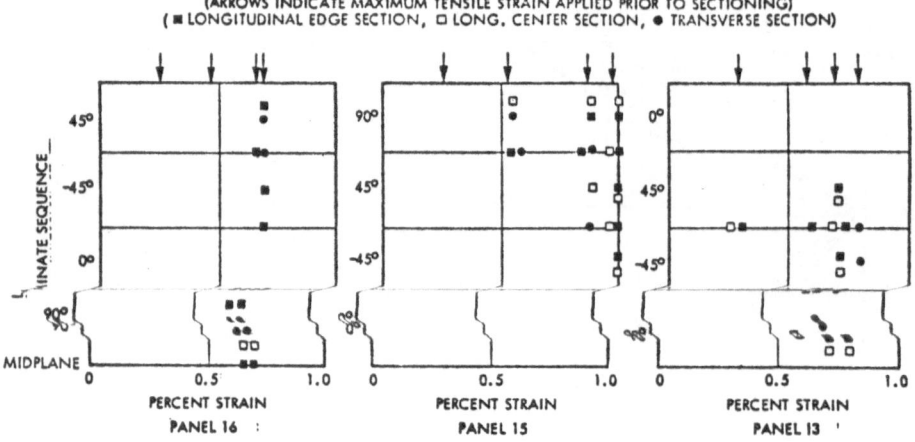

Figure 3. Diagram showing the location of transverse cracking (symbols within each layer) and delamination (symbols at ply interfaces) as a function of laminate stacking sequence and tensile strain.

Figure 4a shows the combination of transverse 90° layer cracks and delamination found at high tensile strain levels in specimen 16-10 loaded to 90% of ultimate tensile strength. The stain patterns give the location of the transverse cracks in the 90° plies. There is some evidence that transverse cracks were formed both before and after the 90° delamination. Near the free edge of specimen 16-10 (±45) and (0/45) interface delaminations are also observed immediately prior to failure. Figure 4b shows the pattern of cracking just prior to failure in specimen 15-3. Transverse cracking and delamination between the 90° and 45° layers occur at strains near 0.5 percent. Further straining leads to 45° transverse cracking and (±45) delamination. No evidence for (0/45) interface delamination was found, leaving the primary load carrying 0° layers relatively unaffected by the damage existing in the outer layers. The lack of damage near 0° layers may account for the particularly high tensile strength of this laminate.

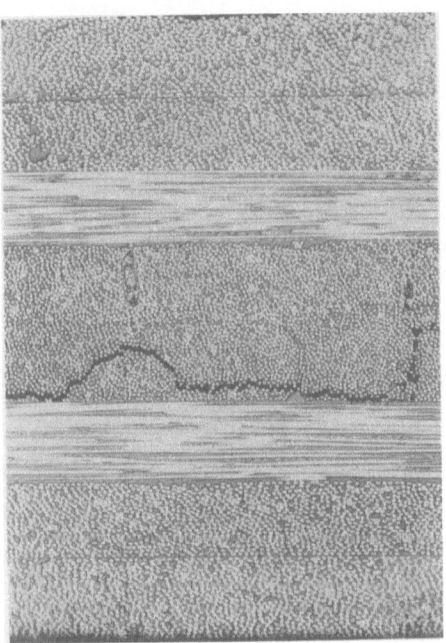

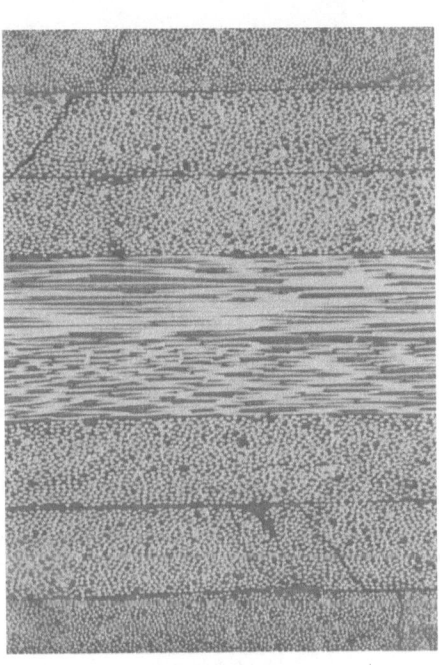

Figure 4a. Longitudinal, x (horizontal)-z (vertical) plane section near the center line of (±45/0/90)$_s$ specimen 16-10.

Figure 4b. Longitudinal x-z plane section at the free edge of specimen 15-3.

Tensile specimens from HMF laminates 1-4 were microscopically examined after loading to ninety percent of ultimate tensile strength. While transverse cracks were found in the 90° oriented fiber bundles, few delaminations were noted. This was surpising since the delamination prone panel 16 showed through thickness normal expansion similar to panels 1 and 3. Figure 5 gives some insight into the lack of extensive delamination in the fabric reinforced laminates. This transverse section of panel 1 shows a short delamination crack extending inward from the free edge at a (0/90) interface. However, because of the weave pattern, the delamination stops near a point where the 0° oriented bundle ends.

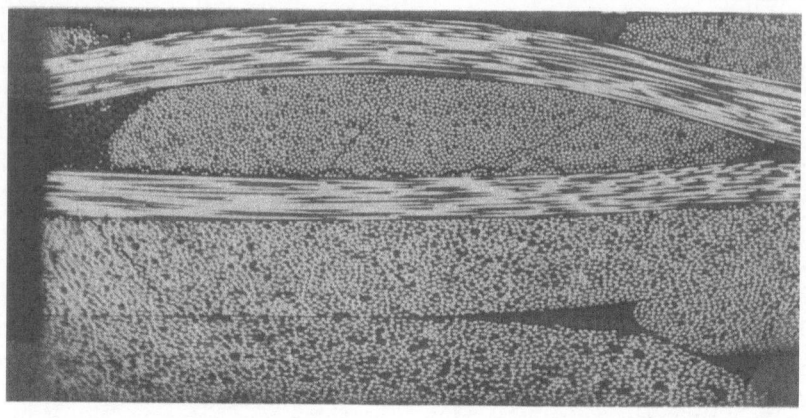

Figure 5. Transverse, y (horizontal)-z (vertical), plane section near the edge of $[(\pm 45)/(0/90)]_s$ HMF330C/34 specimen 1-5. Tracer fiber bundles of Kevlar 49 with larger fiber diameter are visible.

CORRELATION OF FAILURE MODES TO FINITE ELEMENT STRESS ANALYSIS

To determine the three-dimensional stress state in the eight-ply symmetric laminates tested in tension, a generalized plane strain elastic finite element analysis was made of the laminate tensile coupon. Symmetry conditions allowed the analysis to be carried out for one quadrant of the y-z cross-sectional plane of the laminate as shown in Figure 6. A half-width of 16h and a half-thickness of 4h (where h is the ply thickness) was chosen for the model. Elements were concentrated near the free edge at y=b where edge stress gradients were expected. The lamina elastic constants for the unidirectional lamina were based on data for T300/934 [4]. The analysis determines both the stress/strain distribution in the laminates as well as the displacement of each node in the structural model. Table 1 shows the excellent correlation between the measured and calculated out-of-plane Poisson ratio ν_{xz} at the free edge of the quasi-isotropic laminate as a function of stacking sequence.

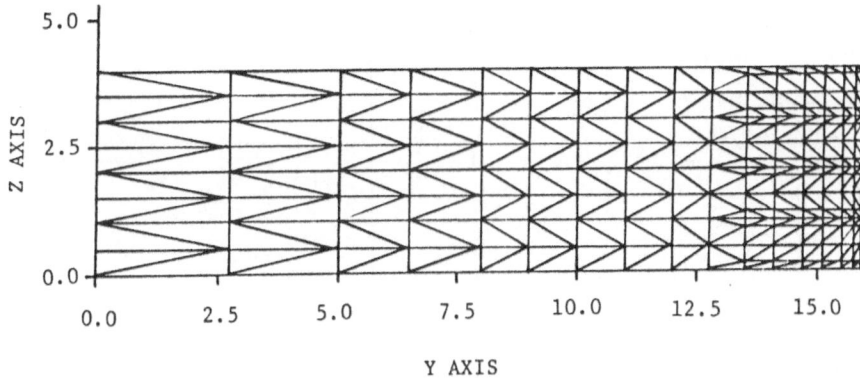

Figure 6. Finite Element Grid

Table 1

STACKING SEQUENCE DEPENDENCE OF THROUGH THICKNESS CONTRACTION
ν_{xz} IN T300/934 QUASI-ISOTROPIC LAMINATES

Panel No.	Stacking Sequence	ν_{xz} Experimental	ν_{xz} Calculated	Ultimate Strength Strength (ksi)
15	$[90/45/-45/0]_s$	0.80	0.72	87
13	$[0/45/-45/90]_s$	-0.05	-0.02	72
16	$[45/-45/0/90]_s$	-0.30	-0.34	63

Figure 7 shows the variation with position y of the out-of-plane normal stress σ_z at each ply interface in the quasi-isotropic laminate (panel 16) under a tensile strain $e_x = 10^{-6}$. Figure 8 shows the variation of selected stress components at several ply interfaces under the same tensile load. The out-of-plane stress τ_{yz} shows a peak value close to the free edge but also drops to zero at the edge. Therefore, the major stress components which are unaccounted for in plane stress laminated plate analysis are σ_z and τ_{xz}, which are observed to reach their maximum value at the free edge. Steep gradients of these stress components are also found in the z direction through the thickness of the laminate. In Figure 9 stress components σ_z and τ_{xz} at y = .999b are plotted as a function of position z through the laminate thickness. Only half of the laminate thickness is shown because of symmetry.

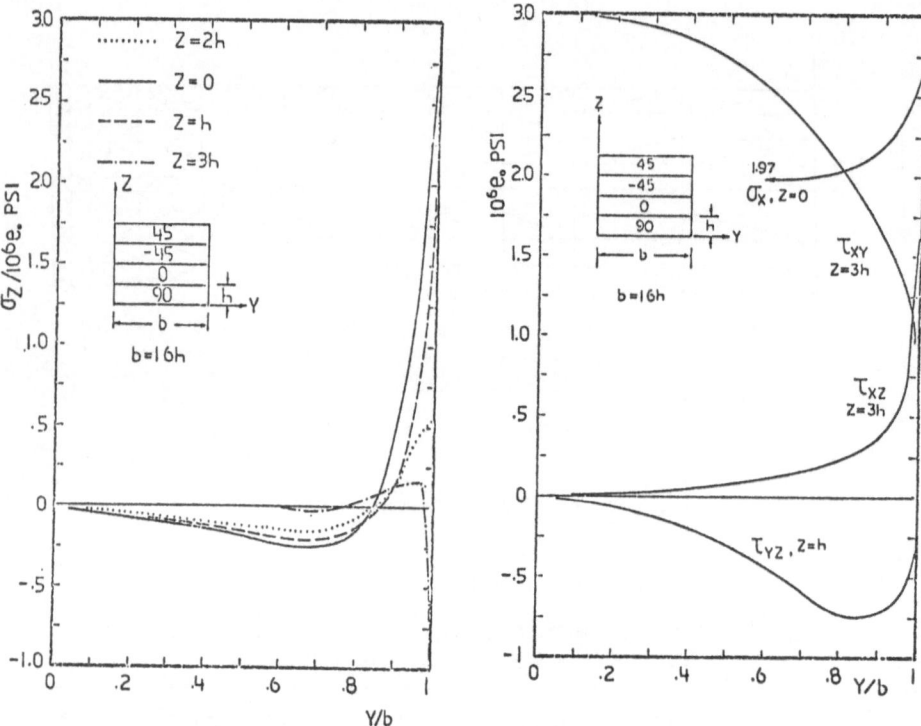

Figure 7. σ_z Distributions along each ply interface for $(\pm 45/0/90)_s$ laminate

Figure 8. Various stress distributions along several ply interfaces in $(\pm 45/0/90)_s$ laminate

The stresses σ_z^T and τ_{xz}^T, induced by a unit temperature increase (°F), and stresses σ_z^m and τ_{xz}^m, induced when a tensile strain $\epsilon_x = 10^{-6}$ is applied, are shown in the figure.

We use the principle of linear superposition to examine combinations of mechanical loading and internal thermal residual stresses which exist due to the cooling of the laminate from its stress-free temperature (the cure temperature of 350°F will be assumed) to the temperature at which it is tensile tested. For example, from Figure 9 the total normal edge stress σ_z at the midplane $z = 0$ in laminate 16 due to residual stresses and loading to 0.5 percent strain is calculated to be:

$$\sigma_z = \sigma_z^{\,m} + \sigma_z^{\,T}$$

$$= [2.6 \text{ psi}/\mu\epsilon] \times [1500 \ \mu\epsilon] + [-15 \text{ psi}/^\circ\text{F}] \times [(75-350)^\circ\text{F}]$$

$$= 17.1 \text{ KSI (118 MPa)}$$

THERMAL EDGE STRESS (PSI/$^\circ$F)

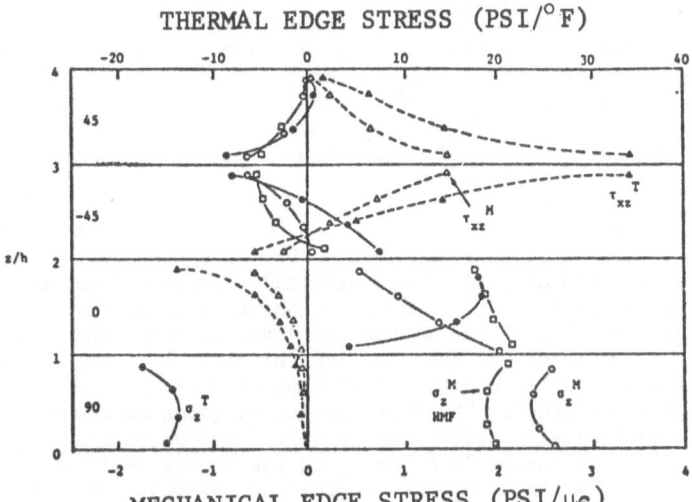

MECHANICAL EDGE STRESS (PSI/$\mu\epsilon$)

Figure 9. Calculated edge normal and shear stresses at y = 0.999b as a function of position z through the laminate thickness in $[45/-45/0/90]_s$ T300/934 Panel 16. Stresses $\sigma_z^{\,M}$, $\tau_{xz}^{\,M}$ due to tensile strain ($\mu\epsilon_x$) and $\sigma_z^{\,T}$, $\tau_{xz}^{\,T}$ due to temperature change ($^\circ$F) are given.

Although the largest free edge stress is the $\tau_{xz}^{\,T}$ component at the ($\pm$45) interface, the general nonlinear and viscoelastic shear stress-strain response is expected to reduce the value calculated here by a purely elastic analysis. Furthermore, the shear contribution due to tensile loading actually operates to reduce the total shear stress τ_{xz}. At 0.5 percent tensile strain the total shear stress near the $\pm$45 interface is:

$$\tau_{xz} = \tau_{xz}^{\,m} + \tau_{xz}^{\,T}$$

$$= [1.5 \text{ psi}/\mu\epsilon] \times [1500 \ \mu\epsilon] + [35 \text{ psi}/^\circ\text{F}] \times [(75-350)^\circ\text{F}]$$

$$= 2125 \text{ psi (14.6 MPa)}$$

While it appears that under tensile loading the out-of-plane normal stress σ_z is primarily responsible for the initiation of free edge delamination, one should note that the shear stresses $\tau_{xz}^{\,T}$ and $\tau_{xz}^{\,m}$ are <u>additive</u> if the uniaxial strain is applied in compression. Therefore, the free edge shear stress τ_{xz}

will play a more significant role in initiation of delamination for static and fatigue testing involving compressive loading.

In Figure 9 the free edge stress σ_z^m for the HMF panel is given. As described in an earlier paper [5], the properties of an HMF lamina (Table 1) are essentially quasi-isotropic; and at the ply level of stress analysis no shear stress exists between a 45° and 0° oriented ply. Furthermore, because the thermal expansion coefficients of a (+45) ply and a (0/90) fabric ply are identical, no thermal stresses are developed because of ply/ply interaction. Calculation of the midplane σ_z stress in panel 1 at 0.5 percent tensile strain gives a value of 10.0 KSI (69 MPa), nearly forty percent lower than that calculated for panel 16, fabricated from tape, which was found to be more susceptible to delamination.

By assuming a hygroscopic expansion $B_y = B_z = 5 \times 10^{-3}/\%M$ in the transverse and through thickness directions due to the absorption of moisture [3], one may examine the combined effects of hygrothermal history and mechanical loading. Figures 10 and 11 from Reference 6 are shown to demonstrate the combined effects of hygrothermal and mechanical loads on free edge stress. Figure 10 depicts the σ_z stress near the edge after a short time during which .04 percent water content is desorbed following absorption to an equilibrium 1.0 percent moisture content. The distribution is given at two temperature extremes. If a mechanical tensile load is superposed as in Figure 11, the free edge stress σ_z, increases by a factor of two in the $(+45/0/90)_s$ laminate, while decreasing nearly to zero in the laminate of opposite stacking sequence.

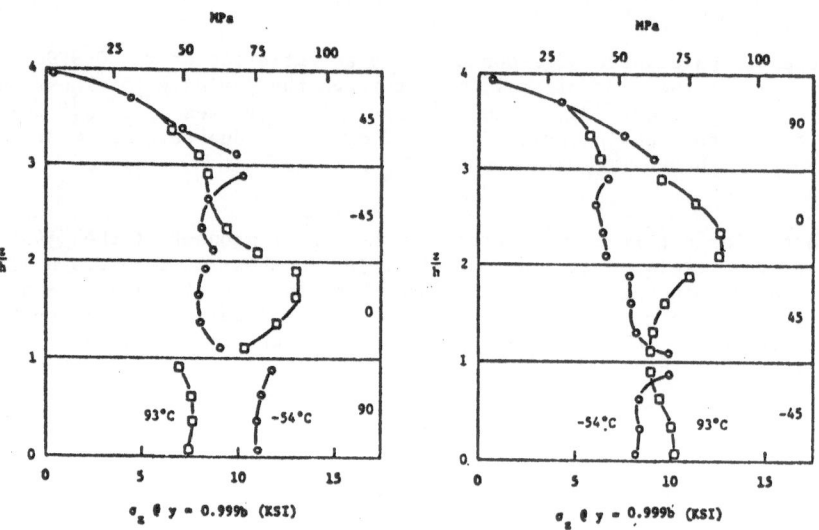

Figure 10. σ_z through thickness distribution in $(+45/0/90)_s$ and $(90/0/+45)_s$ laminates after a desorption time t_{-4} following saturation to M_s and under an applied tensile strain of 5000 $\mu\epsilon$.

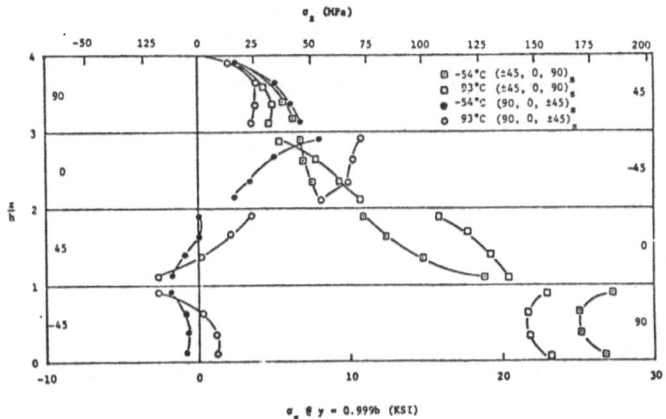

Figure 11. σ_z through thickness distributions in $(\pm45/0/90)_s$ laminates after a desorption time t_{-4} following saturation to M_g and under an applied tensile strain fo 5000 $\mu\varepsilon$.

CONCLUSIONS

Tensile tests of quasi-isotropic tape and fabric-reinforced T300/934 laminates and finite element analysis of these laminates have shown the necessity of determining both the in-plane and out-of-plane stress components near laminate-free edges in order to predict the prevalent fracture modes of these materials. While the elastic finite element analysis has proven useful in predicting the location of the first delamination, it does not predict the subsequent (0/45) and ($\pm$45) delaminations noted in Fig. 3 after application of a tensile stress approaching the ultimate strength of the laminate. Better correlations between stress calculations and all observed failure modes may require a modeling of the delamination process.

By means of linear superposition, stress fields under combined mechanical and hygrothermal loads can be determined within the framework of the theory of elasticity. On the other hand, nonlinear behavior has been observed in resin-matrix composites. It is not clear, however, whether to characterize the composite strictly on the basis of intrinsic nonlinearity or on the basis of material viscoelasticity. Sufficient information on nonlinear and/or viscoelastic composite behavior and the appropriate constitutive relations suitable for three-dimensional stress analysis appears lacking to date. Until these issues are resolved, the elastic description of composite response as presented in this study must be considered a pragmatic means of developing an appreciation of the complex three-dimensional interactions which take place in multidirectionally reinforced laminates.

ACKNOWLEDGMENTS

This study was performed in collaboration with Prof. A. S. D. Wang of Drexel University and Messrs. W. J. Warren and J. B. Bjeletich of the Lockheed Palo Alto Research Laboratory. The work was funded under the Lockheed Independent Research and Development Program on the Behavior of Structural Composites.

REFERENCES

[1] Pipes, R. B. and Pagano, N. J., "Interlaminar Stresses in Composite Laminates Under Uniform Axial Extension", J. Composite Materials, Vol. 4 (1970) p. 538

[2] Reifsnider, K. L., Henneke, E. G., and Stinchcomb, W. W., "Defect Property Relationships in Composite Materials", AFML-TR-76-81 Part II, June 1977

[3] Crossman, F. W. and Wang, A. S. D., "Stress Field Induced by Transient Moisture Sorption in Finite Width Composite Laminates", J. Composite Materials, Vol. 12 (1978) p. 2

[4] Bjeletich, J. G., Crossman, F. W., and Warren, W. J., "The Influence of Stacking Sequence on Failure Modes in Quasi-Isotropic Graphite-Epoxy Laminates", Failure Modes in Composites - IV, Proceedings of Chicago Symposium, October 25-27, 1977. To be published by AIME, 1978

[5] Crossman, F. W. and Wang, A. S. D., "Analysis of Free Edge Stress Induced Fracture of Fiber Composite Laminates", Proceedings of the Symposium on Applications of Computer Methods in Engineering, Univ. S. California, August 23-26, 1977, p. 23

[6] Crossman, F. W., Rothwell, W. S., and Wang, A. S. D., "Alteration of Laminate Free Edge Stresses by Moisture Absorption", Proceedings of the Second International Conference of Composite Materials, Toronto, Canada, April 1978

FRACTURE OF COMPACT BONE TISSUE

I. V. Knets

Institute of Polymer Mechanics
Academy of Sciences of the Latvian SSR

During the last decade, fracture mechanics as applied to different structural materials has become one of the most rapidly developing branches of solid mechanics. In particular, it has played a significant role in practical application and in describing the fracture processes of various man-made composite materials. However, relatively little attention has been given to many of the natural biopolymers such as bone tissues, muscles, tendons, blood vessels, etc.

In this paper, we shall analyze in detail the fracture of compact bone tissue, which is a basic constituent of human and animal skeletons. This tissue forms diaphysis of long tubular bones which carry large mechanical loads and often must sustain the impact of different external traumatic effects. High values of specific strength (ratio of ultimate stress to the material density) of the compact bone tissue have attracted the attention of specialists in biomaterial for many years. This interest lies within the fact that the process of evolution has created many efficient biosystems which are optimal not only with respect to their functional requirements, but also to their mechanical loading bearing behavior.

On the basis of many experimental investigations, it has been found that one of the factors, responsible for high load-carrying capacity of the compact bone tissue is its particular composite structure. It has been assumed that this composite has two main components. The collagen fibers act as the matrix, and the mineral crystals serve as the reinforcement [1,2].

However, complex experiments carried out by destructive and nondestructive testing methods at the Institute of Polymer Mechanics of the Latvian SSR Academy of Sciences have revealed that the compact bone tissue has to be considered as a three-component material with five structural levels [3-6].

The first and lowest structural level is formed by tropocollagen biopolymer macromolecules and inorganic crystals. Each tropocollagen molecule consists of three left-handed helical polypeptide chains (Figure 1) that form a right-handed super helix stabilized by hydrogen bonds.

The second structural level is based on collagen microfibrils and inorganic crystals. Each microfibril is built up by five helical tropocollagen macro-

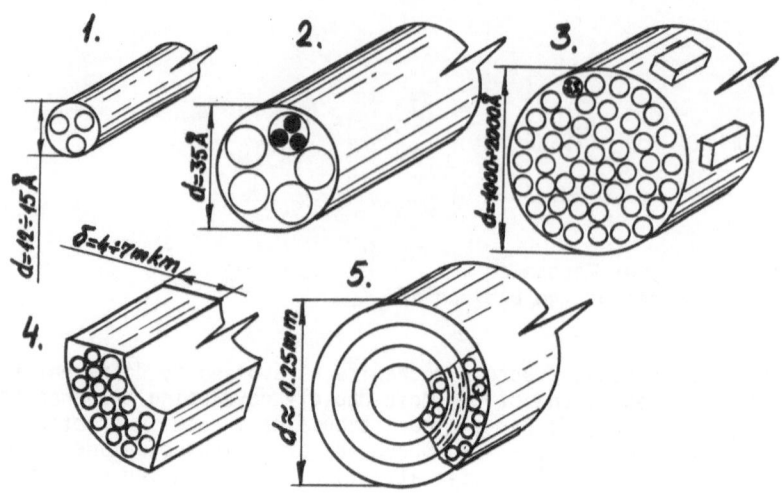

Fig. (1) - Structural levels (1-5) of compact bone tissue

molecules which are laid down in the staggered fashion with a period of 65 nm. The diameter of the microfibril is approximately 3.5 nm.

The third structural level embraces fibers formed by many collagen micro-fibrils and combined with mineral crystals by stereochemical bonds. These crystals (5 x 5 x 50 nm) are located both outside and inside the microfibrils and generally are oriented along the longitudinal axis of the fibers. The main zone of crystal location is between the ends of tropocollagen molecules. The bonds may be formed between separate crystals of neighbouring tropocollagen molecules or microfibrils both in transversal and longitudinal direction. This is confirmed by the fact that after the removal of organic phase the bone tis-sue retains its external shape although the brittleness of the material increases significantly.

It seems to us that this combination of organic fibers and mineral crystals has to be considered as a reinforcement of the compact bone tissue. These col-lagen-mineral fibers are embedded into physically nonlinear interfibrillar sub-stance which consists generally of mucopolysaccharides and glycoproteins and performs the role of a matrix.

It should be mentioned that in the calculation of the elasticity modulus and the ultimate stress of the bone tissue as a composite material the applica-tion of the conventional "law of mixture" is not correct. This law is unable to predict the experimental finding [7]. For example, an increase of mineral con-tent in bone tissue by only 8% causes an increase in the ultimate stress by 350%. In accordance with Lees and Davidson [8], the collagen fibers in calcified bone tissue obtain much higher stiffness as compared with stiffness of collagen fibers in tendon. This hardening, not taken into account by the "law of mixture", is explained by increase of the cross linking density during the process of calcification of collagen-mineral fibers.

The fourth structural level is formed by lamellae, the smallest self-dependent structural element of compact bone tissue. Lamellae can be of different configurations. In the cases of thin-walled plates, curved panels or cylindrical shells, their thickness may change from 4 to 12 mkm. Each lamella is made of the bonding substance (matrix) and collagen-mineral fibers oriented in certain direction depending on the particular lamella.

The fifth structural level comprises osteons which are specific structural elements, generated around the small blood vessels in the bone during its growth. Osteon consists of 5 to 20 concentrically located lamellae. It is significant to mention that the collagen-mineral fibers are oriented in adjacent lamellae which forms angles from 45 to 90°. Lamellae in each osteon and osteons themselves are bonded together by a matrix. The average diameter of osteon varies from 0.15 to 0.29 mm. Over the volume of the middle part of the compact bone tissue layer the osteons form the spatially interlaced system.

The first two structural levels are lowered in scale and considered as structural components at the molecular level. Therefore, the assumption of a homogeneous anisotropic solid at this scale level will cause significant errors. The two last structural levels correspond to the actual structural elements of the bone. Thus, viewing the compact bone tissue as a composite solid we have chosen the third structural level as the basic one.

The development and propagation of cracks in the compact bone tissue depend on the specific multistage structure and other factors such as the type of loading, the orientation of the load with respect to direction of the elastic symmetry axis of the material, the loading or deformation rate and external configuration of the test specimens.

When the bone specimens are stretched along the longitudinal axis, i.e., along the direction of osteons orientation, the fracture surface is inclined at 45 to 90° to this axis. In the case of slow loading, the so-called "pull-out" of osteons from the interosteon matrix and formation of rough fracture surface are observed. In the case of dynamic loading, this surface becomes more flat. The fracture line in the cross-section is generally along the outer surface of osteons or even between the lamellae, but very seldom over the Haversian canals.

During bending the fracture is caused by the combined action of tensile and compressive stresses. The osteons are pulled out in the zone of tension and buckling occurs at the outer surface of the bone specimen where compression prevails. Generally speaking, the origin of fracture is observed in the zone of tensile stress.

Both the specimen and the segment of the bone with an edge notch under bending fail at a considerably lower specific strain energy than the case without the notch. It indicates that the largest part of strain energy causing fracture is used for the creation of the crack. The values of stress-intensity-factor being able to initiate the crack growth are different for the longitudinal and transversal specimens. In accordance with Sih and Berman [9], the average value of stress-intensity-factor for bone tissue along the longitudinal axis of the bone is approximately 56 kg.cm.

During torsion, for a specimen [3] with a rectangular cross-section, the macrofracture begins at the moment when the shear stress on the middle part of the largest edge reaches its ultimate value. For the specimens with a circular cross-section [10] we do not know beforehand the location of the crack. But in experiments with these specimens it is found that the nature of bone tissue fracture depends significantly upon the orientation of load with respect to the axis of material elastic symmetry. The fracture of a longitudinal specimen occurs slowly by generating a great number of microcracks along the osteon system. The specimen tested in torsion becomes split in separate osteons or groups of osteons (Figure 2). Fracture of both transversal and radial specimens is brittle forming the fracture surface at 45° to the direction of specimen axis.

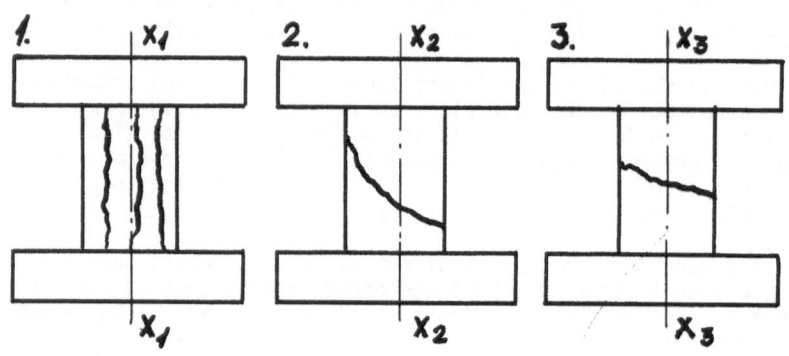

Fig. (2) - Schematic view of fracture lines of the specimens subjected to torsion with respect to longitudinal (1), transversal (2) and radial (3) axis

It should be noted that the nature of bone tissue fracture changes depending upon the age and the possible pathological alterations in the bone. The main features in the fracture of this tissue at senile age are connected with a sharp decrease in the amount of specific strain energy causing fracture, and the formation of a less rough fracture surface. Figure 3 shows the curves due to the change of ultimate shear strain γ_{12}^* and ultimate specific strain energy U_{12}^* with age (index 1 designates the longitudinal axis x_1 of the bone while index 2 refers to the transversal axis x_2). It shows that at the age of 80 the value of U_{12}^* decreases more then twice, when compared with people at an earlier age.

The different kinds of fracture surface formations mentioned above characterize the final state of the solid under ultimate loads. But during the deformation process long before the final fracture the process of formation and accumulation of microcracks begins in the bone tissue. This may be detected by using different methods. It is the author's opinion that the methods of acoustic and photon emission are most promising.

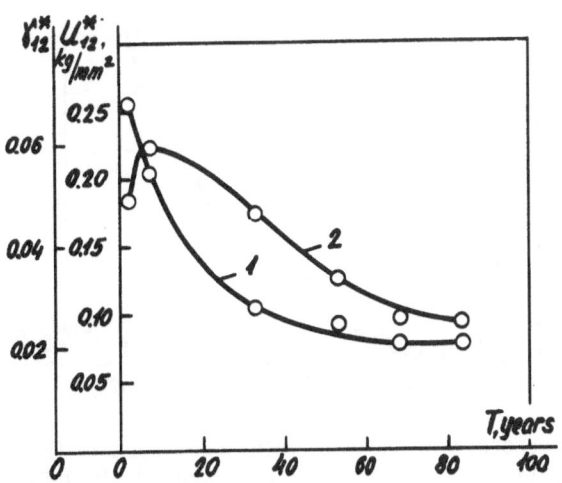

Fig. (3) - Dependence of ultimate shear strain γ_{12}^{*} (1) and ultimate specific shear strain energy U_{12}^{*} (2) upon the age

The method of acoustic emission is based on the detection of acoustic impulses in the material caused by generation and development of structural defects, i.e., dislocations or microcracks formation during the loading process. The main characteristics determined by this method embrace the total number of acoustic impulses N registrated over a certain interval of loading, and the intensity of acoustic impulses Ṅ, i.e., the number of impulses per time unit. As the total number of impulses is a nonlinear function of the stress-intensity-factor, the value of N permits an estimate on the degree of damage corresponding to the moment of fracture.

The measurement of acoustic impulses in the bone tissue specimens subjected to longitudinal tension revealed [11] three distinct regions (Figure 4). In the first region when the relative stress $\sigma_{11}/\sigma_{11}^{*}$ (σ_{11}^{*} = 13.35 kg mm^2 is the ultimate tensile stress along the longitudinal axis of the bone) does not exceed the value of 0.28, there is insignificant manifestation of acoustical effects. Only about 2% from the total number of acoustic impulses N* at the moment of fracture is recorded. The absence of acoustic emission in this stage of loading could be explained as the Kaizer's effect in the bone tissue, i.e., the lack of deformation sounds until the maximum stress level is reached in the previous history of loading. Therefore, this stress could be considered as the physiological level of stress in human tibia.

308

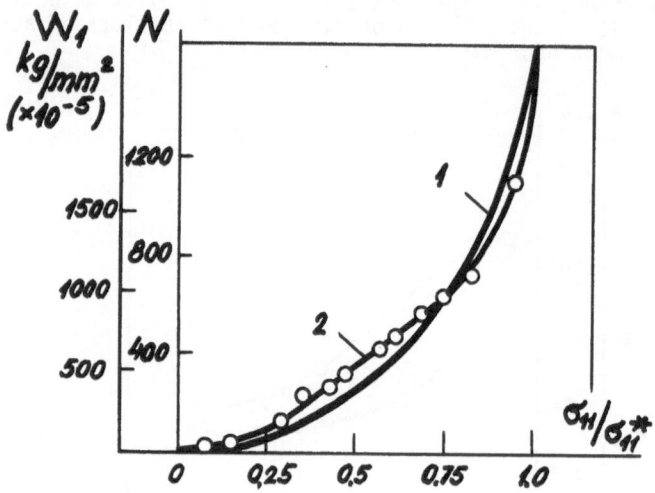

Fig. (4) - Dependence or irreversibly spent specific strain energy
W_1 (1) and the total number of acoustic impulses N (2)
upon relative stress $\sigma_{11}/\sigma_{11}^*$.

In the second region of loading when the value of $\sigma_{11}/\sigma_{11}^*$ changes from 0.28
to 0.91 there is a linear increase in acoustic impulses ($\dot{N}$ = 20 imp./sec
= const). The total number of acoustic impulses over the first two regions of
loading is 0.46 N^*. In the last, third region there is a significant increase
in the intensity of acoustic impulses which correspond to the moment of fracture
reaching the value of 210 imp./sec. In our opinion, this is caused by a sharp
increase in the number of microcracks and of macrocracks.

The comparison of the N versus $\sigma_{11}/\sigma_{11}^*$ and W_1 versus $\sigma_{11}/\sigma_{11}^*$ curves indicate
that there is a high positive correlation (r = 0.95) between them. It confirms
the assumption on the nature of the acoustic emission process.

Besides the method of acoustic emission, we have also used the method of
photon emission [12]. This method is concerned with the recombination of free
radicals generated during the rupture of bonds and the appearance of gaseous discharge at the moment of new surface creation, i.e., formation of microcracks.

The marked effect of photon emission in bone tissue has been found only at
the very last stage of loading ($\sigma_{11}/\sigma_{11}^* \geq 0.96$). The intensity of emission at
this level of loading is 36 specific photon units per second, but at the moment
of final fracture it increased to 58 units per second. At the same time in the
artificial composite materials [13] the photon emission starts at much lower
stress. For example, in the fiberglass reinforced plastic with the orientation

of fibers in adjacent layers at 45° the photon emission begins at $\sigma_{11}/\sigma_{11}^{*} = 0.38$.

The fracture features of compact bone tissue revealed that the high load-bearing capacity of this biopolymer material is ensured by its composite multi-stage structure. The further investigations of compact bone microfracture on different structural levels will lead us not only into a better understanding of this natural material, but also may provide us with some ideas on how to improve on the reinforcement structure of the man-made materials.

REFERENCES

[1] Welch, P. O., "The composite structure of bone and its response to mechanical stress", Recent Advances in Engineering Science, Vol. 5, Part 1, pp. 245-262, 1970.

[2] Herrmann, G. and Liebowitz, H., "Mechanics of bone fracture", Fracture, Vol. 7, H. Liebowitz, ed., Academic Press, New York and London, pp. 771-840, 1971.

[3] Knets, I. V., Pfafrod, G. O., Saulgozis, Yu. Z., Laizan, Ya. B. and Janson, H. A., "Deformability and strength of compact bone tissue in torsion", Mehanika polimerov, No. 5, pp. 911-918, 1973 (in Russian).*

[4] Knets, I. V., Saulgozis, Yu. Z. and Janson, H. A., "Deformability and strength of compact bone tissue in tension", Mehanika polimerov, No. 3, pp. 501-506, 1974 (in Russian).*

[5] Knets, I. and Malmeisters, A., "Deformability and strength of human compact bone tissue", Mechanics of Biological Solids, Proceedings Euromech Colloquium 68 (Varna, Bulgaria, 1975), G. Brankov, ed., Publishing House of the Bulgarian Academy of Science, Sofia, pp. 123-141, 1977.

[6] Knets, I. V., "Mechanics of biological tissues", Mehanika polimerov, No. 3, pp. 510-518, 1977 (in Russian).*

[7] Vose, G. P. and Kubala, A. L., "Bone strength - its relationship to X-ray determined ash content", Hum. Biol., Vol. 31, pp. 262-270, 1959.

[8] Lees, S. and Davidson, C. L., "The role of collagen in the elastic properties of calcified tissue", Journal of Biomechanics, Vol. 10, pp. 473-486, 1977.

[9] Sih, G. C., "Fracture toughness of bones", 27th ACEMB (Philadelphia, 1974), p. 295, 1974.

[10] Pfafrod, G. O., Knets, I. V., Saulgozis, Yu. Z., Kregers, A. F. and Janson, H. A., "Age aspects of compact bone tissue strength in torsion", Mehanika polimerov, No. 3, pp. 493-503, 1975 (in Russian).*

---

*The translation of paper in English is available in "Polymer Mechanics" published by Consultants Bureau, Plenum Publishing Corporation, New York, USA.

[11] Knets, I. V., Krauja, U. E. and Vilks, Yu. K., "Acoustic emission in human bone tissue in longitudinal tension", Mehanika polimerov, No. 4, pp. 685-690, 1975 (in Russian).*

[12] Krauja, U. E., Knets, I. V. and Laizan, Y. B., "Photon emission of human compact bone during fracture", Mehanika polimerov, No. 4, pp. 746-749, 1977 (in Russian).*

[13] Krauja, U. E., Laizan, V. B. Upitis, Z. T. and Tutan, M. F., "Photon emission of fiberglass reinforced plastic in tension", Mehanika polimerov, No. 2, pp. 316-320, 1977 (in Russian).*

---

*The translation of paper in English is available in "Polymer Mechanics" published by Consultants Bureau, Plenum Publishing Corporation, New York, USA.

PREDICTION OF FATIGUE LIFETIME OF FIBERGLASS PLASTICS BASED ON CUMULATIVE DAMAGE

V. M. Parfeyev, P. P. Oldirev and V. P. Tamuzs

Institute of Polymer Mechanics
Latvian SSR Academy of Sciences, Riga

INTRODUCTION

The present paper deals with predicting the lifetime of fiberglass plastics by considering the fatigue fracture under constant deflection, stepwise and programmed loading, and accelerated tests.

The fracture of composites being heterogeneous is mainly of a disperse nature. The irreversibility of the fatigue fracture process of fiberglass plastics and other composites has been discussed briefly in [1] by investigating the accumulation of seismoacoustic emission [2,3]. The change of the physico-chemical properties of the materials is also discussed in [4-9]. Since fiberglass plastics having completely hardened thermosetting binders do not exhibit the strengthening effect during cyclic loading, the data on the kinetics of the cumulative damage can serve as a basis for predicting the lifetime of these materials.

Two types of specimens of fiberglass plastics with smooth surface were tested under symmetric cyclic loads. Polyester unidirectional fiberglass plastic was loaded in bending by rotation at a frequency of 50 Hz while the epoxy woven fiberglass plastic was tested in pure bending at a frequency of 18 Hz. The elastic moduli of both of these plastics are 3,030 and 5,380 kg/mm^2 and the limit strength in static bending are 54.9 and 74.8 kg/mm^2, respectively.

Fiberglass plastics give rise to heat during fatigue which can be cooled by intense blowing. This effect should be taken into consideration when considering the kinetics of cumulative damage in the material. This paper considers the experimental data related to the accumulation of damage under different load conditions: stationary, stepwise and programmed, isothermic and non-isothermic.

STATIONARY AND ACCELERATED FATIGUE TESTS

Figure 1 shows the fatigue curves for the fiberglass plastics and Figure 2 the data on damage growth under isothermic (T = const) and non-isothermic (T ≠ const) conditions with constant amplitude loading, i.e., σ_α = const. The damage was evaluated by change of the deflection f, residual static strength σ_b and the sample rigidity EI in bending.

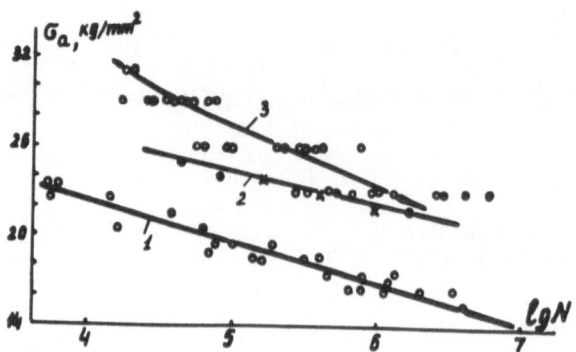

Fig. (1) - Fatigue curves of unidirectional polyester fiberglass
reinforced plastics subjected to bending introduced by
rotation at T = 25°C: curve 1 for constant amplitude
load and curve 2 for fixed displacement load. Curve 3
is for epoxy fiberglass reinforced plastic under constant amplitude fatigue and non-isothermic condition.
The points (x) represent calculated values and (o) experimental values for f = const.

The process of fatigue fracture consists of three stages: a decreasing,
constant and increasing rate of damage accumulation as illustrated in Figure 2a.
It is clear from Figures 2a and 2d that the relative lifetimes are connected
with the change of the properties and, consequently, with the damage level. For
the same relative lifetime, the damage of the fiberglass plastics increases with
decreasing σ_α. The same is valid for the critical damage n_b corresponding to
the beginning of the third stage of fracture.

The foregoing features of fatigue fracture are observed as a result of the
accelerated fatigue test technique [6]. The method gives rise to accelerated
accumulation of damage caused by brief overstress. The critical damage n_b tends
to increase with decreasing stress amplitude. The damage that occurs in the
specimen is measured by parameters related to the area enclosed by the hysteresis loop or the temperature difference due to self-heating of the specimen. The
recommended sequence of loading and the self-heating curve are given in Figure 3.
The lifetime under the applied stress σ_H can be determined by extrapolating the
self-heating curve up to the critical temperature determined from the overstress
tests. The accelerated test method involving 10^7 cycles reduces the experimental
test time by as much as 80 times.

It should be emphasized that the accelerated test method can be justified experimentally as follows:

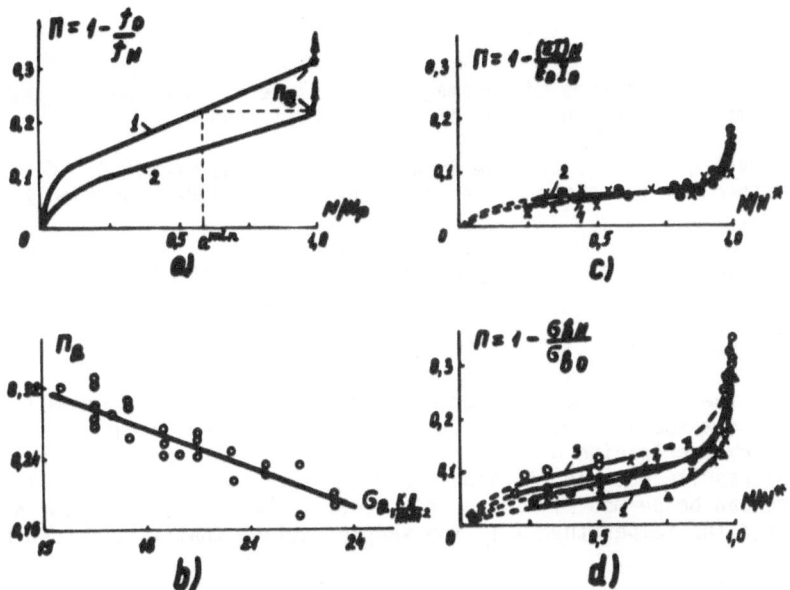

Fig. (2) - Functions characterizing the cumulative damage: (a) and (b) for polyester, and (c) and (d) for epoxy fiberglass reinforced plastics. (a) $\Pi(N/N_p)$ at T = 25°: Curve 1 for σ_α = 16.5 kg/mm² and curve 2 for σ_α = 21.5 kg/mm². (b) $\Pi_B(\sigma_\alpha)$ at T = 25°C. (c) and (d) $\Pi(N/N^*)$ with self-heating: Curve 1 with data points (Δ) for σ_α = 29 kg/mm², curve 2 with data points (.) for σ_α = 26 kg/mm², curve 3 with data points (o) for σ_α = 23 kg/mm² and curve 4 with data points (x) for programmed loading

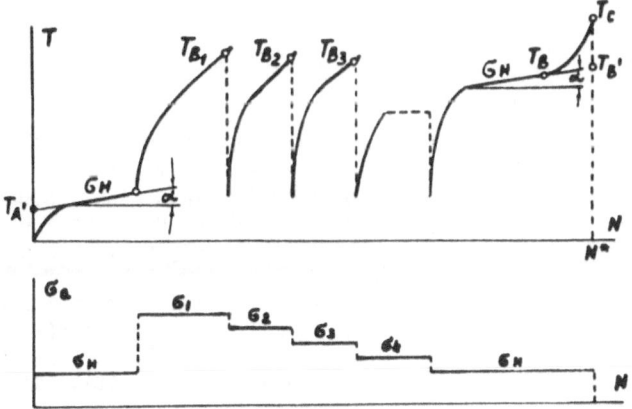

Fig. (3) - Sequence of loading in accelerated test for determining of T_B for low stress level σ_H

(1) Cyclic loading causes microdamage in a volume of material and leads to changes in the macroscopic properties of the material; and

(2) the level of critical damage is a function of the magnitude of the stress amplitude σ_α at the onset of macrofracture and is independent on the loading history.

Similar observations have also been made on plastics with the Kevlar type of reinforcement.

NON-STATIONARY AND ISOTHERMIC CONDITIONS ($\sigma_\alpha \neq$ const, T = const)

At present, there is no general agreement on the hypothesis of summation of relative lifetime associated with the fracture of composites. Self-heating arising in the cyclic loading of fiberglass plastics further complicates a comparison of the results between stationary and non-stationary loadings. The relative lifetimes can be summed by taking into account the variation of the stress amplitude σ_α and the temperature T in the sample section that is stressed most severely:

$$\int_0^{N^*} \frac{dN}{N_p[\sigma_\alpha(N), \, T(N)]} = a \qquad (1)$$

where N_p is the lifetime at the constant T and σ_α level N* is the lifetime with σ_α and T varying in accordance with $\sigma_\alpha(N)$, T(N) while a is the summation constant of the relative lifetimes.

Let us first examine the isothermic conditions T = const and $\sigma_\alpha \neq$ const. For the lifetime calculation of non-stationary loading, we use the empirical relationships:

$$\Pi = f(\sigma_\alpha, N), \quad _B = f(\sigma_\alpha) \qquad (2)$$

which describe the damage accumulation process of the material (polyester fiberglass plastics) under stationary and isothermic conditions. At first, we consider the method for lifetime prediction of fiberglass plastics under the fixed displacement condition (f = const). To this end, we use equations (2) where the damage has been determined from the change in the rigidity when σ_α is constant.

Let the deflection or initial stress and the small damage increment $\Delta\Pi$ be given. From equations (2) we obtain the number of cycles ΔN necessary to find $\Delta\Pi$ and the stress due to variation in the rigidity. The same procedure can then be repeated. The relations $\sigma_\alpha(N)$ and $\Pi(N)$ for the two initial stresses are shown in Figure 4. To determine the life of a sample from these curves, we use the fracture criterion by assuming that the lifetime of the material terminates upon reaching the critical damage which depends only on the stress level at the moment of failure.

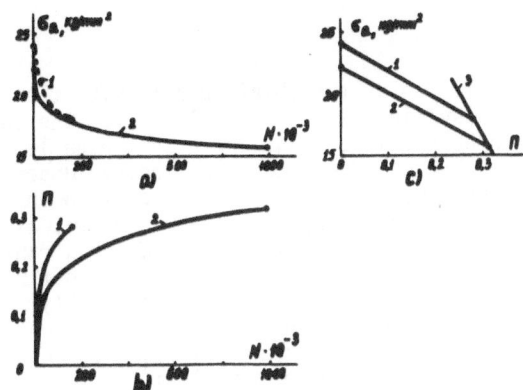

Fig. (4) - Calculated relations with fixed displacement: (a) for $\sigma_\alpha(N)$, (b) for $\cap(N)$ and (c) for $\sigma_\alpha(\cap)$. Curve 1 for an initial stress of 24 kg/mm² and curve 2 for an initial stress of 22 kg/mm²

Lifetime estimate can be made by excluding N from $\sigma_\alpha(N)$ and $\cap(N)$. Curves 1 and 2 in Figure 4 give σ_α and $\cap$ whose intersection with the straight line 3 gives the critical damage under stationary loading, Figure 2b, from which life-time estimate can be made. The experimental data on lifetime, Figure 1, as in the case of damage and stress relaxation under fixed displacement loading, are in good agreement with the calculations. Hence, the lifetime under the condi-tion of fixed displacement has been obtained from the results of constant load test.

Next, we examine a stepwise load increase shown in Figure 5. From the load-ing pattern of curve 1 and the use of equations (2), we calculate the damage accumulation curve. The intersection of the latter with the curve of critical damage, labelled 3 in Figure 5, will define the failure point and the lifetime. The life for the case of programmed isothermic loading can be calculated in the same way.

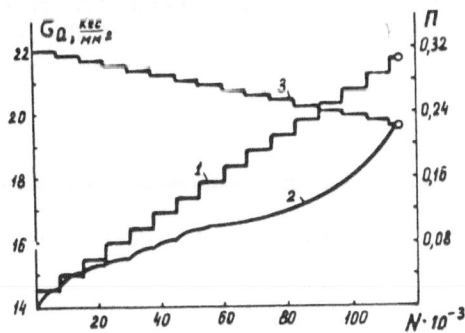

Fig. (5) - Calculated relation $\cap(N)$ for multi-stress level loading given by curve 2

Let us examine a correlation of the obtained data with the linear summation law of relative lifetime, equation (1). It is clear from the cumulative damage curves in Figure 2a that the summation constant a should be less than one for increasing loads and more than one for decreasing loads. Under increasing loads, the damage can be accumulated at the start. An increase in the load causes the specimen to fail within several hundreds of cycles. For a two-stress level loading applied to the fiberglass plastics, the mean value of $\bar{a}$ ranges from $0 < \bar{a} < 1$ for increasing load and $1 < \bar{a} < 2$ for decreasing load. These effects due to the change in the sequence of loading was verified experimentally. At the last level of loading, the critical damage was the same as that of continuous loading with the same stress amplitude. For the fixed displacement condition with decreasing loads, the values of the constant a fall within the range of 1.3 to 1.7.

For programmed block loading, the summation constant a should be less than one since failure takes place at the high stress level when n_b is the smallest. The value of a depends on the magnitude of the stress drop and the duration of high and low stress levels the variation of which yields a within the limits of $a^{min} < a < 1$. The upper limit is reached when the high stress dominates. In the opposite case, a tends to the lower limit a^{min} whose magnitude can be estimated by a two-stress level test (Figure 2a) with the stress increasing from the minimum to the maximum value. All of the above conclusions have been verified experimentally. Thus, when high stresses dominate in the programmed loading block, the lifetime can be calculated approximately by means of the hypothesis proposed earlier.

The given results show that the damage level and the lifetime of fiberglass plastics under different non-stationary and isothermic conditions can be predicted with sufficient reliability from a knowledge of the cumulative damage obtained under the stationary and isothermic conditions.

NON-STATIONARY AND NON-ISOTHERMIC CONDITIONS ($\sigma_\alpha \neq$ const, $T \neq$ const)

The lifetime prediction under non-stationary and non-isothermic conditions is made for a two-stress level loading of the fiberglass reinforced plastics [8] and PMMA [9]. The sequence of loading again influences the value of the summation constant a which can be determined by replacing equation (1) with the expression

$$\int_0^{N*} \frac{dN}{N*(\sigma_\alpha)} = a \tag{3}$$

where $N*(\sigma_\alpha)$ is the lifetime for stationary and non-isothermic loading condi-

For tests involving multiple periods of rest and programmed loading, the constant a in equation (3) is often above one which may not necessary be connected with strengthening. In three-stress level programmed tests with the loading program and self-heating curve given in Figure 6, $a \approx 2.8$.

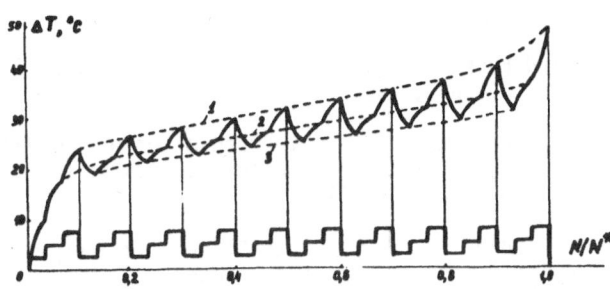

Fig. (6) - Self-heating curve of epoxy fiberglass reinforced
plastics in programmed loading. Stress block duration of 9.10^3 cycles. The stress amplitudes 23, 26
and 29 kg/mm² characterized by the lifetimes of
$1.1.10^6$, 2.10^5, $4.1.10^4$ cycles (curve 3 in Figure 1)

The overall kinetics of self-heating in programmed tests are of the same
nature as in stationary loading. Curves 1, 2 and 3 in Figure 6 drawn through
the points at which load variations take place are similar to those of self-
heating. The slope of the curves is associated with the change in the intensity
of self-heating due to the cumulative damage level. Unlike the stationary loading under programmed loads, different damage level may refer to the same temperature. Curves 1, 2 and 3 can control the behavior of the specimens during cyclic
programmed loading as in the case of σ = const.

For non-isothermic programmed loading, the epoxy fiberglass plastic softens
owing to the accumulation of damage. Refer to curve 4 in Figures 2c and 2d.
The softening process is approximately the same as that of stationary loading
with the mean stress determined from the block loading. In the programmed tests,
the failure occurring under maximum stress is more severe than that occurred under the same stress for the stationary and non-isothermic test conditions. The
short duration of maximum load was not able to warm up the material to the temperature causing damage. The material was able to cool down during unloading.
For this reason, evaluation of lifetime in the programmed tests by using equation (3) is not always accurate and equation (1) is far better although it involves labor-consuming isothermic tests and theoretical calculations of the self-
heating curves which are rather complicated even in static loading [1].

The present study shows that in stationary and non-stationary cyclic loading
of fiberglass reinforced plastics, the cumulative damage can be described by the
same relationships and are determined by the amount of damage, stress and temperature. On the basis of the experimental data on cumulative damage, the lifetime under a variety of non-stationary isothermic load conditions was calculated
and the method of accelerated fatigue tests has been worked out.

REFERENCES

[1] Tamuzs, V. P. and Kuksenko, V. S., "The fracture micromechanics of polymer materials", Riga, 294 pp., 1978.

[2] Tutans, M. J. and Urzumtsev, Yu. S., "Prediction of the fracture process of fiberglass reinforced plastics by the seismoacoustic method", Mekhanika Polimerov, No. 3, pp. 421-429, 1971.

[3] Williams, R. S. and Reifsnider, K. L., "Investigations of acoustic emission during fatigue loading of composite specimens", J. Comp. Mat., Vol. 8, No. 4, pp. 340-345, 1974.

[4] Oldirev, P. P. and Tamuzs, V. P., "Variation of the properties of glass laminate in cyclic tension-compression", Mekhanika Polimerov, N. 5, pp. 864-872, 1967.

[5] Oldirev, P. P., "Investigation of the deformation properties, energy dissipation and fracture of rigid polymer materials in long-term cyclic loading", Diss. of Candidate of Technical Sciences, Riga, 174 pp., 1968.

[6] Oldirev, P. P., Parfeyev, V. M. and Komar, V. I., "Improvement of the method for the determination of the lifetime of polymer materials by the warmup temperature", Mekhanika Polimerov, N. 5, pp. 906-913, 1977.

[7] Owen, M. J., "Fatigue damage of fiberglass reinforced plastics", In: Composite Materials, Vol. 5, Fracture and Fatigue, edited by L. Broutman, Moscow, pp. 333-363, 1978.

[8] Broutman, L. I. and Sahu, S. A., "A new theory to predict cumulative fatigue damage in fiberglass reinforced plastics", Comp. Mater. Test. and Des. 2nd Conf., Phil., pp. 170-188, 1972.

[9] Oldirev, P. P. and Parfeyev, V. M., "The lifetime of PMMA in stationary and stepwise non-isothermic loading", Mekhanika Polimerov, N. 5, pp. 795-803, 1975.

Section V

EXPERIMENTAL METHODS

METHODS AND MEANS FOR NONDESTRUCTIVE STUDY OF THE DAMAGEABILITY OF COMPOSITE MATERIALS

V. A. Latishenko and I. G. Matiss

Institute of Polymer Mechanics, L.S.S.R.
Academy of Sciences

ABSTRACT

Changes in the properties of composite materials and their structural imperfections are closely associated with physical parameters dealing with acoustic,
thermal, electric, etc. To disclose these interrelations, the method of nondestructive testing of composite materials may be used. A number of these methods
will be discussed.

INTRODUCTION

Conventional methods of studying the formation and accumulation of microcracks and flaws include X-ray dissipation, infrared spectroscopy, electromicros-
copy and acoustic emission. However, damageability in the broader sense is connected with the variations of material structure due to loading, aging, technological imperfections, etc. Our studies of less known methods used in fracture
mechanics associated with the emission of acoustic, thermal and electric energy
show that in some cases these methods are more effective than the conventional
ones. Their main advantage lies in the nondestructive evaluation of the integral
characteristics and variations of numerous composite materials.

A. Basic Studies of the Physical and Mechanical Properties

Experience has shown that all composites are inhomogeneous and vary significantly in their composition and properties even in the absence of gross defects. The range of variability of mechanical properties is illustrated in Table
1. The quantities in Table 1 are defined as:

σ_{11}, E_{11} - longitudinal strength and modulus of elasticity along the fiber
direction;

σ_{22}, E_{22} - longitudinal strength and modulus of elasticity transverse to
the fibers;

τ_{31} - interlaminar shear strength along the fiber direction;

τ_{32} - interlaminar shear strength transverse to the fibers.

TABLE 1 - THE VARIABILITY OF MECHANICAL PROPERTIES OF
REINFORCED PLASTICS (max value/min value)

Material	Mechanical Properties					
	σ_{11}	σ_{22}	E_{11}	E_{22}	τ_{31}	τ_{32}
Unidirectional GFRP AG-4-S	2.3	2.3	1.3	2.0	2.6	-
33-18-S	1.9	1.5	1.7	2.4	1.5	2.0
27-63-S	2.0	-	1.7	-	-	3.1
P-2-1S	1.9	-	2.6	-	-	-
GFRP with two-dimensional reinforcement PPN	1.4	1.3	1.6	1.7	1.6	1.6
GFRP with fabric reinforcement	2.0	2.7	1.8	1.3	1.3	1.4

Evaluation of the in-service performance reliability of composite materials during loading requires nondestructive testing (NDT). Careful study on the detailed physical and mechanical properties of typical reinforced plastics provided by M. Z. Medvedev and A. V. Sandalov shows a highly definite correlation between the physical parameters and mechanical properties. We could therefore set up an equation as follows:

P and F = f(S,K,T)

P and S = f(F,K,T)

where P is the parameter of strength or reliability; F is the physical parameter which can stand for the velocity of ultrasonic waves, thermal conductivity, dielectric constant, dissipation factor, ratio of light penetration, etc., S is the parameter of structure such as fiber volume fraction, fiber orientation, distribution of reinforcement, porosity content, interlaminar adhesive strength; K is the mechanical characteristics of components; and T is the processing characteristics such as winding tension moulding pressure, temperature of polymerization and others. The equation shows that to assure high durability the NDT physical methods must be combined with technological control.

Investigations conducted by V. D. Shtrauss and J. Matison disclose correlation between the strength and some physical parameter during the process of artificial aging of GFRP. Figure 1 shows the correlationship between bending strength and the physical parameter. Both parameters are presented in relative units:

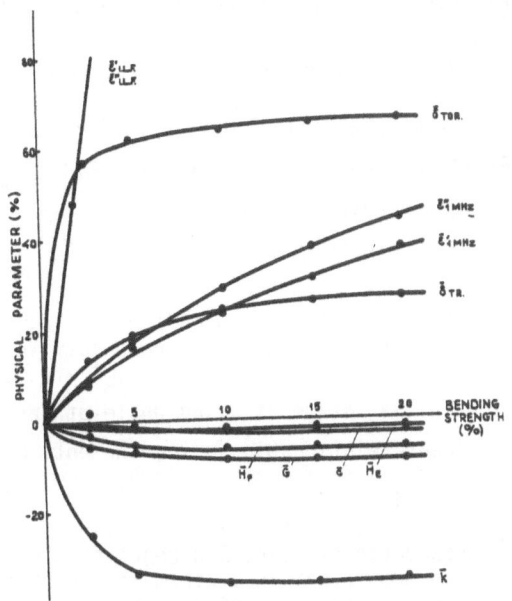

Fig. (1) - Correlation between the physical parameter and
bending strength during the composite aging

$$\bar{\sigma} = \left| \frac{\sigma_{b,s}}{\sigma_{o,b,s}} \right| \cdot 100\% \qquad \text{(bending strength)}$$

and the physical parameter

$$\bar{y} = \left| \frac{y}{y_o} \right| \cdot 100\%$$

where $\sigma_{o,b,s}$ is the bending strength of the initial specimen; $\sigma_{b,s}$ is the bending strength of the specimen under aging; y_o is the physical parameter of the initial specimen; y is the physical parameter of the specimen under aging.

In our investigation the emphasis was placed on the following physical parameters:

1. Dielectric constant (ε'_{ILF}) and dissipation factor (ε''_{ILF}) at the infra-low frequency ranging from 0.001 up to 1 Hz.

2. Dielectric constant (ε'_{1MHz}) and dissipation factor (ε''_{1MHz}) at 1 MHz frequency.

3. Dynamic modulus of elongation (He) and dynamic modulus of flexibility (H_F).

4. Dynamic shear modulus (G).

5. Decrement of torsional (δ_T) and transversal (δ_{TR}) oscillations.

6. Ratio of light penetration (K).

7. Velocity of ultrasonic waves.

Figure 1 shows the correlation between the bending strength and different physical parameters (except the velocity of ultrasonic waves and dynamic modulus of flexibility). Some physical parameters (K, δ_T, δ_{TR}) are more sensitive to the bending strength in the initial stage of aging while others ($\varepsilon'_{ILF}, \varepsilon''_{ILF}, \varepsilon'_{1MHz}$, ε''_{1MHz}) have an almost linear relationship during the entire aging process.

B. Thermal Characteristics

Investigations conducted by V. F. Zinchenko have noted high sensitivity of the thermal activity through a combined parameter which incorporates thermal conductivity, capacity and density that arise in the study of the damageability of composites. The thermal activity was determined from the experimentally determined thermal flux diagram in which the difference between the thermal flux in the material under investigation and in the standard material is plotted. The relationship obtained from the flux diagram provides information with regard to material inhomogeneity owing to imperfections, microcracks, unbonded areas of the laminated structure, etc. The cracks size and their location depth can be determined as well. Figure 2 exhibits the thermal flux diagram obtained for materials with different microporosity and for materials with macro imperfections. It can be seen from the graphs in Figure 2 that the quantity of microporosity determines the amplitude of the curve, while the presence of imperfections determines the curves variation rate.

C. Acoustic Emission Study

The approach proposed by M. J. Tutans complies formation of the output signal by summing the amplitudes of individual acoustic impulses during the loading test. The experiments show that this approach is very helpful in studying the microcrack formation and accumulation. The relationship between the output signal and the continuous load (Figure 3), and the relationship between the output signal and the tensile load (Figure 4) indicate that the fracture of the composite material starts from the very onset of loading and continues during the entire loading process. It should be noted that correlations in Figure 4 correspond to the exponential function. It is attractive to assume that under recurrent loads of identical magnitudes acoustic emission is decreasing after each cycle (Figure 5). A further increase in load leads to the exponential correlation as in the previous experiment.

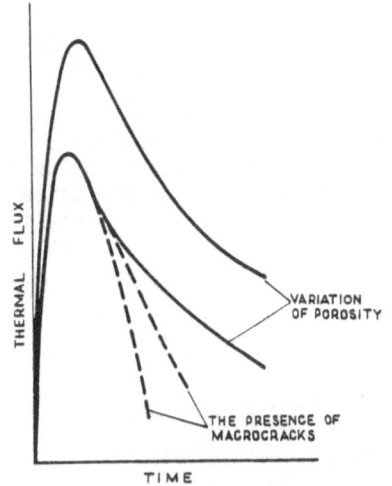

Fig. (2) - Thermal flux diagram

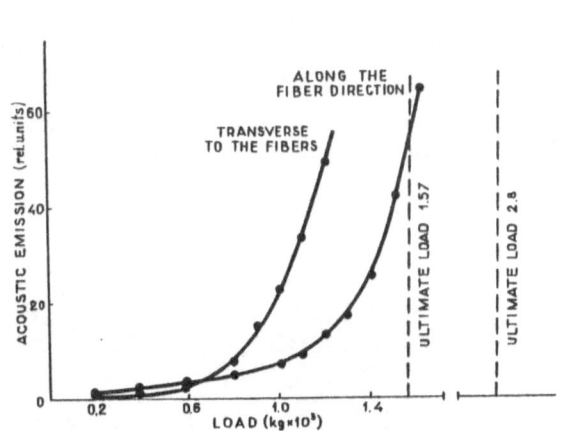

Fig. (3) - Relationship between the acoustic
emission and a continuous load

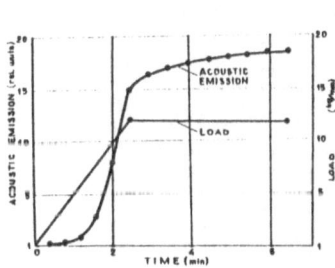

Fig. (4) - Relationship between the
acoustic emission and a
tensile load

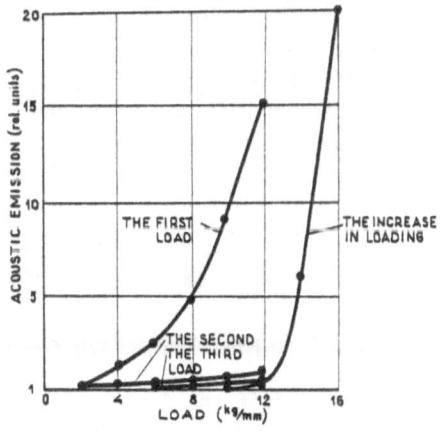

Fig. (5) - The effect of recurrent loads
on acoustic emission

D. NDT Technique

In order to characterise damageability by the above mentioned physical parameters, a special test equipment was designed. The dynamic modulus and decrement were measured by ICHZ-9 designed by A. A. Balodis (see Figure 6). Comparison between our device and the conventional technique shows the advantage of using ICHZ-9 which provides a direct reading of the measuring parameters, separation of the parameters of different waves and contactless determination of the magnitude of oscillations. The velocity and damping ratio of the ultrasonic waves were determined by means of ISZU designed by A. A. Balodis (see Figure 7) which is patented in Canada - Nr. 896138, Great Britain - Nr. 1277065,

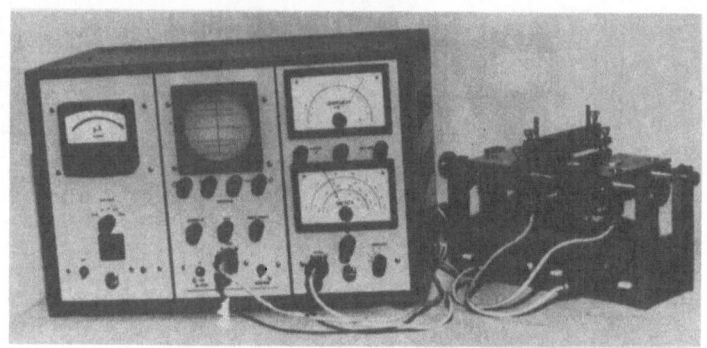

Fig. (6) - ICHZ-9 for vibration inspection of composites

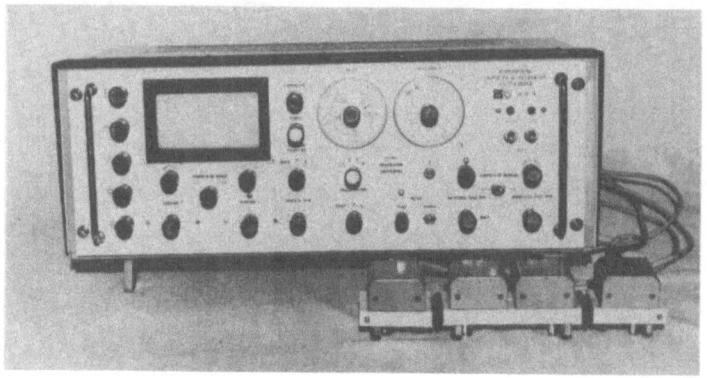

Fig. (7) - ISZU for ultrasonic inspection

the FRG - Nr. 1947646 and Japan - Nr. 4712828. The scanning is performed by two external exciters and two internal receivers of ultrasonic waves disposed symmetrically on one line. Each of the receivers receives the signal which covers different distances in the material. The analysis of these signals from both receivers provides high accuracy of measurement due to the exclusion of the main disturbing factors which take place in the measuring process.

NDT of electric characteristics of composites is based on the application of unilateral capacitance transducers. In our development program great attention is given to the multiparameter measuring technique. Experiments have shown that variations of the distribution of electric field energy over the active area of the transducers face offer one of the greatest potentials for forming the multidimensional signal. In order to analyze the multidimensional signal and separate the variables, special mathematical blocks were used. According to the algebraic interpretation of multiparameter control, each component of the multidimensional signal represents an equation. In the simplest case, the mathematical block must provide a solution of a system of linear equations. The more complicated cases (high accuracy, great measuring range) involve nonlinear equations which can be more easily handled by combining the measuring equipment with a computer.

Multiparameter control provides simultaneous measuring of one, two or more parameters or the measuring of one parameter under conditions of significant disturbing factors such as uneven or rough surfaces of the material, various thickness of thin plates, etc. The variation of the electrical parameters in the bulk of the material can be determined as well.

The general construction of a capacitance transducer consists of a set of coplanar electrodes connected in different combinations with respect to the given field distribution.

On the bases of the above described principles, several two-dimensional measuring equipment was designed: IDP-7 for controlling the dielectric constant of objects with uneven surface (see Figure 8) and DP-201 for controlling the dielectric constant of thin plates (see Figure 9). The basic principles of these

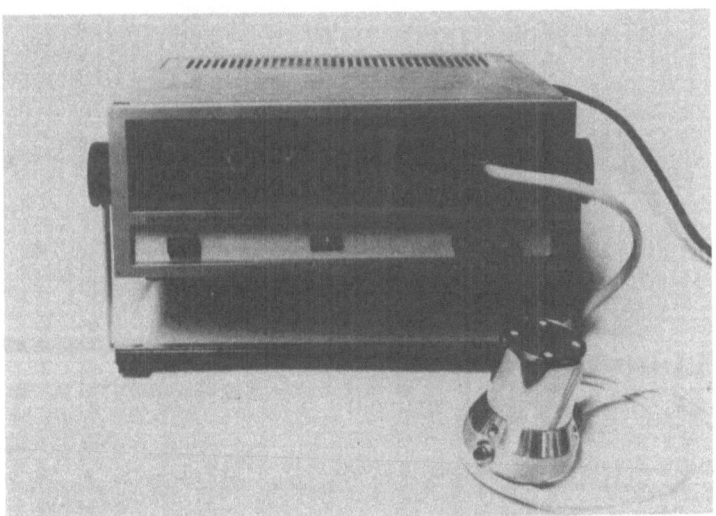

Fig. (8) - IDP-7 for measuring the dielectric constant
of objects with uneven surface

Fig. (9) - DP-201 for measuring the dielectric constant
of thin plates

devices have patented in the USA - Nrs. 3671857, 3694742), Great Britain - Nrs. 1260359, 1275324, 1277334 and France - Nrs. 2047332, 2067519).

IMPETUS OF COMPOSITE MECHANICS ON TEST METHODS FOR FIBER COMPOSITES

C. C. Chamis

National Aeronautics and Space Administration, Lewis Research Center
Cleveland, Ohio 44135

ABSTRACT

The impetus of composite mechanics on the composite test methods and/or on interpreting test results is described using examples from the three major areas of composite mechanics: composite micromechanics, composite macromechanics and laminate theory. The specific examples described occurred over the last 12 years and include contributions such as criteria for selecting resin matrices for improved composite strength, the 10° off-axis tensile test, criteria for configuring hybrids and superhybrids for improved impact resistance and the "reduced bending rigidities" concept for buckling and vibration analyses.

INTRODUCTION

Over the last twelve years composite mechanics has contributed significantly to the development of test methods for composites, to the interpretation of test results and, thereby, to the immense progress of the whole composites technology. In this paper, significant contribution of the three major areas of composite mechanics (composite micromechanics, composite macromechanics and laminate theory) to the development of test methods are illustrated with selected examples. The selected examples are limited to those with which the author was personally involved. However, these examples cover contributions over a time span of about ten years and can be considered as being representative of the contribution of composite mechanics to the development of composite test methods.

The specific examples describe contributions such as criteria for selecting resin matrices for improved composite strength, the 10° off-axis tensile test, procedures for configuring hybrids and the concept of "reduced bending rigidities". The pertinent composite mechanics equations associated with each contribution are given and are supplemented by tabular and/or graphical data which illustrate the significance of the contribution. The symbols are defined when they are first used and are summarized in the Appendix for convenience.

COMPOSITE MICROMECHANICS

The impetus of composite micromechanics in identifying constituent properties which influence composite strength is illustrated herein using two examples: (1) matrix properties influencing composite transverse tensile, compressive and intralaminar shear strengths [1] and major constituent contributors to composite impact resistance [2].

Matrix Properties Influencing Composite Transverse Tensile, Compressive and Intralaminar Shear Strengths

Stress-strain curves for high and low modulus matrix resins are shown in Figures 1(a) and (b), respectively. Also included in the figures is the stress-strain curve for a unidirectional composite tested in the transverse direction.

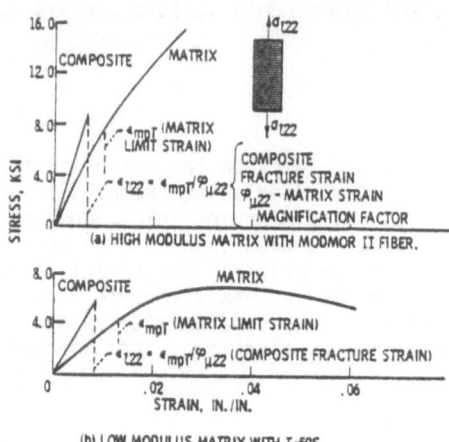

Fig. (1) - Transverse composite and matrix stress/strain curves [1]

It can be seen from Figure 1 that only the initial portion of the matrix stress/strain curve is utilized in the composite. The notation to be used in subsequent discussion is defined in Figure 1. Note that the matrix limit strain, ε_{mpT}, is taken to be the point at which the matrix stress/strain curve exhibits a pronounced nonlinearity.

The governing micromechanics equations are from [1]:

Transverse tensile strength $(S_{\ell 22T})$

$$S_{\ell 22T} = \beta_{22T} \frac{\varepsilon_{mpT}}{\beta_v \phi_{\mu 22}} E_{\ell 22} \tag{1}$$

Transverse compressive strength $(S_{\ell 22C})$

$$S_{\ell 22C}^{'} = \beta_{22C} \frac{\varepsilon_{mpC}}{\beta_v \phi_{\mu 22}} E_{\ell 22} \tag{2}$$

Intralaminar shear strength $(S_{\ell 12S})$

$$S_{\ell 12S} = \beta_{12} \frac{\varepsilon_{mpS}}{\beta_v \phi_{\mu 12}} G_{\ell 12} \tag{3}$$

The undefined notation in equations (1), (2), and (3) is as follows: β denotes the theory experiment correlation coefficient reflecting the fabrication process; β_v denotes the void influence; ϕ_μ is the matrix-strain-magnification factor; the subscripts T, C, and S denote tension, compression, and shear, respectively; ε_{mpT} is the matrix limit strain as defined from Figure 1; and correspondingly for compression and shear; $E_{\ell 22}$ and $G_{\ell 12}$ are the composite transverse and shear moduli, respectively.

There are three groups of variables with distinct physical meaning in equations (1), (2), and (3). These groups can be easily identified by writing equation (1) in the following form:

$$S_{\ell 22T} = (\frac{\beta_{22T}}{\beta_v})(\frac{E_{\ell 22}}{\phi_{\mu 22}}) \varepsilon_{mpT} \tag{1a}$$

where (β_{22T}/β_v) represents the particular fabrication process and depends only on the fabrication process; $(E_{\ell 22}/\phi_{\mu 22})$ is defined herein as the "strength parameter" which depends on the local and average composite geometry and on the elastic properties of the constituents; and ε_{mpT} is the matrix limiting strain as defined previously. Corresponding variables in equations (2) and (3) can be grouped in the same fashion with analogous physical interpretations.

The matrix variables influencing $S_{\ell 22T}$ enter through either $(E_{\ell 22}/\phi_{\mu 22})$ or ε_{mpT}. The group (β_{22T}/β_v) does not depend (at least not explicitly) on the matrix elastic or strength properties.

The variation of $(E_{\ell 22}/\phi_{\mu 22})$ and $(G_{\ell 12}/\phi_{\mu 12})$ with matrix modulus for a Thornel-50/epoxy composite with a 0.5 fiber volume fraction and zero voids is shown in Figures 2 and 3, respectively. As can be seen in Figures 2 and 3, the matrix modulus markedly affects the transverse and shear strength parameters.

The results in Figures 2 and 3 suggest that transverse and intralaminar shear strength tests should be sensitive to matrix modulus.

The variation of the transverse strength parameter $(E_{\ell 22}/\phi_{\mu 22})$ with fiber volume ratio is shown in Figure 4 for three matrix moduli, zero voids and 10 per-

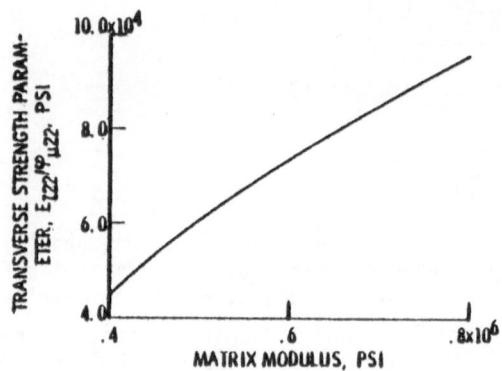

Fig. (2) - Effect of matrix modulus on transverse strength parameter.
TH-50/epoxy with 0.5 fiber volume ration and zero voids [1]

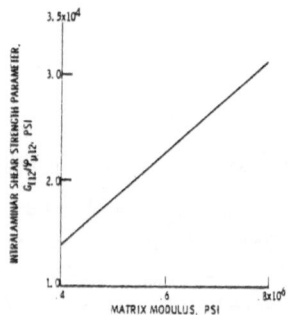

Fig. (3) - Effect of matrix modulus on intralaminar shear strength
parameter. TH-50/epoxy with 0.5 fiber volume ration and
zero voids [1]

cent voids. The curves in Figure 4 show that the transverse strength parameter
is sensitive to both matrix modulus and void content. However, it is not as sensitive to fiber volume ratio. These observations also apply to the intralaminar
shear strength parameter.

The above composite micromechanics results guided the experimental investigation described in [1]. The combined results led to a simple criterion for selecting resin matrices for improved composite strength. This criterion as stated
in [1] is: "Of the various simple matrix properties, the area under the matrix
stress/strain curve up to the proportional limit strain (initial area) is the
best index for assessing matrix influence on composite strength and overall com-

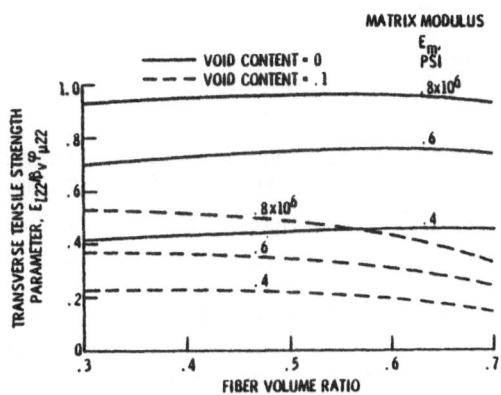

Fig. (4) - Effects of Thornel-50/resin unidirectional composite
transverse tensile strength limited by in situ matrix
tensile strain (elongation) [1]

posite structural behavior". An even simpler version of this criterion is:
"The initial modulus of the resin stress-strain curve is a good index in assessing the contribution of the resin matrix to composite strength". The higher the
initial modulus the higher the composite strength. It is interesting to note
that the total elongation-to-fracture of the resin does not influence composite
strength [1].

Major Constituent Contributors to Composite Impact Resistance

The total energy stored in a uniformly stressed unidirection composite under
uniaxial tension along the fiber direction is simply

$$U = \frac{1}{2} \varepsilon^*_{\ell11T} S_{\ell11T} V \tag{4}$$

or

$$U = \left(\frac{S^2_{\ell11T}}{2E_{\ell11}}\right) V \tag{4a}$$

where U is the strain energy, ε^* is the fracture strain, S is the fracture
strength, V is the volume, and E is the modulus. The subscript group $\ell11T$ is
defined as follows: ℓ refers to unidirectional properties, 11 identify outward
normal to the plane and stress directions in that order, and T identifies the
sense of the stress. Using composite micromechanics $S_{\ell}11T$ and $\bar{\varepsilon}_{\ell11}$ are expressed

in terms of fiber and matrix properties [2]. The impact energy density (IED) equals the strain energy divided by the volume. The IED of composites with an E_f/E_m ratio greater than 20 is approximated by

$$IED = \frac{(1-k_v)k_f\beta_{fT}^2 S_{fT}^2}{2E_f} \tag{5}$$

with an approximation error of less than 5 percent. The variables in equation (5) are as follows: k_v and k_f denote void and fiber volume ratios, respectively; β_{fT} represents the in-situ fiber strength efficiency which reflects the fabrication process. The subscript f refers to fiber property. The important points to be noted in equation (5) are the quadratic dependence of the strain energy density on the fiber strength S_{fT}^2 and the fabrication process variable β_{fT}^2. For a high impact resistance composite, equation (5) imposes the following requirements: a high strength low modulus fiber, approximately 100 percent fiber properties translation efficiency, high fiber volume ratio, and low void volume ratio.

The transverse IED is given by

$$IED = \frac{1}{2}\left(\beta_{22}\frac{\varepsilon_{mPT}}{\beta_v\phi_{\mu22}}\right)^2 E_{\ell22} \tag{5a}$$

where the notation has been defined in equations (1) to (3). As was the case for equation (1), the important resin property for transverse IED is the modulus (E_m).

The ranking of IED of various composites predicted using equations (5) and (5a) are compared with measured data in Table 1. As can be seen the comparison is excellent.

TABLE 1 - MINIATURE IZOD IMPACT DATA FOR FIBER/EPOXY COMPOSITES [2]

Fiber	Type	Fiber volume ratio	Average impact energy				Rank			
			Longitudinal		Transverse		Longitudinal		Transverse	
			cm-N	in.-lb	cm-N	in.-lb	Measured	Predicted	Measured	Predicted
Graphite	Thornel 50S	0.532	85.9	7.6	7.9	0.7	5	5	3	3
	Thornel-50	.583	208.0	18.4	3.4	.3	4	4	5	5
	HTS	.523	56.5	5.0	14.7	1.3	6	6	2	2
	Modmor-I	.542	215.0	19.0	4.5	.4	3	3	4	4
Glass	S	0.486	757.0	67.0	15.8	1.4	1	1	1	1
Kev-49	-	-	280.0	24.8	3.4	0.3	2	2	5	-

COMPOSITE MACROMECHANICS

The impetus of composite macromechanics in developing test methods for characterizing unidirectional composites is illustrated herein using the 10° off-axis test method [3]. A schematic of the test specimen for this method is shown in Figure 5.

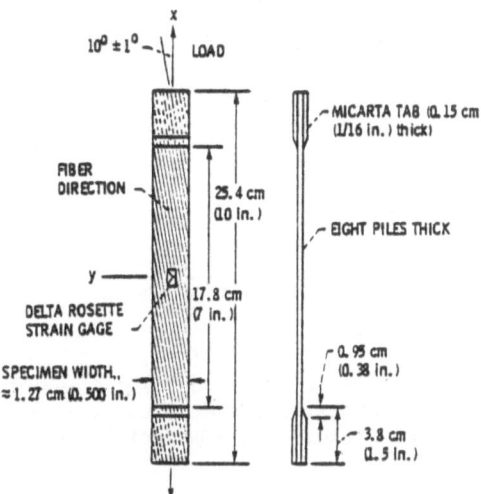

Fig. (5) - Schematic showing geometry and instrumentation of proposed 10° off-axis tensile specimen for fiber composite intralaminar shear characterization [3]

The composite macromechanics equations used in developing the 10° off-axis test method for intralaminar shear characterization are respectively (refer to Figure 6): plain stress transformation

$$
\begin{Bmatrix} \varepsilon_{\ell 11} \\ \varepsilon_{\ell 22} \\ \varepsilon_{\ell 12} \end{Bmatrix} =
\begin{bmatrix}
\cos^2\theta & \sin^2\theta & \frac{1}{2}\sin 2\theta \\
\sin^2\theta & \cos^2\theta & -\frac{1}{2}\sin 2\theta \\
-\sin 2\theta & \sin 2\theta & \cos 2\theta
\end{bmatrix}
\begin{Bmatrix} \varepsilon_{cxx} \\ \varepsilon_{cyy} \\ \varepsilon_{cxy} \end{Bmatrix}
\tag{6}
$$

or in matrix form

$$\{\varepsilon_\ell\} = [R_\ell]\{\varepsilon_c\} \tag{6a}$$

plain stress transformation for uniaxial loading (Figure 6)

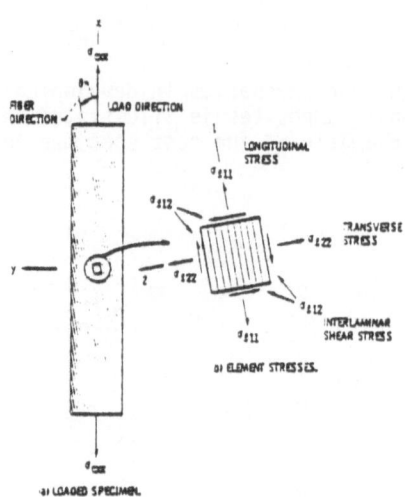

Fig. (6) - Schematic depicting loaded 10° off-axis tensile test
specimen and stresses at element at 10°-plane (x,y -
structural axes; 1,2 - material axes) [3]

$$\sigma_{\ell 11} = \sigma_{cxx} \cos^2\theta; \quad \sigma_{\ell 22} = \sigma_{cxx} \sin^2\theta; \quad \sigma_{\ell 12} = \frac{1}{2} \sigma_{cxx} \sin 2\theta \qquad (7)$$

and the two dimensional failure criterion

$$1 - \left[\left(\frac{\sigma_{\ell 11}}{S_{\ell 11T}}\right)^2 + \left(\frac{\sigma_{\ell 22}}{S_{\ell 22T}}\right)^2 - K_{\ell 12}\left(\frac{\sigma_{\ell 11}}{S_{\ell 11T}}\right)\left(\frac{\sigma_{\ell 22}}{S_{\ell 22T}}\right) + \left(\frac{\sigma_{\ell 12}}{S_{\ell 12S}}\right)\right] \geq 0 \qquad (8)$$

The undefined notation in equations (6) to (8) is as follows: ε denotes strain;
σ denotes stress; θ is the angle between load or composite axis and material axis
(along fiber) (Figure 6); S denotes strength and $K_{\ell 12}$ is a function of the elastic
properties of the composite [4]. The subscript ℓ denotes material axis property
and c denotes composite axis property. The subscript T and S denote tension and
shear, respectively.

The variation of the material-axes strains as a function of load angle is
plotted in Figure 7 for a Mod-I/epoxy unidirectional composite. As can be observed in this figure, the material-axes shear strain (intralaminar shear strain
$\varepsilon_{\ell 12}$) is maximum at about a 10° load orientation angle and appears to be insensitive to small errors about this angle. These are significant results that led to
the recommendation of the 10° off-axis tensile test specimen to measure the intralaminar shear modulus and fracture shear stress. Other composites, for example,

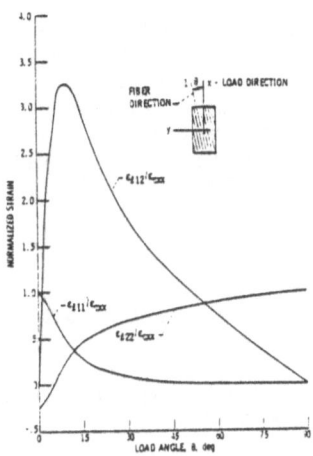

Fig. (7) - Variation of material axes strains in unidirectional
composite (Mod-I/epoxy) plotted against load direction
[3]

approach their peaks at about 11° for T-300/epoxy (PR288) and 15° for S-glass/epoxy (PR288).

The fracture stress of a 10° off-axis specimen [3] is 343 mPa (49.8 ksi) which is equal to σ_{cxx} in equation (7). The material axes stresses from equation (7) are, respectively: $\sigma_{\ell 11}$ = 333 MPa (48.3 ksi); $\sigma_{\ell 22}$ = 10 MPa (1.5 ksi); and $\sigma_{\ell 12}$ = 59 MPa (8.5 ksi). The strengths are: $S_{\ell 11T}$ = 563 MPa (81.7 ksi); $S_{\ell 22T}$ = 28 MPa (4 ksi) and $S_{\ell 12S}$ = 52 MPa (7.6 ksi). The parameter $K_{\ell 12}$ = 1.44 from [3]. Using these numerical values in equation (8) yields:

$$1 - [(\frac{48.3}{81.7})^2 + (\frac{1.5}{4.0})^2 - 1.44 \frac{48.3 \times 1.5}{81.7 \times 4.0} + (\frac{8.5}{7.6})^2]$$

which reduces to

$$1 - [0.350 + 0.141 - 0.319 + 1.25] = -0.421$$

Since this value is less than zero, according to the failure criterion, fracture has occurred. The important observation to be noted here is that the major stress contribution to fracture is from the intralaminar shear stress which is the last term in the brackets. The contribution from the longitudinal and transverse stresses (first three terms in the brackets) tend to cancel each other. It is worth noting that the cancellation tendency observed here is not exhibited when the relative magnitudes are compared on an individual stress basis. The numerical results from the combined-stress failure criterion just discussed lead to the conclusion that fracture of the 10° off-axis tensile specimen is initiated by the intralaminar shear stress.

Therefore, the use of composite macromechanics helped identify the two important features, peak shear strain and shear stress, that induced fracture at the 10° plane. These features led to the recommendation of the 10° off-axis tensile test method for intralaminar shear characterization. Comparisons of intralaminar shear modulus and strength as measured using the 10° off-axis tensile test with literature data are shown in Table II [3]. As can be seen the data from the 10° off-axis tensile specimen are within the range of the literature data.

TABLE 2 - COMPARISON OF MEASURED INTRALAMINAR SHEAR PROPERTIES FROM 10° OFF-AXIS TENSILE SPECIMEN WITH THOSE REPORTED ELSEWHERE [3]

Composite	10° Off-axis tensile specimen		Reported elsewhere				10° Off-axis tensile specimen		Reported elsewhere			
			Low		High				Low		High	
	Modulus						Fracture stress					
	N/cm^2	psi	N/cm^2	psi	N/cm^2	psi	N/cm^2	ksi	N/cm^2	ksi	N/cm^2	ksi
Mod-I/epoxy	0.61×10^6	0.88×10^6	0.44×10^6	0.64×10^6	0.62×10^6	0.90×10^6	5.9×10^3	8.6	4.7×10^3	6.8	6.1×10^3	8.9
T-300/epoxy	.43	.63	.42	.61	.69	1.00	8.3	12.1	6.2	9.0	9.2	13.3
S-glass/epoxy	.65	.94	.57	.83	1.2	1.74	7.1	10.3	4.5	6.5	12	17.1

LAMINATE THEORY

Several examples are described in this section in order to illustrate the impetus of linear laminate theory (LLT) or testing composites and on interpreting composite behavior. These examples include configuring hybrid composite laminates, lamination residual stresses, laminate warpage, and the quasi-isotropic laminate analogy for planar randomly reinforced fiber composites.

Criteria for Configuring Hybrid Composite Laminates

The influence of the constituent plies on the section properties and thermal forces of hybrid composite laminates is best illustrated by briefly examining the general LLT equations for determining these properties:

$$[A], [C], [D] = \sum_{i=1}^{N_\ell} [\int_{Z_{i-1}}^{Z_i} (1,Z,Z^2) [R]^T[E]^{-1}[R]dz]_i \qquad (9)$$

$$\{N_T\}, \{M_T\} = \sum_{i=1}^{N_\ell} [\int_{Z_{i-1}}^{Z_i} (1,Z)\Delta T[R]T[E]^{-1}\{\alpha\}dZ]_i \qquad (10)$$

The notation in equations (9) and (10) is as follows: $[A]$, $[C]$, and $[D]$ denote membrane, coupling and flexural (bending) stiffness matrices, respectively; these matrices are $[3\times3]$ for plane problems and $[5\times5]$ in cases where the transverse (through the thickness) shear deformations are taken into account. The term Z denotes the laminate thickness coordinate referred to some convenient plane; the index i denotes the ith ply in the stacking sequence of the laminate; $[R]_i$ denotes the transformation matrix locating the ith ply material axes (parallel to and transverse to the fiber direction) from the laminate structural axes (coincident with the principal load direction (equation (6a)); $[E]_i$ denotes the ith ply strain-stress relations; $\{N_T\}$ and $\{M_T\}$ denote the thermal forces and moments; ΔT_i denotes the difference between ply and reference temperature; and $\{\alpha\}_i$ denotes the ply thermal expansion coefficients.

Referring to equation (9), it is seen that the constituent plies influence the hybrid section properties (1) through the ply-strain stress relations $[E]_i$, (2) the ply orientation relative to the hybrid structural axes $[R]_i$, and (3) the ply location in the stacking sequence Z_i. Laminate configuration concepts such as the core/shell hybrid and the super-hybrid are readily deduced from equation (9). The ply properties used in equation (9) for interply hybrids are obtained either by measurement or by the use of micromechanics. The ply properties for intraply hybrids are presently obtained by measurement.

The force deformation relationships for a composite laminate are given by

$$\begin{Bmatrix} \{N_c\} \\ \{M_c\} \end{Bmatrix} = \begin{bmatrix} [A] & [C] \\ [C] & [D] \end{bmatrix} \begin{Bmatrix} \{\varepsilon_{co}\} \\ \{\kappa_c\} \end{Bmatrix} + \begin{Bmatrix} \{N_T\} \\ \{M_T\} \end{Bmatrix} \qquad (11)$$

The undefined notation in equation (11) is as follows: $\{N_c\}$ denotes force, or stress resultant at the section, $\{M_c\}$ is the corresponding moment, $\{\varepsilon_{co}\}$ denotes the reference plane strains, $\{\kappa_c\}$ denotes the corresponding curvature and $\{N_T\}$ and $\{M_T\}$ the thermal forces and moments, respectively.

The LLT equation that has been used to predict ply strains in laminates and in hybrid composite laminates may be expressed in matrix form as follows:

$$\{\varepsilon\}_i = [R]_i \, [A]^{-1} \, <\{N_c\} + \{N_T\} + [C] \, \{\kappa_c\}> - Z_i [R]_i \, \{\kappa_c\} \qquad (12)$$

where $\{\varepsilon\}_i$ denotes the strains in the ith ply. The other symbols have been de-fined previously. Note that the thermal moments are included in $\{\kappa_c\}$.

The equation to predict ply stress is obtained by multiplying equation (12) with the ply stress-strain relations and accounting for the free thermal strains. The resulting matrix equation may be expressed as follows:

$$\{\sigma\}_i = [E]_i^{-1} \, <\{\varepsilon\}_i - \Delta T_i \{\alpha\}_i> \qquad (13)$$

where $\{\sigma\}_i$ denotes the stresses in the ith ply of the hybrid, $\{\varepsilon\}_i$ is determined from equation (12), and the other symbols have been defined previously.

Equations (9) to (13) were used to configure the superhybrid composites shown in Figure 8 [5]. Briefly, the concept of superhybrid composites involves the

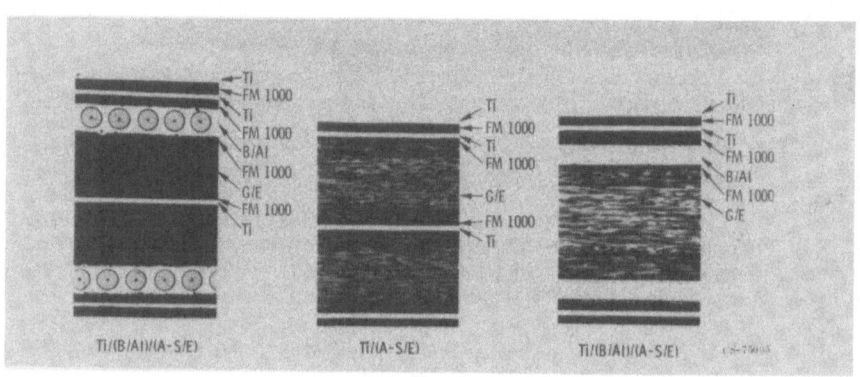

Fig. (8) - Superhybrids. Composite specimen cross sections; [5]

strategic location of the titanium foil and B/Al plies to provide maximum re-sistance to transverse and shear forces. A direct way to assess whether this is achieved in superhybrids is to compute the ply stress influence coefficients due to uniaxial membrane and bending composite stresses. These influence co-efficients are computed using the LLT equations (9) to (13). Selected results obtained for a superhybrid are summarized in Table 3. These results are for a particular ply type as it is first encountered progressing inward from the sur-face. Note that to obtain the ply stress, the influence coefficients must be multiplied by the membrane (bending) stress taken with the correct sign.

TABLE 3 - PREDICTED PLY STRESS INFLUENCE COEFFICIENTS DUE
TO UNIT UNIAXIAL COMPOSITE STRESS FOR SUPER-HYBRID
(Ti, B/Al, Gr/Ep), [5]

| Ply | Uniaxial Membrane Stress | | | | | Uniaxial Flexural Stress [a] | | | | |
| | Longitudinal | | Transverse | | Shear | Longitudinal | | Transverse | | Shear |
	Longitudinal	Transverse	Longitudinal	Transverse		Longitudinal	Transverse	Longitudinal	Transverse	
Top Titanium alloy	0.824	0.032	0.373	1.95	2.08	0.768	0.026	0.161	1.22	1.28
Adhesive	0.011	0.002	0.009	0.028	0.024	0.010	0.002	0.004	0.016	0.014
B/Al	1.63	−0.014	0.184	2.45	2.43	1.12	−0.013	−0.017	1.13	1.10
Gr/Ep	0.912	−0.002	−0.184	0.226	0.206	0.420	0	−0.095	0.070	0.063
Center Gr/Ep	0.912	−0.002	−0.184	0.226	0.206	−0.420	0	0.095	−0.070	−0.063
B/Al	1.63	−0.014	0.184	2.45	2.43	−1.12	0.013	0.017	−1.13	−1.10
Adhesive	0.011	0.002	0.009	0.028	0.024	−0.010	−0.002	−0.004	−0.016	−0.014
Titanium alloy Bottom	0.824	0.032	0.373	1.95	2.08	−0.768	−0.026	−0.161	−1.22	−1.28

[a] To obtain ply stress, multiply influence coefficient by the flexural stress with the correct sign.

As can be observed from the data in Table 3, the titanium foil and B/Al plies have large ply stress influence coefficients for uniaxial transverse and shear composite stresses. Therefore, the titanium foils and the B/Al in the superhybrids provide practically all the resistance for transverse and shear forces. This verifies their role in the superhybrid concept. Note in Table 3 that the ply stress influence coefficients of the adhesive are negligible for all uniaxial composite stresses. Therefore, fracture will occur first in one of the nonadhesive constituents as desired in the superhybrid concept.

Laminates Residual Stresses

The LLT equation for predicting lamination residual stresses in angle-plied laminates is given in [6]

$$\{\sigma\}_i = [E]_i^{-1} \, <[R]_i \, \{\varepsilon_{co}\} - Z_i [R]_i \, \{\kappa_c\} - \Delta T_i \, \{\alpha\}_i> \qquad (14)$$

where $\{\varepsilon_{co}\}$ and $\{\kappa_c\}$ are obtained from equation (11) with $\{N_c\} = \{M_c\} = 0$. Equation (14) in conjunction with composite micromechanics can be used to predict the effects of laminate configuration, fiber volume ratio and void volume ratio on the ply residual stresses. Ply residual transverse stress for two angle-plied laminates from high-modulus/polyimide-matrix composite system versus fiber volume ratio are plotted in Figure 9. Corresponding results versus void

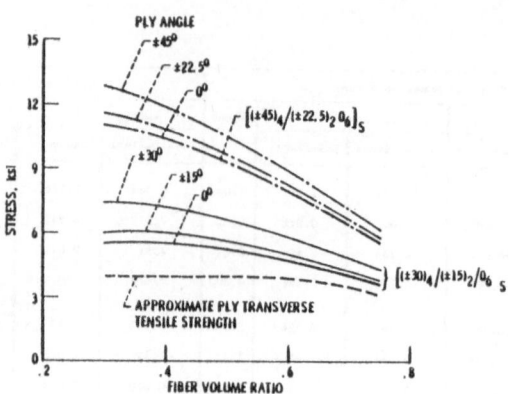

Fig. (9) - Ply residual transverse stress for Modmor-I/polyimide
composites. Temperature difference = -600°F [6]

volume ratio are plotted in Figure 10. The plots in Figure 9 show that the transverse ply stresses are relatively high compared to corresponding strength and will, therefore, cause the type of transply cracks shown in Figure 11. Also,

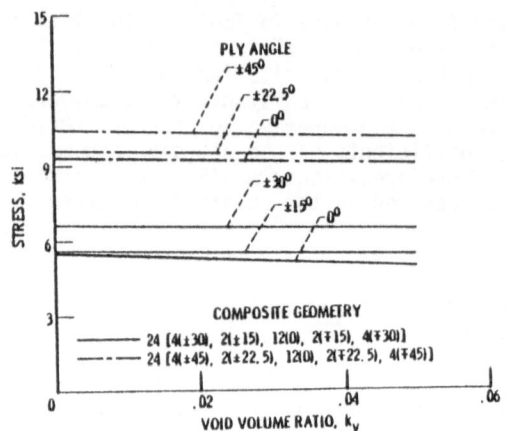

Fig. (10) - Effects of voids on ply transverse residual stress.
Modmor-I/polyimide composites. Fiber volume ratio
= 0.50. Temperature difference = -600°F [6]

the laminate configuration and the fiber volume ratio have a strong effect on the ply residual stress while the void volume ratio has negligible effect.

These results were confirmed by the experimental data of [7 and 8]. The results were also used to recommend laminate configurations for jet engine compressor fan blades to avoid transply cracks [9]

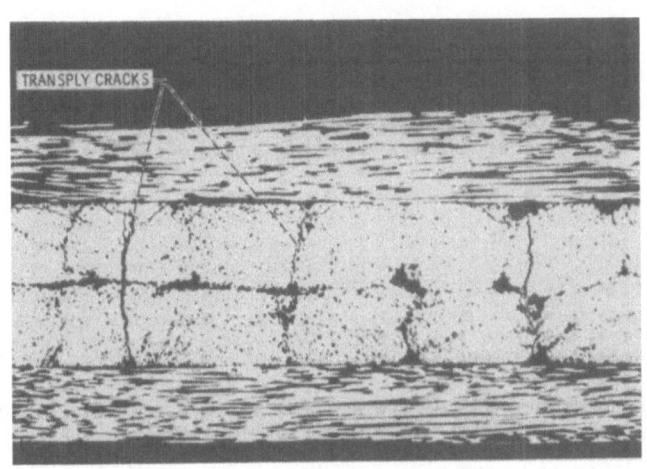

Fig. (11) - Photomicrograph showing transply cracks $(0/90)_S$ high-modulus/epoxy

Laminate Warpage Due to Thermal Stress

Unsymmetric angle-plied laminates will warp when subjected to changes in temperature. Unsymmetries caused by ply misorientation produce warpage in flat laminates upon removal from the mold [10]. A schematic of a warped laminate is depicted in Figure 12. The corner deflection (at point C, Figure 12) is given by

$$w(x,y) = \frac{1}{2} \kappa_{yy} b^2 + \kappa_{xy} ab \tag{15}$$

where the curvatures are determined from LLT equation (11) with $N_C = M_C = 0$. The required equation is

$$\{\frac{\{\epsilon_{co}\}}{\{\kappa_C\}}\} = [\frac{[A]}{[C]} \frac{[C]}{[D]}]^{-1} \{\frac{\{N_T\}}{\{M_T\}}\} \tag{16}$$

where $\{\kappa_C\} = [\kappa_{xx}, \kappa_{yy}, \kappa_{xy}]$, [A], [C], and [D] are given by equation (9) and $\{N_T\}$ and $\{M_T\}$ are given by the LLT equation (10). Equations (15), (16), (10) and (9) can be used to determine the possible degree of ply misorientation in angle-plied laminates which warp due to temperature changes.

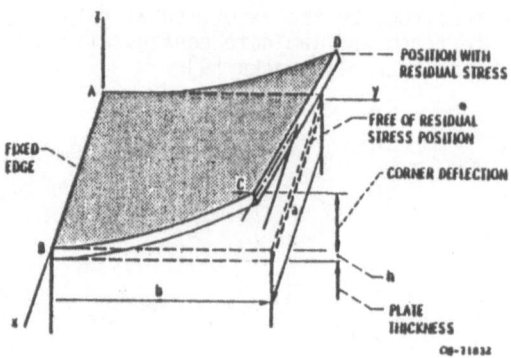

Fig. (12) - Schematic depicting corner deflection due to warpage [10]

This procedure was used to determine the possible ply misorientation in two warped laminates: $[0_2/\pm30]_S$ and $[0_2\pm45]_S$ [10]. These laminates were 30.5 cm (12 in.) square plates and were made from Modmor I/epoxy [10]. These laminates warped when they were cooled from cure temperature (about 461 K (370°F)) to room temperature (294 K (70°F)). The corner displacements measured at point C, Figure 12, were 0.56 cm (0.22 in.) for the $[0_2/\pm30]_S$ angle-plied laminate and 3.05 cm (1.2 in.) for the $[0_2/\pm45]_S$ angle-plied laminate. Note that these warpage corner deflections are relatively large when compared to the laminate thickness of 0.15 cm (0.06 in.). Possible ply misorientations which will yield comparable corner deflections using equations (9), (10), (16), and (15), were as follows: $[0_2/30.4/-30/-30/30/0_2]$ and $[0_2/\pm45/\pm39/0_2]$. As can be seen, the perturbations were 0.4° for the $[0_2/\pm30]_S$ laminate and 6.0° for the $[0_2/\pm45]_S$. These perturbations are relatively small and can be caused inadvertently during the fabrication process. A large number of other possible combinations of ply misorientations exist which will produce comparable corner deflections.

The important point from the above discussion is that LLT can be used effectively to identify problems resulting from the fabrication process.

Quasi-Isotropic Analogy

Linear laminate theory (LLT) can be used to determine the influence of ply misorientation on the modulus and Poisson's ratio (elastic properties) of quasi-isotropic (π/n) laminates. The elastic properties are determined from the array [A], equation (9). The results can then be used to assess the elastic behavior of planar randomly reinforced composites (PRRC) because of the elastic properties equivalence that exists between quasi-isotropic laminates and PRRC [11].

The influence of 5° ply misorientations in the 0° plies of π/n (n = 3,4,6, and 8) quasi-isotropic laminates on composite modulus is shown in Figure 13 and

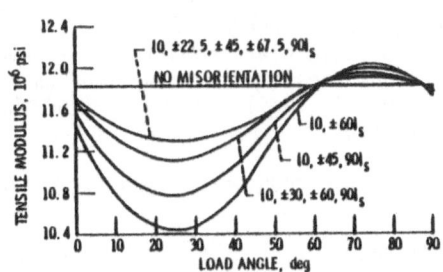

Fig. (13) - Effect of 5° misorientation of 0° plies on the tensile
modulus of quasi-isotropic laminates [12]

for Poisson's ratio in Figure 14 [12]. In these figures the modulus and Poisson's ratio are plotted versus load angle (between load and 0° ply directions) for all four laminates. As can be seen both modulus and Poisson's ratio approach their respective "No Misorientation" straight line as n becomes progressively larger.

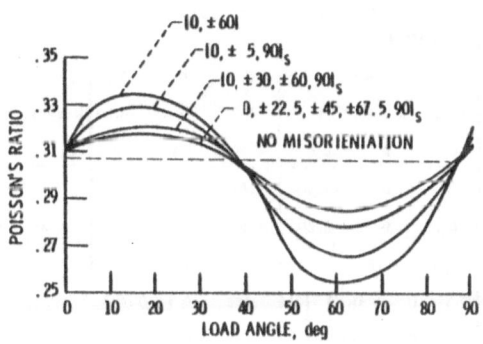

Fig. (14) - Effect of 5° misorientation of 0° plies on the Poisson's
ratio of quasi-isotropic laminates [12]

The important conclusion from the plots in Figures 13 and 14 is that PRRC would require at least 8 different fiber directions at any section through the laminate thickness to achieve isotropic elastic behavior.

Reduced Bending Rigidities for Buckling and Vibration Analysis of Laminates with Coupled Responses

The buckling and vibration analysis of composite laminates having coupled responses such as bending-stretching and/or twisting-stretching requires the solution of the nonlinear anisotropic plate equations. However, approximate buckling loads and vibration frequencies can be determined using the "reduced bending rigidities" method. The method is easily derivable from the LLT equation (11). The details are described in [13].

The governing equation for the reduced bending rigidities is given by

$$[D_R] = [D] - [C]^T [A]^{-1} [C] \tag{17}$$

where $[D_R]$ is the array of the "reduced bending rigidities". The other arrays were defined in equation (11). The values of these arrays are given in Table 4 for a specific laminate. The buckling load obtained using the "reduced bending rigidities" from Table 4 in the computational procedure described in [13] is

TABLE 4 - FORCE DEFORMATION RELATIONSHIPS FOR A $[45_{10}/-45_{10}]$ BORON EPOXY LAMINATE (27.9 cm (11.0 in) x 24.8 cm (9.75 in) x 0.28 cm (0.11 in) [13]

$$\begin{Bmatrix} \{N_c\} \\ --- \\ \{M_c\} \end{Bmatrix} = \begin{bmatrix} [A] & \vdots & [C] \\ --- & \vdots & --- \\ [C] & \vdots & [D] \end{bmatrix} \begin{Bmatrix} \{\epsilon_{co}\} \\ --- \\ \{X_c\} \end{Bmatrix}$$

$$\begin{bmatrix} [A] & \vdots & [C] \\ --- & \vdots & --- \\ [C] & \vdots & [D] \end{bmatrix} = \begin{bmatrix} 10.21 & 8.50 & 0 & \vdots & 0 & 0 & -0.20 \\ 8.50 & 10.21 & 0 & \vdots & 0 & 0 & -0.20 \\ 0 & 0 & 8.62 & \vdots & -0.20 & -0.20 & 0 \\ 0 & 0 & -0.20 & \vdots & 1.01 & 0.84 & 0 \\ 0 & 0 & -0.20 & \vdots & 0.84 & 1.01 & 0 \\ -0.20 & 0.20 & 0 & \vdots & 0 & 0 & 0.85 \end{bmatrix} \begin{array}{l} \times (10^5 \text{ lb/in}^2) \\ \\ \text{or} \\ \\ \times (6.9 \times 10^3 \text{ MPa}) \end{array}$$

$$[D_R] = \begin{bmatrix} D_{11} & D_{12} & D_{13} \\ D_{21} & D_{22} & D_{23} \\ D_{31} & D_{32} & D_{33} \end{bmatrix} = \begin{bmatrix} 552 & 383 & 0 \\ 383 & 552 & 0 \\ 0 & 0 & 431 \end{bmatrix} \begin{array}{l} \times \text{(lb/in)} \\ \text{or} \\ \times (0.0069 \text{ MPa}) \end{array}$$

65.8 kN/m (376 lb/in). This value is in very good agreement with the measured value 65.0 kN/m (371 lb/in) and with that from nonlinear finite element analysis

69.0 kN/m (394 lb/in). The buckling load obtained using orthotropic plate buckling equations such as those in [14] is 154 kN/m (680 lb/in) which is 80 percent higher than the measured value. Vibration frequencies are treated in a similar fashion.

The important conclusion from the above discussion is that LLT was effectively used to obtain a good solution to a complex buckling problem and, therefore, was essential in interpreting properly the experimental results.

SUMMARY

The impetus of composite mechanics on test methods and on the proper interpretation of test results has been reviewed using selected examples. The examples include composite micromechanics, composite macromechanics, and laminate theory. The examples selected demonstrate the following:

1. Composite micromechanics was the essential ingredient required to identify simple tests for identifying resin matrix properties that contribute to improved composite strength and for identifying the major constituent contributors to impact resistance.

2. Composite macromechanics was necessary in the development of the 10° off-axis tensile test for intralaminar shear characterization. Three aspects of composite mechanics that were necessary are: strain transformation, stress transformation and combined-stress failure.

3. Laminate theory is essential for configuring hybrids for improved impact resistance, for assessing lamination residual stress on laminate strength, for identifying possible or inadvertent ply misorientations, for interpreting what may be thought to be low buckling loads of composite plates which exhibit coupling and for identifying sensitive tests to experimentally measure the effects of all of these.

4. Composite mechanics, in general, has contributed significantly to the advancement of composite technology through its impetus on the development of discriminating test methods and through its extensive usage in interpreting test results.

REFERENCES

[1]   Chamis, C. C., Hanson, M. P. and Serafini, T. T., "Criteria for selecting resin matrices for improved composite strength", Modern Plastics, May 1973. Also NASA TM X-68166, 1973.

[2]   Chamis, C. C., Hanson, M. P. and Serafini, T. T., "Impact resistance of unidirectional composites", Composite Materials Testing and Design (Second Conference), ASTM STP 497, American Society for Testing and Materials, 1972, pp. 324-349. Also, NASA TN D-6463, 1971.

[3]   Chamis, C. C. and Sinclair, J. H., "Ten-deg off-axis test for shear properties in fiber composites", Experimental Mechanics, Vol. 17, No. 9, 1977, pp. 339-346. Also NASA TN D-8215, 1976.

[4]   Chamis, C. C., "Failure criteria for filamentary composites", Composite Materials Testing and Design, ASTM STP 460, American Society for Testing and Materials, 1969, pp. 336-351.

[5]   Chamis, C. C., Lark, R. F. and Sullivan, T. L., "Boron/aluminum-graphite/ resin advanced fiber composite hybrids", Materials on the Move, National SAMPE Technical Conference Series, Vol. 6, 1974, pp. 369-385. Also NASA TM X-71836, 1975.

[6]   Chamis, C. C., "Lamination residual stresses in cross-plied fiber composites", Proceedings of the 26th Annual Conference of the SPI Reinforced Plastics/Composite Institute, Section 9-D, Society of the Plastics Industry Incorporated, New York, 1971. Also NASA TM X-52881, 1971.

[7]   Daniel, I. M. and Liber, T., "Measurement of lamination residual strains in graphite fiber laminates", Second International Conference on Mechanical Behavior of Composite Materials, ICM-II, Boston, Mass, August 16-20, 1976.

[8]   Daniel, I. M. and Liber, T., "Effects of laminate construction on residual properties of composites", Society of Experimental Stress Analysis, 1976 SESA Spring Meeting, Paper No. WR-45-1975.

[9]   Hanson, M. P. and Chamis, C. C., "Graphite-polyimide composite for application to aircraft engines", Proceedings of the 29th Annual Conference of the Reinforced Plastics/Composite Institute, Section 16-C, 10, Society of the Plastics Industry, Inc., 1974. Also NASA TN D-7698, 1974.

[10]   Chamis, C. C., "A theory for predicting composite laminate warpage resulting from fabrication", Proceedings of the 30th Annual Conference of the SPI Reinforced Plastics/Composite Institute, Section 18-C, 9, Society of the Plastics Industry, Inc., New York, 1975. Also NASA TM X-71616, 1975.

[11]   Chamis, C. C., "Design properties of randomly reinforced fiber/resin composites", Proceedings of the 27th Annual Conference of the SPI Reinforced Plastics/Composites Institute, Section 9-D, 10, Society of the Plastics Industry, Inc., New York, 1972. Also NASA TM X-67948, 1971.

[12]   Sullivan, T. L., "Elastic properties and fracture strength of quasi-isotropic graphite/epoxy composites", Paper presented at NASA TM X-73592, 1977.

[13]   Chamis, C. C., "Buckling of anisotropic composite plates", Journal of the Structural Division, ASCE, Vol. 95, No. ST10, Proc. Paper 6779, pp. 2119-2139, 1969.

[14]   Lekhnitskii, S. G., Anisotropic plates, Gordon Beach, 1968.

THE NATURE OF CRACK GROWTH IN COMPOSITE MATERIALS

A.R. BUNSELL

Ecole Nationale Supérieure des Mines de Paris

ABSTRACT

Whilst it is often convenient to consider a fibre composite material as a continuous medium which can contain developing cracks as in homogeneous bodies the actual mechanisms of crack growth can be very different. The applications of the ideas of fracture mechanics to this type of material is all too often taken to imply the progressive advance of a crack tip through the material and the creation of two continous fracture surfaces. When failure of the composite is controlled by the matrix material this type of behaviour can be seen but fibre dominated composites do not necessarily fail by simple crack growth across the fibres.

INTRODUCTION

In explaining the failure of a material it is natural to consider existing theories of failure which have been successfully applied to other materials. This has been the case where linear elastic fracture mechanics based on an explication of crack propagation in glass has been adapted to explain failure criteria for metals. Adaptions to the the original theory for the relatively homogeneous and amorphous glass have had to take into account plastic deformation and restrictions imposed by a crystalline structure. In considering the failure of composite materials it is tempting to extend the theories and ideas which have proved useful for glass and for metals to this type of material. This has been done by many authors(1-3) with varying degrees of success. Difficulties have been experienced in determining material parameters which are independant of specimen shape or type of loading and also because fibres lead very often to crack deflection. The application of theories based on the cracking of single phase materials to composites can be profitable but they can also be confusing as they imply a certain mode of crack propagation which for certain composites is far from what happens in reality. It is for this reason that the physical mechanisms of failure of serveral composites will be examined here.

FAILURE OF ASBESTOS CEMENT

Cement reinforced by randomly distributed asbestos fibres is a type of composite material which predates by centuries composite material such as glass fibre reinforced resin. Cement is a brittle material which has very little tensile strength, no ductility and one which cracks easily, it is also a suitable material to which the theories of fracture mechanics can be applied (4). Asbestos fibre are added to cement in order to improve its properties and in particular to inhibit crack propagation. The fibre volume fraction typically used is of the order of six per cent.

If a CT specimen shape is used, Figure 1, it is possible to observe slow crack growth and study the stages of the load-displacement curve shown in Figure 2.

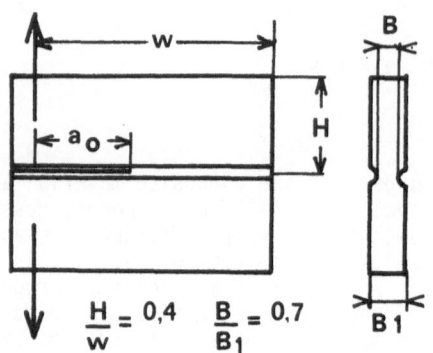

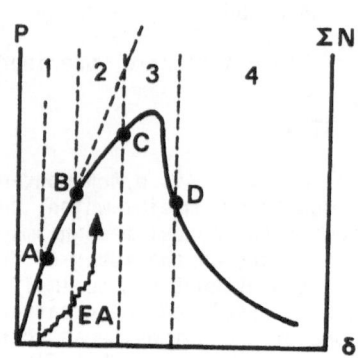

Fig. (1) - Compact tension form of specimen

Fig. (2) - Load/C.O.D. for asbestos cement showing the point at which acoutic emissions begin

Acoustic Emission has revealed that there are four stages to crack propagation in asbestos cement (6). Acoustic emission activity starts at the point A in the linear elastic region of the load displacement curve and shows that at this point a zone of microcracks is created ahead of the notch. The stress intensity factor associated with the creation of this zone we shall designate as Ko. At the point of non-linearity (B) the zone of microcracks has attained a certain size and the crack begin to propagate slowly. During this period of stable propagation the zone of microcracking increases in size until the point C is reached after which it remain constant. During the propagation of the crack it is possible to observe the pull-out of asbestos fibres from the cement matrix across the major crack (Figure 3).

The straining of fibres in the zone of microcracking in front of the crack and the pull out of fibres which bridge it act as energy absorbing mechanisms and increase the toughness and the resistance of the material to crack propagation. These two mechanisms supplement each other at the beginning of slow crack growth but after the point C it is the increasing crack length over which fibre bridging occurs which accounts for a contributing increase in resistance. At the point D the opening of the crack corresponds to a maximum length of fibre bridged crack giving a maximum resistance to propagation. From this point the crack resistance of the material remains constant and as the specimen is tested in an increasing displacement mode it quickly fails.

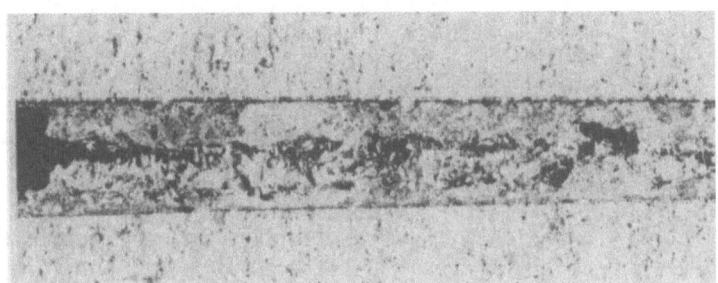

Fig. (3) - Pull out of asbestos bridging a major crack in cement.

As crack growth is occuring in the cement matrix, although modified by the presence of the fibres, this type of composite seems suitable for an approach using linear elastic fracture mechanics which tells us that fracture toughness is related to

$$K_c = \sqrt{\sigma_c^2 a}$$

where σ_c is the breaking stress of the material and "a" the crack length.

Irwin (5) was able to relate the rate of release of energy from the body (G) to K_c and show that

$$K_c^2 = E \, G_c \qquad \text{where E is the modulus of the material.}$$

Unstable crack propagation occurs when the rate of energy released from the body is equal to or greater than the energy required to create the new fracture surfaces.

The presence of the fibres are an inhibiting factor (K_r) as they act to prevent microcracks and the major crack from opening. We can therefore write for the stress intensity factor (K_R) necessary to cause propagation

$$K_R = K_0 + K_r$$

where K_0 is the stress intensity factor of the matrix in the presence of the fibres.

This situation is analogous to the model proposed by Dugdale (7) in which the zone of plastic deformation ahead of a crack in a metal is replaced by a closing pressure over an imaginary crack tip extended into the region ahead of the real crack. For a metal this closing pressure is introduced to retain an elastic model and it accounts for the reduced stresses due to plastic yielding at the crack tip but for asbestos cement the closing pressure (p) is real and arises from the fibre bridges across the major crack and in the region of microcracks.

Paris and Sih (8) here proposed an analytical expression to calculate the additional stress intensity factor introduced by Dugdale so that equation 4 can be rewritten as

$$K_R = K_0 + \frac{P}{\pi} (a + \alpha Z)^{1/2} \left[\frac{\pi}{2} - \sin^{-1} \frac{a_0}{(a + \alpha Z)} + \left[1 + \frac{a_0^2}{(a + \alpha Z)^2} \right]^{1/2} \right]$$

in which a_0 is the original crack length, $a = a_0 + \Delta a$ in which Δa is the measurable crack extension and the zone of microcracks of size Z is represented by an imaginary linear increase of crack length of $Z = \alpha Z$, where $\alpha < 1$, to which the closing pressure P is uniformly applied.

It has been shown that this expression and an approach based on linear elastic fracture mechanics correctly explains crack propagation in asbestos reinforced cement (9). This success of linear elastic fracture mechanics is because crack growth is still similar to that found in the unreinforced brittle matrix but the short asbestos fibres provide means of energy absorption as they inhibit crack opening. Failure of the composite is controlled by the cracking of the cement matrix during which the short asbestos fibres are pulled out of the matrix at the fracture surface.

BORON ALUMINIUM

In contrast to the previous material which is produced at low cost for such applications as corrugated roof coverings and pipes boron aluminium is almost exclusively of interest to the aerospace industry and is far from being cheap. Another difference is in the fibre volume fraction used which is typically 40-50%. The properties of B.Al are therefore dominated by those of the fibres which are continuous and arranged in unidirectional layers in the aluminium matrix. To break the composite it is necessary to break the fibres although it has been found that the mechanics of failure depend on the type of aluminium alloy used as the matrix.

Three types of alloy have been studied, 1200, 6061 and 2024, in ascending order of elastic limit. The importance of cyclic work hardening also increases in this order so that the 1200 matrix shows almost no work hardening where as the 2024 matrix steadily work hardens during cyclic loading.

It has proved extremely difficult to conduct slow crack propagation tests on unidirectional specimens of B.Al in which the fibres lie across the crack path. It is found that the specimen fails either catastrophically or the crack does not propagate or that the crack is deviated and cracking parallel to the reinforcing boron fibres occurs. Tests which simulate impact conditions which might be encountered in service show that cracking is not the immediate result but rather the failure of the boron fibres near the point of impact (10). The matrix which can deform plastically does not fail but the brittle reinforcing fibre fails in bending. Once the fibre has failed the composite is weakened and complete failure can follow.

In order to study crack propagation under cyclic loading B.Al specimens have been tested in repeated circular bending tests. Fatigue tests of notched and unnotched B.Al specimens having the softest matrix (1200) produced cracking parallel to the fibres and originating at a fibre break. Some boron fibres may be broken during the hot pressing manufacturing process and particularly weak fibres at the lower end of the strength distribution curve break during the initial loading of the specimen. This crack propagation is seen normally to occur in the matrix and not necessarily at the fibre matrix interface (11). In this way individual fibre breaks can be isolated and do not lead to points of stress concentration in the composite.

Cracking in the B 6061 specimens also originates at breaks in fibres, but, as Figure 4 shows, thay run at right angles to the fibre direction. When the crack front reach an intact fibre it passes around without producing

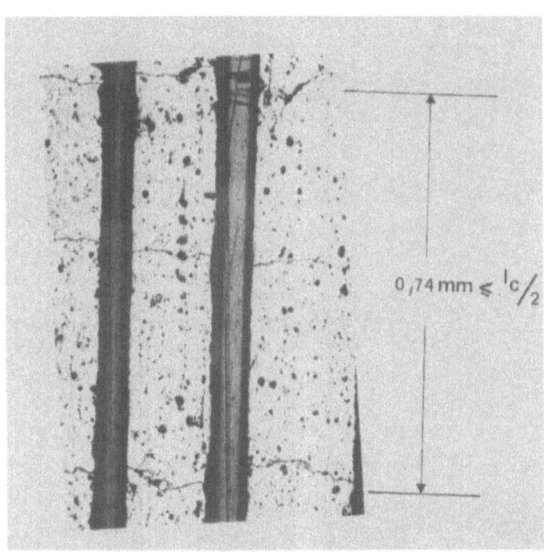

$$0{,}74\,mm \leqslant {}^{l_c}\!/_2$$

Fig. (4) - Cracks in B 6061 which have originate at fibre breaks and
pass around intact fibres.

its failure. In a limiting case one fibre failure could lead to cracking of
the specimen and removal of the matrix contribution over a section of the
specimen through the break. This reduction in strength would be confined to
a section having a width which was some function of the critical load
transfer length (l_c) along the fibre so that we can write

$$\sigma_c \propto \sigma_f \, V_f \Big]_{l_c}$$

where σ_c is the breaking stress of the composite, V_f the fibre volume fraction and σ_f the strength of a bundle of boron fibres of length l_c.
The crack length or the stress concentration produced by the crack does
not come into the expression.

The behaviour of B 2024 is greatly influenced by the higher yield
point of the matrix which results in load concentrations, in fibres neighbouring a fibre break, occuring by load transfer through the matrix. In a
simple bend test it has been observed (11) that a series of fibre failures
can occur in the outer layer which is put most into tension, without
cracking of the matrix. Under cyclic conditions B 2024 is found to deterio-
rate and fibres are found to break. To facilitate the study of fatigue failure mechanisms in B-2024 the material has been further heat treated to
weaken the fibres. Figure 5 shows a series of fibre failures in the outer-
most layer of a specimen which had been subjected to 4×10^4 cycles. Inspection of the fractures reveals that the origin of this damage is one or
two broken fibres and that their neighbours have progressively broken in a
direction radiating outward from the original failure. A second fracture
of one of the fibres illustrates the possibility of multiple failure of
fibres in a composite at a distance equal to or greater than half the critical load transfer length. The development of fatigue damage in heat treated B-2024 is then shown to be

i) fibre failure

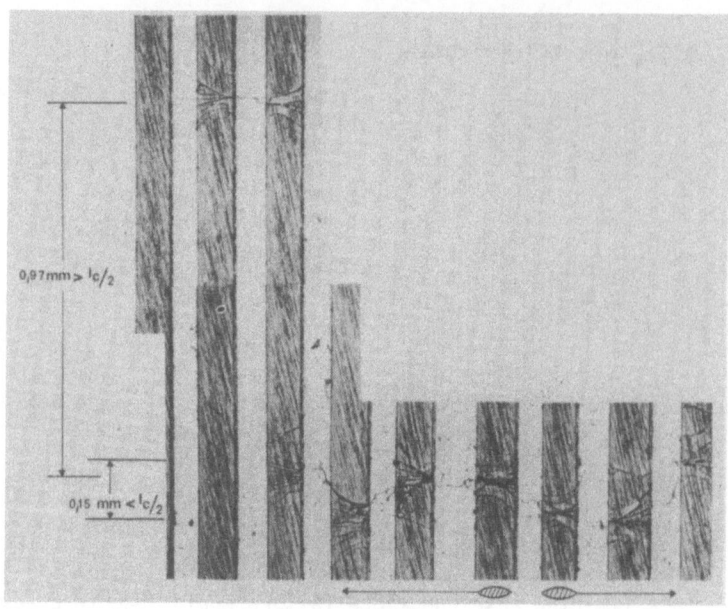

Fig. (5) - Series of fibre breaks in B-2024 produced during 4×10^4
cycles and originating at one of two fibre breaks

ii) neighbouring fibre broken by load transfer through the matrix

iii) the matrix bridge is broken

The fibres do not break because of the influence of a crack in the
matrix but because of load transfer through the matrix resulting from an
initial fibre break. Cyclic work hardening of the 2024 matrix accounts for
delayed fibre failure as it results in increasing load concentrations. It
has been shown (II) that a series of n fibre breaks in a given section
of the material leads to a load concentration in the next intact fibre of

$$(K_n)_e = \frac{4.6.8... \ (2n+2)}{3.5.7... \ (2n+1)}$$

This relationship is for the elastic (e) case and for the plastic case
(p) such as for B-2024 the load concentrations are reduced. Work hardening
of the 2024 matrix gradualy increases these concentrations and results in
further fibre failures.

When "n" adjacent fibres are broken by a defect or a notch of length
"2a" in a total specimen width "w" the stress threshold (σ_{net}) in the

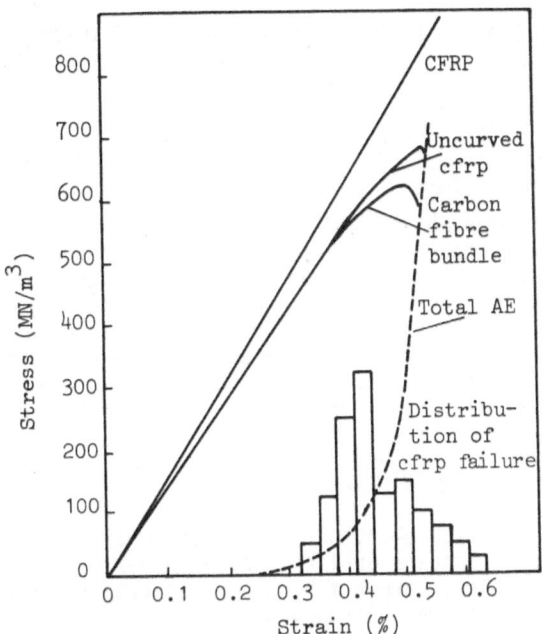

Fig. (6) - Typical stress/strain curves for curved and uncurved cfrp together with a similar specimen from which the matrix has been removed. The dotted AE curve indicates how activity increases with the straining of cfrp specimens.

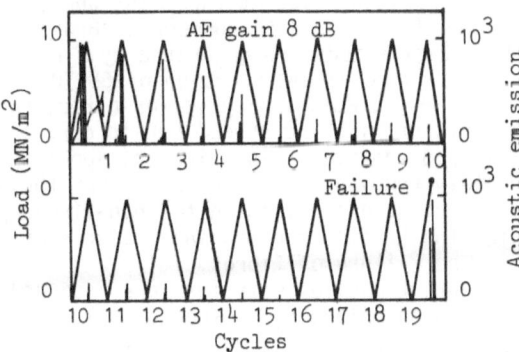

Fig. (7) - AE output from a polar wound cfrp pressure vessel taken through 19 pressure cycles and then burst.

reduced cross section (w-2a) below which damage is not produced in controlled by

$$\sigma_{net} < \frac{\sigma_c^*}{(Kn)_p} \cdot \frac{w}{(w-2a)} \tag{10}$$

where σ_c^* is the failure stress of the composite.

Work hardening produces an increasing elastic limit during cyclic loading of 2024 and the load concentrations produced tend increasingly towards
the fully elastic case $(Kn)_p$ $(Kn)_e$ and so the damage threshold falls.

We have seen then that for the B.Al specimens considered here crack
propagation is very different from that which is found in more homogeneous
materials. For all three types of aluminium alloy failure has been seen to
originate at a fibre failure and for B.2024 the resulting crack propagation
in the matrix can be locally in an opposite direction to the general direction of damage propagation.

CARBON REINFORCED EPOXY RESIN

This material has been welcomed, with the occasional misgiving, by the
aerospace industry and is finding applications in other areas because of
its high specific properties. Carbon fibres possess properties which are
similar to those of boron fibres but they are very much finer having diameter of 7μm as against 140μm for the boron fibres. Because of the far greater number of fibres in carbon reinforced epoxy resin it is not possible to
examine their failure in the same direct way as is possible for boron aluminium although it has been shown that isolated broken fibres can be found
all over the specimen after loading (12). Simple crack propagation through
the matrix and fibres seems unlikely in this material. An indirect technique for following the internal damage in cfrp is acoustic emission which,with
the aid of suitable equipment, allows the stress waves created by the damage mechanisms such as fibre failure to be recorded. It has been shown
that for an unnotched unidirectional specimen the emissions produced during
a monotonic test start to be generated at the point where the weakest
fibres begin to break and increase as the number of broken fibres increases. This means that most emissions are generated by fibre failures. The
composite typically has a fibre volume fraction of 60-70%.

Figure 6 shows the acoustic emission curve for a rectangular unidirectional cfrp specimen together with the stress strain curves of the specimen
and those of similar shaped specimens in which the specimen has left un-
cured and also completely removed. It can be seen that the major part of
the emissions are generated at extensions in the non-linear part of the
stress strain curve for the bundle specimen indicating that the probable
source of emissions in the failure of fibres.

In cyclic loading a great deal of emissions are recorded during the
first loading although non on unloading. A second cycle procudes no emissions until just before the previous maximum load is attained when the
material starts again to emit. Repeated cycling leads to fewer and fewer
emissions each cycle until they almost completely cease (13).

Figure 7 shows the emissions recorded during cyclic pressurisation of a
polar wound cfrp pressure vessel in which the internal pressure resulted
in almost exclusive tensile loads being transmitted to the fibres. Even at
loadings to 90% of the failure load the amount of AE was found to diminish

and eventually to almost cease with cycling. The emission have been shown to be the result of irreversible damage which accumulates during cycling as a monotonic test of the material, once the emissions have almost ceased during cycling, produces no emissions until a load is reached which is approximately four per cent greater than the maximum load experienced during cycling, Figure 8. After this point the rate of emission rapidly

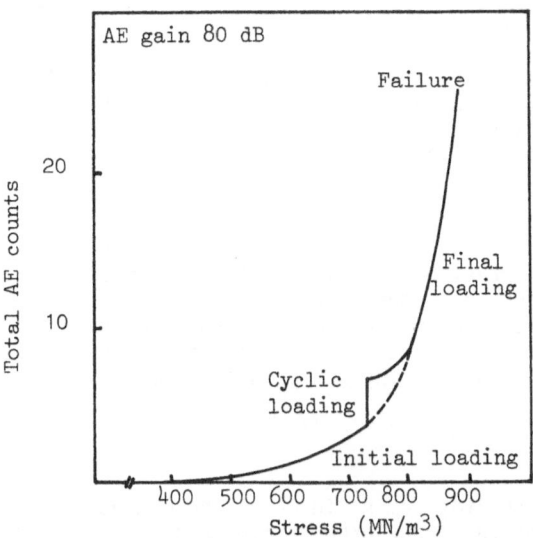

Fig. (8) - A cumulative AE count obtained during a cyclic test followed by a monotonic loading to failure shows that damage occurs during cycling which in a simple tensile test would have occurred at loads slightly higher than the maximum cyclic load.

increases to attain that which would have been recorded at that load if the speciment had simply been monotonically loaded and had experienced no cyclic loads. The conclusion is that if the maximum load experienced during a cyclic test is within about four per cent of the breaking load of the speciment eventual failure will occur because of the accumulation of damage. Maximum loads which are less than this limit will not lead to failure and the specimen will tend to stabilise. This behavour agrees with SN curves which have been obtained for this type of material as they are almost horizontal and lie within the scatter band for the simple tensile strength.

Crack growth does not account for the observed behaviour and the damage which is produced must be the failure of fibres, at first randomly and then in more concentrated regions as the stresses in the specimen sections and around breaks increase and produce further failures. Final specimen failure occurs when a given section is sufficiently weakened by the interaction of different fibre breaks.

The overloading of fibres in the composite occurs in a cyclic test because of the anelastic behaviour of the matrix which results in it supporting less of the load as cycling continues. The time dependant properties of the matrix also result in the increase of the critical load transfer length around each fibre break and the subsequent interaction of fibres breaks with the further weakening of the material.

Under constant load conditions it is found that although no creep can be detected if the specimen is loaded in the direction of reinforcement acoustic emission activity continues long after the loading. The rate of emission reduces with time and the total number of counts may be described by the relation

$$N = A \ Ln \ ((t+\tau)/\tau)$$

where N   is the total number of emissions
      A   a dimensionless parameter, given a constant load
      t the time
      τ a time constant

This formula shows that complete stabilisation and hence cessation of emissions and progressive damage does not occur but fortunately their rate rapidly diminishes. If a specimen is subjected to a brief overloading and then returned to its previous load level it is found that an aging of the material has been induced as the rate of emissions is less. Given further information on the residual properties of cfrp when it is subjected to loading over a period of time the monitoring of acoustic emission activity provides a means of proof testing and predicting minimum lifetimes for this type of material.

CONCLUSION

In the case of asbestos cement in which the cement matrix controls the failure of the composite, crack propagation occurs in a similar manner to that which is found in more homogeneous materials. The use of the theories of linear elastic fracture mechanics has been shown to be appropriate for asbestos cement and the assumptions which this implies have been found to be reasonable. The failure of boron aluminium and carbon reinforced epoxy resine is found to be controlled by the fibres and that, although the matrixes influence the failure of the fibres, it is by transfering load to fibres neighbouring fibre breaks and not by crack propagation through the matrix. Although the accumulation of fibre breaks is analogous to a criterial crack length the mechanisms by which this situation is produced are unlike those found in homogeneous materials.

REFERENCES

1   Piggott, M.R. ; J. Mat. Sci. 5 (1970) 669.

2   Harris, B., Beaumont, R.W.R. and Mon Cunill de Ferran, E. ; J. Mat. Sci. 6 (1971) 238.

3   Sih, G.C., Chen, E.P., Huang, S.L. ; Eng. Fracture Mech. 6(1974) 343.

4   Brown, J.H. ; Mag. of Conc. Res. 12 (1972) 185.

5   Irwin, G.R. ; Appl. Mat. Res. (April 1964) 65.

6   Lenain, J.C. ; Doctorat Thesis, Univ. de Tech. de Compiègne, France (1976).

7   G.I. Barenblatt ; Adv. Appl. Mech. 7, (1962) 55.

8   Paris, B.C., Sih, G.C. ; ASTM STP 381 (1964) 67.

9   Lenain, J.C., Bunsell, A.R. ; to be published in J. Mat. Sci.

10   Rich, G., Bunsell, A.R. ; Proceedings of ICCM-II, Toronto (1978).

11   Leddet, I, Bunsell, A.R. ; Proceedings of ASTM Spring Meeting,
     New Orleans (1978).

12   Fuwa, M., Bunsell, A.R., Harris, B. ; J. Mat. Sci. 10 (1975) 2062.

13   Fuwa, m., Bunsell, A.R., Harris, B., J. Phys. D. 8 (1975) 1460.

APPLICATION OF EXPERIMENTAL METHODS TO FRACTURE OF COMPOSITES

R. E. Rowlands and E. L. Stone
University of Wisconsin, Madison, Wisconsin 53706, USA.

ABSTRACT

The applications of experimental methods, particularly photomechanics and fatigue, to fracture of composites at room and cryogenic temperatures are discussed. Static and impact loadings are included.

INTRODUCTION

Optimum use of composite materials requires that confident design and analysis information be available under extremes of temperature and loading. While much of the approach to composite materials is based on experience with metals, the heterogeneity and anisotropy of composites introduces significantly different behaviors and mechanisms. These additional complexities underscore the need for parallel or combined experimental, numerical and theoretical analyses in composites at both the micro- and macro-level. Certainly fracture, together with its initiation, propagation and controlling parameters, represents a most significant problem of composites, and one that must be well understood and documented.

This paper discusses the applications of experimental methods to the measurement and evaluation of stresses, strains, displacements, wave motion, crack propagation, fracture and fatigue of composites at room and cryogenic temperatures. Conventional and photomechanical experimental methods of high-speed photography, holography and moiré are employed. Glass-reinforced epoxy and polyester man-made composites, and wood, are discussed.

The text draws upon several examples, rather than the detailed analysis of a single problem. Capabilities and limitations of the various experimental methods are emphasized relative to fracture analysis. Compatible with the objectives, many aspects reported herein represent current research at the authors' institute.

EXPERIMENTAL STRESS ANALYSIS

The response of a loaded composite can be represented in terms of the deformations. With respect to the principal material axes of an orthotropic composite under plane-stress, the stresses are given by (1)

$$\left\{ \begin{array}{c} \sigma_1 \\ \sigma_2 \\ \sigma_{12} \end{array} \right\} = \left[\begin{array}{ccc} Q_{11} & Q_{12} & 0 \\ Q_{12} & Q_{22} & 0 \\ 0 & 0 & Q_{66} \end{array} \right] \left\{ \begin{array}{c} \varepsilon_1 \\ \varepsilon_2 \\ \varepsilon_{12} \end{array} \right\} \qquad (1)$$

where the coefficients of the reduced stiffness matrix are

$$Q_{11} = E_{11}/(1 - \nu_{12}\nu_{21})$$

$$Q_{22} = E_{22}/(1 - \nu_{12}\nu_{21})$$

$$Q_{16} = \nu_{21}E_{11}/(1 - \nu_{12}\nu_{21}) \qquad (2)$$

$$Q_{66} = G_{12}$$

and

$$\nu_{12}E_{22} = \nu_{21}E_{11}.$$

In terms of any general orthogonal coordinate system (ξ,η) in the plane, Eq. (1) becomes

$$\{\sigma_{\xi\eta}\} = [\bar{Q}_{ij}]\{\varepsilon_{\xi\eta}\} \qquad (3)$$

where the components of the stiffness matrix $\bar{Q}$ are related to those in terms of the principal material axes and hence the engineering constants of Eq. (2). The strains of Eqs. (1) and (3) are the appropriate spatial derivatives of the in-plane displacements. These equations are valid irrespective of whether the event is static or dynamic, and apply to each ply of a stacked laminate. If discrete fibers and matrix are modelled, the isotropic form of Eqs. (1) through (3) apply to the individual components. The equations obviously become more complicated for other than Hookean response.

Moiré and holography are two optical methods of recording full-field displacements of loaded composites, including those in the neighborhood of cracks. Figures 1A and 1B are moiré fringe photographs of portions of a vertically tensile loaded $[0/\pm45/0/\overline{90}]_s$ glass-epoxy plate containing a central circular hole 2.54 cm. in diameter [2]. The applied stresses were 205.8 and 209.6 MPa, respectively. The photographs were taken with through-the-plate illumination. Individual fringes represent relative vertical displacements of 0.0254 mm. Cracking initiated on the right boundary of the hole in Fig. 1A. It is accompanied by a discontinuous displacement (fringe) field and a fringe concentration at the crack tip. One could similarly record the other in-plane displacement field, differentiate the displacements and thereby compute the strains and stresses in the vicinity of the crack tip. If the strains were assumed uniform through the laminate,

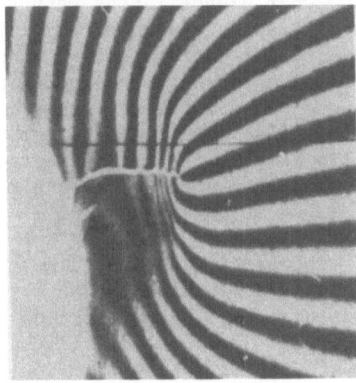

A (205.8 MPa)

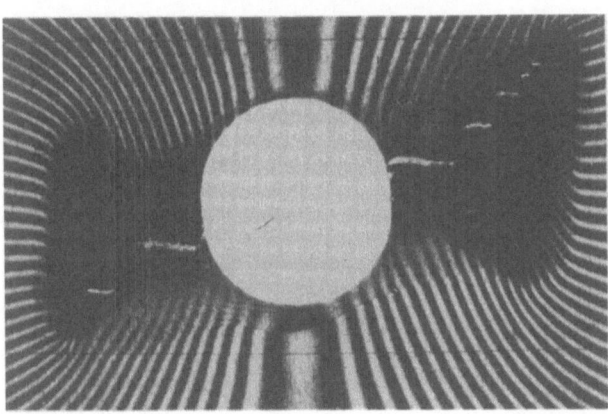

B (209.6 MPa)

Fig. 1 Crack Initiation and Propagation

in $[0/\pm45/0/\overline{90}]_s$ Laminate

then the stresses near the crack could be evaluated in each lamina from the measured surface displacements and application of Eqs. (1) through (3) on a ply-by-ply basis. In view of probable interlaminar stresses generated at the crack and a possible non-normal crack front, the validity of assuming uniform strain through the laminate may be questionable. On the other hand, recognition of the difficulty of measuring the strains of individual plies in the immediate vicinity of the crack during actual fracture may render surface measurements quite acceptable.

Upon increasing the applied load to 209.6 MPa (Fig. 1B), fracture pro-
gressed outward from the hole in a somewhat discontinuous manner. While
individual normal crack propagation occurred fairly horizontally, gross
fracture proceeded in a 45-degree direction. The sharpness of the white
images of Fig. 1 suggests these were indeed through-the-laminate fractures.
Interply fracture may well have occurred between the seemingly discrete,
discontinuous normal cracks of Fig. 1B. Classical fracture mechanics
assumes crack propagation occurs co-linear with the original crack, which
is not the case in Fig. 1. Discontinuous crack propagation in composites
has been reported previously by Wu [3].

Figure 2 illustrates the moiré recorded displacements in the neighborhood of a machined edge crack in a tensile strip of sitka spruce (vertical
grain). The vertical strip was 6.4 cm wide by 0.6 cm thick and the edge
crack was 1.6 cm long and 0.15 cm wide. As identical 197 lines per mm
(5000 ℓpi) horizontal model and analyzing rulings were utilized, the fringes

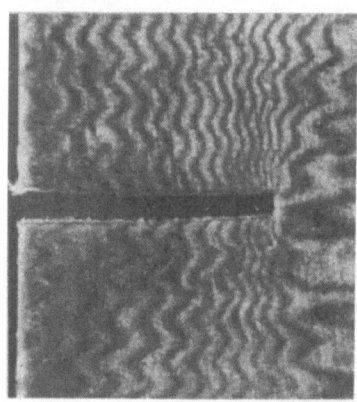

Fig. 2 Moiré (197 ℓpmm) Measured

Displacements in the Neighbor-

hood of an Edge Crack in a

Spruce Tensile Plate

represent a relative vertical displacement of 5×10^{-3} mm. Similar 5000 ℓpi
moiré rulings have been used to determine the strains associated with bolted
joints in wood [4]. This appears to be the first time such dense moiré
rulings have been employed directly for stress analysis. Chiang and Slepetz
previously used moiré to record the displacements in the neighborhood of a
crack in a double-cantilever fiberglass fracture specimen [5].

Although numerous dynamic photomechanical fracture and crack propagation
studies of homogeneous materials have been reported in the literature [6],
applications of dynamic photomechanics to composites are extremely limited.
Daniel recorded transient isochromatic fringes associated with a propagating
crack in a simulated composite tensile model consisting of bonded parallel
strips of glass and plastic [7]. Dally, Link and Prabhakaran [8], and

Rowlands, Daniel and Prabhakaran [9] have recorded transient photoelastic
fringe patterns of impacted composite plates fabricated from glass reinforced polyester. References [9,10] employed moiré and birefringent coatings
to monitor the transient displacements and fracture of impacted glass-,boron-
and graphite-composites and rock. The latter is quite anisotropic and
heterogeneous and, like bone and wood, constitutes a natural composite.

While Kobayashi has recorded, with moiré, the transient strains in the
vicinity of running cracks in metal [11], such has not yet been done for
propagating cracks in composites. Certainly the technology is available,
as demonstrated by Fig. 3. This moiré pattern is one from a sequence of
25 images recorded (250,000 frames per second) of the wave propagation in
a $[0/90/0/0/0/\bar{0}]_s$ glass-epoxy half-plane containing a circular hole. The

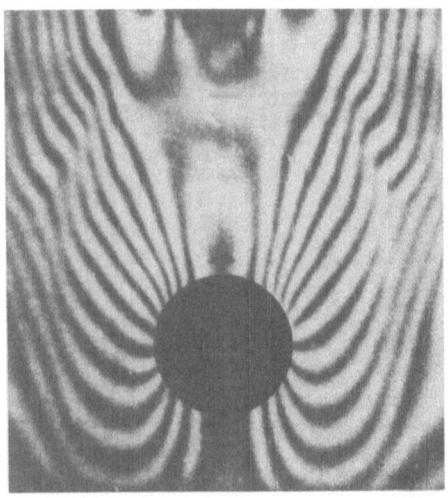

Fig. 3 Dynamic Moiré (39.4 ℓpmm)

for an Impacted

$[0/90/0/0/0/\bar{0}]_s$ Gℓ-E

Composite Containing a Hole

plate was supported along the bottom and explosively impacted at the top.
The fringes again represent relative vertical displacements of 0.0254 mm
and the hole diameter was 3.8 cm. The wave front of Fig. 3 has engulfed
the hole while a reflection from the top of the hole has propagated back
into the plate. High modulus, light-weight composites can possess wave and
crack velocities appreciably higher than metals, necessitating very high-
speed photography. Reference [9] utilized photographic framing rates beyond
10^6 pictures per second to record wave propagation in graphite-and boron-
epoxy composites. It should be noted that running cracks in laminates may
also display complicated situations such as in Fig. 1.

The holographic method of optically recording surface displacements of
actual opaque structural components enjoys the advantage over moiré in that

no surface ruling or pattern must be applied. The method is also extremely
sensitive. However, as suggested by Fig. 4, the quality of holographically
recorded in-plane displacements currently tends to be marginal for strain
determinations around sharp cracks. This figure is a holographic fringe
photograph of the vertical displacements in the vicinity of a pin-loaded
hole in a vertically loaded tensile balsa plate. The hole was 2.54 cm in
diameter. Each fringe represents a relative displacement of 4.5×10^{-5} cm.
This holographic fringe pattern was recorded according to the details contained in Ref. [4]. High-speed, multiple-frame holography is not yet
sufficiently developed for dynamic fracture studies.

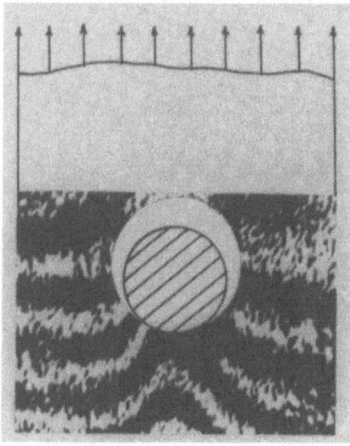

Fig. 4 Holographically Recorded

Displacements in the Vicinity

of a Pin-Loaded Hole in a

Balsa Tensile Plate

Anisotropic transmission photoelasticity requires further development
for studying fracture or crack propagation in composites, although
Rowlands et al. have recorded the isopachics and isochromatics in the
neighborhood of a stationary crack in a glass composite [12]. Classical 2-
and 3-dimensional isotropic photoelasticity has been used to stress analyze
composites at the micro-mechanical level. However, such investigations
involve the inherent uncertainties associated with modelling of the actual
situation. Propagating cracks in a carbon composite have been studied using
a birefringent coating by Green and Pratt [13].

FATIGUE AT CRYOGENIC TEMPERATURES

The high strength and stiffness of fiber reinforced plastics, coupled
with low thermal and electrical conductivity, renders such materials advantageous for superconducting machinery and cryogenic structures. A major

review article on the performance of structural composites at cryogenic temperatures was published by Kasen in 1975 [14]. The existing cyclic compressive fatigue data on unidirectional glass composites are for an epoxy matrix at room temperature [15,16]. At cryogenic temperatures, only one cyclic tensile fatigue study on unidirectionally reinforced glass-epoxy is known to the authors [17].

Compressive fatigue data (cyclic stress normalized with respect to the ultimate compressive strength) for a unidirectionally reinforced glass poly-ester are plotted in Fig. 5 for three temperatures [18]. To simulate the anticipate loading of a superconducting magnet, haversine loading was super-imposed upon a constant baseline to produce a minimum to maximum stress ratio

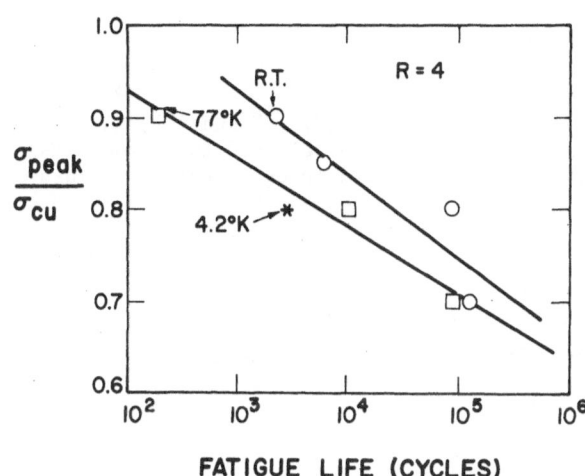

Fig. 5 Compressive Fatigue Data for Unidirectionally Reinforced Glass-Polyester at Room, Liquid Nitrogen and Liquid Helium Temperatures

of R = 4. At least five specimens were tested at each of the different stress levels at room temperature and at 77°K (liquid nitrogen), while three specimens were cyclically loaded at one stress level at 4.2°K (liquid helium). Many of the fatigue specimens failed by mushrooming towards one end of the specimen; others resulted in an approximately 45° fracture line, Fig. 6. This is taken as evidence of compressive failure in shear. Berg and Salama described similar shear-type failure in a unidirectional graphite composite under compressive cyclic loading and discuss the compressive resistance in terms of longitudinal cracking and fiber buckling [19]. The present room temperature fatigue data is intermediate between the compressive cyclic results for S-glass-epoxy of Refs. [15] and [16], Fig. 7. The

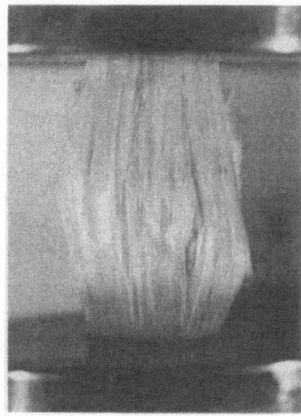

Fig. 6   Failed Glass-Polyester

Compressive Fatigue Specimen

fiber content of the composites of Fig. 7 were 83%, 75% and 72%, in order of decreasing strength. The data of Fig. 5 are believed to be the first 77°K and 4.2°K compressive fatigue data reported on unidirectional glass-reinforced polyester.

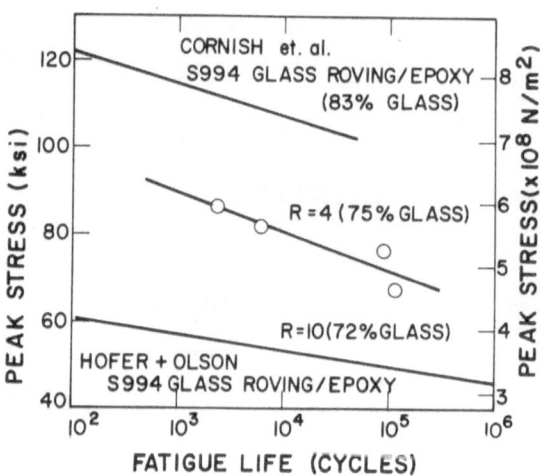

Fig. 7   Comparison of Compressive Cyclic Response

of Glass Composites at Room Temperature

SUMMARY, CONCLUSIONS AND DISCUSSION

Comtemporary technological demands on materials necessitate that fracture information of composites be available under extremes of temperature and loading. Photomechanical techniques are well suited for determining stresses, strains and displacements associated with crack initiation, propagation and arrest. While structural applications of composites typically involve room or elevated temperatures, such materials also find use at cryogenic environments where their mechanical response is less well understood. The compressive fatigue behavior at cryogenic temperatures of unidirectionally reinforced glass polyester is significant because of the material's potential usefulness for structures such as superconducting magnets.

This work suggests at least three areas worthy of future pursuit: (1) extension of photomechanical techniques to cryogenic environments, (2) application of dynamic moiré, and perhaps holography, to crack initiation, propagation and arrest in composites, and (3) evaluation of fatigue and fracture of other composites under cryogenic conditions.

ACKNOWLEDGEMENTS

Aspects of the described work have been sponsored by the U.S. Air Force, the U.S. Energy Research and Development Administration (now DOT), the IIT Research Institute, the U.S. Department of Agriculture and the University of Wisconsin. The paper reflects the authors' associations with I. M. Daniel, L. O. El-Marazki, T. D. Dudderar, E. A. Fuchs, J. Jung, P. Lemens, R. Prabhakaran, T. Richard, J. B. Whiteside, T. L. Wilkinson, and W. C. Young. The manuscript was typed by Mrs. M. Lynch.

REFERENCES

[1] Vinson, J. R. and Chou, T. W., Composite Materials and their Use in Structures, J. Wiley & Sons, 1975.

[2] Rowlands, R. E., Daniel, I. M. and Whiteside, J. B., "Stress and Failure Analysis of a Glass-Epoxy Composite Plate with a Circular Hole," Exper. Mech., 13(1), pp. 31-37, 1973.

[3] Wu, E. M., "Application of Fracture Mechanics to Anisotropic Plates," Jour. Appl. Mech., 34(4), pp. 967-974, 1967.

[4] Wilkinson, T. L., Fuchs, E. A. and Rowlands, R. E., "Photomechanical Determination of Stresses in the Neighborhood of Loaded Holes in Anisotropic Media," Proc. 6th Int'l. Conf. on Exper. Stress Anal., Munich, September, 1978.

[5] Chiang, F. and Slepetz, J., "Crack Measurements in Composite Materials," Jour. Comp. Mat'ls, 7(1), pp. 134-137, 1973.

[6] Experimental Techniques in Fracture Mechanics, Vols. I and II,
     ed. by A. S. Kobayashi, Iowa State Univ. Press, Ames, Iowa, 1973,
     1975.

[7] Daniel, I. M., "Photoelastic Study of Crack Propagation in
     Composite Models," Jour. Comp. Mat'ls, 4(2), pp. 178-190, 1970.

[8] Dally, J. W., Link, J. A. and Prabhakaran, R., "A Photoelastic
     Study of Stress Waves in Fiber Reinforced Composites,"
     Proc. 12th Midwestern Mech. Conf., pp. 937-949, 1971.

[9] Rowlands, R. E., Daniel, I. M. and Prabhakaran, R., "Wave Motion
     in Anisotropic Media by Dynamic Photomechanics," Exper. Mech., 14(11),
     pp. 433-439, 1974.

[10] Daniel, J. M. and Rowlands, R. E., "On Wave and Fracture Propagation
     in Rock Media," Exper. Mech. 15(12), pp. 449-457, 1975.

[11] Kobayashi, A. S., Harris, D. O., Engstrom, W. L., "Transient Analysis
     in a Fracturing Magnesium Plate," Exper. Mech. 7(10), pp. 434-440,
     1967.

[12] Rowlands, R. E., Dudderar, T. D., Prabhakaran, R. and Daniel, I. M.,
     "Holographically Determined Isopachics and Isochromatics in the
     Neighborhood of a Crack in a Glass-Composite," to be published.

[13] Green, A. K. and Pratt, P. L., "The Measurement of the Dynamic Stress-
     Intensity Factor at the Tip of a Running Crack in Some Modern
     Engineering Materials," Proc. 5th Int'l Conf. on Exp. Stress Anal.,
     Italy, pp. 213-220, 1974.

[14] Kasen, M. B., "Mechanical and Thermal Properties of Filamentary -
     Reinforced Structural Composites at Cryogenic Temperatures,"
     Cryogenics, 15(6), pp. 327-349, 1975.

[15] Cornish, R. H., Nelson, H. R. and Dally, J. W., "Compressive Fatigue
     and Stress Rupture Performance of Fiber Reinforced Plastics,"
     Proc. SPI, 19, Section 9-E, pp. 1-22, 1964.

[16] Hofer, K. E. and Olson, E. M., "An Investigation of the Fatigue and
     Creep Properties of Glass Reinforced Plastics for Primary Aircraft
     Structures," N67-32662 or AD652415, 1967.

[17] Tobler, R. L. and Read, D. T., "Fatigue Resistance of Uniaxial
     S-Glass/Epoxy Composite at Room and Liquid Helium Temperatures,"
     Jour. Comp. Mat'ls., 10, pp. 32-43, 1976.

[18] Stone, E. L., El-Marazki, L. O. and Young, W. C., "Compressive Fatigue
      Tests on Unidirectional Glass-Polyester Composite at Cryogenic
      Temperatures," presented at Int'l Cryogenic Mat'ls. Conf., Munich,
      July, 1978.

[19] Berg, C. A. and Salama, M., "Fatigue of Graphite Fiber-Reinforced
      Epoxy in Compression," Fiber Sci. and Tech., 6, pp. 79-118, 1973.

CARBON FIBER SURFACES - CHARACTERIZATION, MODIFICATION AND EFFECT ON THE FRACTURE OF HIGH MODULUS FIBER - POLYMER COMPOSITES

I. L. Kalnin

Celanese Research Company
Summit, New Jersey 07901

ABSTRACT

The types of existing fiber surface treatments, the resulting changes in the fiber surface properties, and the effects of the surface condition on the microfracture of oriented carbon fiber - polymer composites are discussed.

INTRODUCTION

The development of strong and stiff carbon fiber (CF) and its use as the reinforcing element in lightweight structural parts is one of the major technological achievements of the past decade, and the production volume of CF is projected to expand as new applications arise [1]. As the usage of the various CF types increases, so does the understanding of those CF properties that are needed for a good composite performance. This performance often depends on the degree of adhesion between the fiber and the binder phase that in turn depends considerably on the state of the CF surface. Already the early CF development work in Great Britain indicated that an adhesion promoting surface treatment is necessary to realize an adequate composite strength [2]. Since then the number of different surface treatments has increased greatly, although the understanding of their action is still far from complete.

DISCUSSION

Surface Treatment

The finding that surface oxidation of carbon blacks, while changing their initially hydrophobic surfaces into hydrophylic ones, greatly increases their adhesion strength to polymeric materials [3], has been found equally applicable to CF, as a consequence of which the treatments most commonly described and used are oxidative. In addition to being oxidative or non-oxidative, they can be classified further as being carried out in 1) gas phase or liquid phase; 2) at relatively low temperatures (<150°C) or at elevated temperatures, typically >400°C; 3) by chemical, electro-chemical or other means, and finally, 4) whether the treatment results in an overall fiber weight gain or loss. The oxidative gas phase treatments are typically carried out by heating the CF in air at 400-800°C [4], although other oxygen-containing gases, such as CO, CO_2, H_2O, etc., may be used or added to the air to modify the oxidation kinetics and mechanism. The high temperature gaseous treatments will result in surface etching the extent of which must be strictly controlled, lest the filament strength be degraded by excessive pitting [5]. Treatments at lower temperatures, 120-150°C, by ozonized air [6] or at room temperature by plasma excited oxygen or ammonia [7] with a lesser tendency toward pitting are known. Likewise, oxidative liquid phase treatments using chemical or electrochemical means have been explored widely. Of the chemical ones, many aqueous solutions of oxidants, such as nitric acid, hypochlorite [8,9], chlorate or dichromate [10] were found useful. Also, the electrochemical (anodic) surface oxidation being particularly adaptable to continuous operation has been utilized [11,12]. Other surface treatments that improve the fiber resin adhesion by depositing more active forms of carbon have been reported, such as the highly effective, but expensive whiskerization treatment [13], the deposition of pyrolytic carbon [14] or the grafting of adhesion promoting polymeric species [15].

To summarize, then, the main principles for effective surface treatment were well established by 1973. The most successful ones involve controlled oxidation, either from the gas or the liquid phase. These treatments have the advantages that the oxidation 1) can be carried out by means of inexpensive commodity chemicals, 2) removes the original weakly bonded layer of surface carbon, about 15-50 nm thick, and 3) reacts with the surface carbon to form several types of oxygen-containing groups, some of which are strongly chemisorbed to the surface carbon atoms and serve as the bonding sites for the polymeric binder.

Internal and Surface Structure

In order to better understand the interaction of the surface with the binder, one ought to be familiar with the surface structure which is basically similar to the internal structure of the CF. The early model, which represented the CF as a conglomerate of highly oriented interconnected undulating submicron size ribbons [16], has been modified substantially in view of the extensive studies of Diefendorf and coworkers [17], the plasma etching work of Barnet and Norr [18], and the identification of phase heterogeneity, including the possible presence of a three-dimensional graphite phase [19,20]. According to this, the high modulus CF contains at least two phases, both comprised of turbostratic graphite ribbons, plus three-dimensional graphite inclusions when the fiber is heat treated above 2000°C. One of the turbostratic phases consists of wider, thicker, more axially oriented ribbons, and the other one of narrower, less oriented, intertwined ones (fibrils). Both phases contain numerous microvoids, microcracks, and other lattice faults. The less oriented material is present primarily in the central portion of the fiber ("core"), while the more perfect ribbons are located on the outside ("sheath") with a mixed region in between. A current model of such a CF is shown in Figure 1-A, B. Although the surface is the most highly oriented part of the fiber, it contains exposed edge planes as well as basal planes. The sp^2 carbons comprising the high energy edge planes have a free bond and, therefore, will chemisorb oxygen eagerly to form a very strong carbon-oxygen complex [21] that can bond further with the binder resin. The low-energy basal planes that interact by the much weaker π bonding will not form strong bonds at

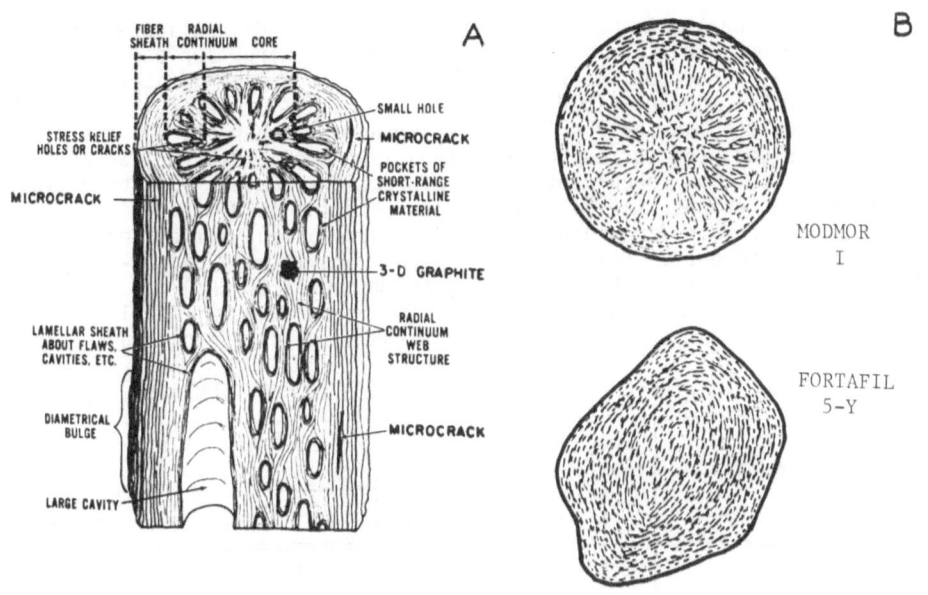

Fig. (1) - (A) Model of axial section of a high modulus ex-PAN fiber, [18]; (B) Model of cross sections of some ex-PAN fibers, [17]

the surface. In addition, the surface exhibits a more or less fibrillar micro-texture, microporosity, cleavage cracks, crystallite boundaries, foreign inclusions or impurities, and fracture-inducing flaws. All of these features will affect the fiber-resin adhesion, and therefore, require characterization in order to assess their significance.

The Surface Properties of CF

A complete characterization of the CF surface would require the knowledge of: 1) the surface composition, i.e., the quantity and nature of the surface atoms and their reactivities, 2) the surface energy and its change upon sorption of the gas or liquid species of interest, 3) the specific surface area, rugosity (roughness), and accessible microporosity, 4) the surface flaws and their distribution and 5) the fiber shape, size, surface texture and anisotropy. Unfortunately, much of the published characterization has been quite limited in scope, the pertinent information being summarized below.

After an oxidative treatment, the surface oxygen content increases slightly, to ca. 100-300 µg/g of C. Part of the oxygen is chemisorbed very strongly onto the most energetic edge carbon atoms to give the before-mentioned surface complexes, the rest forming both acidic groups, or neutral groups, most likely lactone and hydroxyl ones. As mentioned, both -C=O and -C-O- configurations have been identified by ESCA. The portion of the highest energy surface, likely to give the strongest adhesion, decreases sharply with increasing fiber modulus. Thus, for a 250 GPa modulus fiber the values are ca. 4% of the total SSA increasing to 10% after the manufacturer's surface treatment, whereas, for a similar 400 GPa modulus one, the corresponding change is from ca. 0.5 to 1-3% of the SSA, [22]. Judging from the concurrent liquid adsorption studies [23], both types of oxygen-containing groups, the "active complex" and the acidic one, are capable of bonding further to polar organic molecules, such as butylamine and presumably to other epoxy hardeners. However, these groups occupy only ca. 10-12% and 20-25% of the total SSA of the high (400 GPa) and medium (250 GPa) modulus ex-PAN fibers, respectively. For the ex-rayon CF, the percentage may be slightly larger, Table 1. It is evident that the majority of the CF surface atoms must bond to

TABLE 1 - THE CHEMICAL CHARACTERIZATION OF CF SURFACES

Fiber Trademark & Type	Specific Surface Area (SSA) (m^2/g)	"Active" Surface Area* (% of SSA)	Acid Group Content		Oxygen Content ($\mu g/g$)
			(% of SSA)	(μ equiv./g)	
Torayca T-300 [24]	0.55		20	13	400
Torayca M-40A	0.4		10	7	220
Rigilor AC (no S.T.) [25]				7	
Rigilor AC (S.T.)				16-20	
Grafil A-U [22]	0.48	4 ± 1			
Grafil A-S	0.56	10 ± 4			
Grafil HM-U	0.40	0.5			
Grafil HMS	0.42	2 ± 1			
Thornel 50-S [26]	1.1	7	26		
Celion®GY-70	0.75				200

*Surface area occupied by oxygen adsorbed at 340°C on samples after vacuum degassing at 950°C.

the resin through low energy interactions, such as, dispersion (London), dipole-dipole, dipole-induced dipole electrostatic, π-bonding, hydrogen bonding, or donor-acceptor interactions [27] which for CF have not been analyzed. A method based on contact angle measurements of the solids against a series of reference liquids, permitting the separation of the total surface energy into a polar and dispersive component [28], has been applied to type A (both treated and untreated) HTS, HMS, and Modmor II CF [29]. The resulting total surface energies were rather low and about the same, 53-57 mJ/M², with the polar and dispersive contributions divided about evenly. Since these values were practically the same for surface treated and untreated fiber, they probably represent the surface energy associated with the basal planes.

Dimensional properties, primarily the SSA, and the micropore volume, are normally measured by low temperature sorption of nitrogen, argon, or krypton. Unless the CF has been vacuum degassed at ca. 900°C, the micropore volume is small, but can increase substantially when treated with some oxidants, like HNO_3 [22]. For most commercial ex-PAN fibers, the SSA ranges from 0.3 to 0.7 m^2g; for the ex-rayon ones - 1 to 2 m^2/g. Finally, the combined effect of the SSA and the chemical functionality can be assessed by measuring of the surface charge density [30] or surface capacitance [31] of CF immersed in aqueous electrolyte solutions.

On basis of the surface properties, the surface treatments can be divided into 1) those that increase the oxygen content and chemical functionality with little or no increase in the SSA, and 2) those that increase both the SSA and the oxygen content. The second kind of treatment is typically applied to the high modulus fiber when a relatively large, 3-4 fold, increase in their strength is desired.

Flaws

Since the strength of CF has been correlated with its flaw content and severity, the appearance of the flaws has been studied extensively, primarily by electron and optical microscopy, e.g., [32], also references cited therein. The flaws are found both in the interior and at the surface. The interior flaws that are mostly particulate inclusions, round or needle-like voids, and a combination of both, are related to the quality of the precursor fiber [33]. The surface flaws on the other hand are cracks and grooves, apparently incurred during the fiber manufacture, small highly graphitic "microflakes", as well as the above mentioned inclusions and microvoids located at the surface. Surface treatments, capable of reducing the severity of the surface flaws to raise the mean fiber tensile strength by up to 50%, have been reported [34]. The effect of these flaws manifests as a distribution of CF fracture strength that can be described statistically, as has been done for fiberglass [35].

The Effect of CF Surface Condition on Composites' Fracture

The interfacial fracture behavior is determined not only by the surface structure of both components, but also by any additional structural features introduced during the interface formation. Although the surface of the epoxy resin binder, after contact with the CF surface, apparently has not been investigated, studies on other polymers, nylon or PE, show that these form epitaxial, highly oriented, strongly adherent coatings on CF surfaces [36]. Such coatings are likely to have mechanical and chemical properties different from those of the bulk resin to change the expected residual stress pattern as well as the adhesive strength. Furthermore, proprietary binder systems are often modified with reactive additives to increase the fracture toughness, to change the Tg, improve the wettability, etc. Also, many CF grades sold have been prewetted with proprietary sizing agents by the manufacturer in order to facilitate the fabrication of preimpregnated goods ("prepregs"). In addition, the interface may contain numerous microvoids (1-3 fiber diameters in size), caused by incomplete wetting as well as large microvoids formed by coalescence of the former. Any of the above phenomena will affect the interfacial adhesion strength and make the contribution of the original CF surface more difficult to assess.

Shear

In composites, the effect of the surface treatment manifests itself most
dramatically as a greatly enhanced shear strength. About 30-40% of the total increase appears to be due to the removal of the original weak boundary layer on
the as-manufactured surface, and the rest is caused by a) the strong bonding of
the acidic oxygen-containing groups and the weaker bonding of the basal oxygen-
containing groups and atoms with the monomers of the resin, and b) increase surface rugosity to give a larger SSA to bond with that arises from preferential
oxidation of the edge crystallites, exposing additional edge and basal planes.
For the medium modulus (250 GPa) CF, having a relatively large edge carbon content, 20-30% of total surface, a mainly adhesion-promoting surface treatment may
be adequate, whereas for the higher modulus (>400 GPa) CF a treatment increasing
the rugosity is usually necessary in order to obtain acceptable shear strength.
A larger than 2-3 fold increase in the SSA, however, is likely to result in reduced tensile strength due to increased severity of the etched-out surface flaws.

As the rugosity increases, the interfacial stress which is pure shear on a
perfectly smooth surface becomes a combined stress of shear and tension, or shear
and compression when acting on the edgewise or basally exposed crystallites inclined at different angles with respect to the fiber axis. Consequently, the
difference between the shear adhesion and tensile adhesion strength ought to decrease with increasing rugosity of the fiber surface. Examination of manufacturer's data indicates that the difference between the composite short beam shear
strength and transverse tensile strength is indeed the least for the more rugous
high modulus CF, such as Celion®GY-70 (30-40 MPa), becoming larger for the less
rugous medium modulus type II CF (40-55 MPa) and still larger for the S-glass reinforced epoxy, (50-70 MPa). Although these data are obtained on composites
comprised of different epoxy systems and tested under different conditions, they
may nevertheless be indicative of the trend. Consequently, the differentiation
between the bond strengths in tension and in shear, maintained in micromechanical
models [37], is less meaningful for CF with high rugosity. Among the techniques
that have been proposed and used to measure the interfacial shear strength, the
so-called "horizontal shear strength" test, ASTM D 2344, also called "short beam
shear" test (SBS), is by far the most popular because of the extreme simplicity
in specimen preparation and testing. Unfortunately, it is highly unsatisfactory
theoretically [38], and is recommended for quality control purposes only. Nevertheless, its use persists, and shear data on other than short beam shear specimens are rather scarce in the literature. Other techniques, utilizing oriented
composite test specimens, Figure 2, are available, but none has been unequivocally
accepted. New methods are being proposed continually, e.g., [39], and may eventually replace the SBS method. Typical SBS strength values range from 55-70 MPa
for the high modulus (>400 GPa) fibers, and 100-125 MPa for the medium modulus
(250 GPa) ones, as compared to ca. 15 and 50 MPa, respectively, before the surface treatment. Intermediate values may be obtained, if desired, by manipulating
the surface treatment process conditions.

Methods to measure the shear adhesion of single filament interfaces, such as
the shear debond method [40], or the fiber pullout [41], Figure 3, used successfully on fiberglass, have been less successful with the CF. The first method caused
premature fiber fracture [42], while the second one gave shear strengths of 5 and

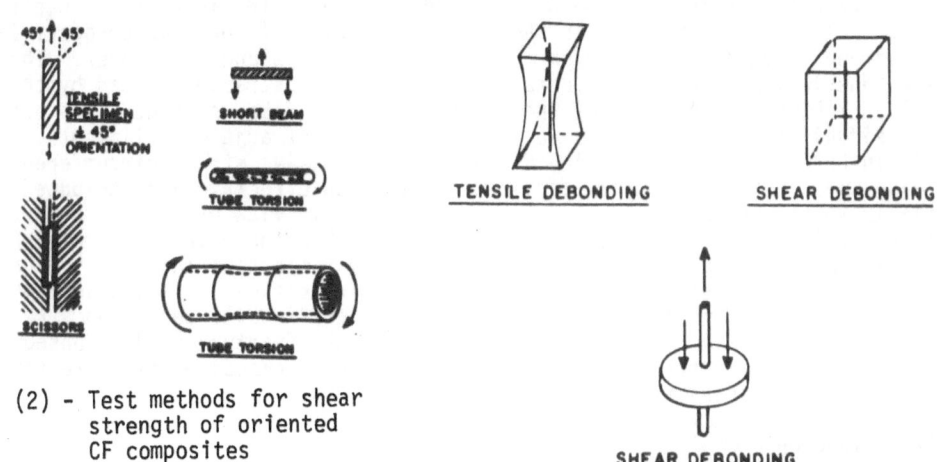

Fig. (2) - Test methods for shear
strength of oriented
CF composites

Fig. (3) - Test methods for filament-resin
adhesion strength

17 MPa for untreated HM and HT fibers, respectively, and 31 and 57 MPa for the
treated ones [43] - about half of their respective SBS shear strength values.
Taking into account the inverse dependence of the pullout strength on the imbedded
length might possibly give higher shear values [44]. Another approach utilizes
measurement of the fragment lengths of fractured resin-imbedded filaments to determine the so-called "critical length" (ineffective length, shear transfer
length), L_c, to calculate the maximum shear adhesion stress τ [45]. Since the CF
filament strengths show a broad distribution, the fragmented lengths will also
reflect this distribution, so that the determination of the L_c and hence the τ becomes quite complicated. Despite that, two estimates of L_c are available, both
for the RAE type I CF (now discontinued). For the treated fiber the L_c was ~0.1
-0.14 mm [46], or ~0.14 mm, as compared to ca. 1.2 mm for an untreated one [47].
The calculated shear adhesion strengths were 47 MPa and 6 MPa respectively, again
substantially lower than similar SBS values. When comparing the shear data obtained on single filament and composited filament bundles, it should be remembered,
however, that the axial shear adhesion depends also on the filament misalignment
[48], the void content [49] and the chemical nature of the binder. The misalignment is expected to increase the apparent shear, the voids will decrease it, whereas the effect of the binder is often quite specific and may be either beneficial
or detrimental.

Transverse Tension

The ultimate tensile strength of unidirectional composites increases after a
typical (manufacturer's) surface treatment from ca. 15 to 30 MPa (to 0.35-0.40%
ultimate strain) for the >400 GPa modulus fiber composites and from ca. 20 to 60

(to 0.5-0.6% ult. strain) for the 250 GPa modulus ones. The increase is due to the same causes as the increase in shear, i.e., removal of the weak boundary layer and strengthened bonding between the functional groups and the binder. The tensile adhesion, however, is affected by residual microstresses caused by the thermal shrinkage and the cure of the resin. For high fiber contents, the estimated residual stress is compressive in the direction of adjacent filaments, and tensile in between, [50], imparting a non-uniform adhesive stress circumferentially. Also, microvoids and matrix strain magnification effects can contribute substantially to the transverse failure [51]. However, in view of the low observed fracture strain values that are 6-10 times smaller than the bulk fracture strain of typical epoxy resins, the transverse fracture must start at interfaces. Unfortunately, detailed studies of the nature of the interfacial tensile adhesion in CF composites are lacking. Attempts to measure the tensile debonding strength of imbedded single filaments failed owing to premature fracture [42]. Consequently, the transverse tensile strength data of unidirectional composites is currently the only means to assess the tensile adhesion strength.

Axial Tension

Although surface treatments have been reported to increase the fiber tensile strength, ordinarily the effect is small, 10% or less, and has little significance for increasing the composite axial tensile strength, S_a. Without a surface treatment, however, the fracture strength is relatively low, because as the applied stress increases the weak fiber-resin interface debonds and fails prematurely. When the composite shear strength is at or above a certain fraction, ca. 5% of the S_a, the latter is independent of the shear level. At still higher shears, 8-10% of the maximum S_a, the strength decreases, not only because of possible surface damage by over treatment, but also by the inability of the fibers to fully dissipate their fracture strain energy into the pullout, resin cleavage, or debonding modes [45], resulting in more extensive microfracturing and subsequent failure at a lower applied stress.

Compression

Axial compression studies on untreated unidirectional CF composites indicated a need for increased interfacial bonding, since the loaded filaments were observed to undergo transverse tensile debonding simultaneously with filament fracture [52], but the optimal adhesion strength level is not known. Voids, residual stresses, fiber misalignment and the intrinsic fiber structure, however, exert a substantial effect on the compressive strength [53], and may negate attempts to assess the effects of increased adhesion. No experimental studies were found in the literature concerning the effect of interface on the transverse compression.

Impact and Fracture Toughness

Charpy impact resistance studies of uniaxial CF specimens showed that high shear (manufacturer's surface treated) specimens have very low impact resistances, and a SBS strength of 50-60 MPa (intermediate adhesion) gives the best balance between the overall failure characteristics [54]. Also important for the impact toughness were - high tensile strength and fiber content. A study of fracture toughness of uniaxial composites, made with type I RAE CF (now discontinued), showed that at a 60 vol. % fiber content the K_{Ic} of surface treated fiber composites was about half of that of untreated ones, ca. 25 versus 45 $MN.m^{-3/2}$, but that at lower fiber contents the difference was greater [46]. Since for the CF composites the fiber pullout energy appears to be the major contributor to the fracture toughness [55], the detrimental effect of strong shear adhesion is probably due to decreased shear transfer length.

SUMMARY AND CONCLUSIONS

A survey of surface treatments for carbon fibers shows that the predominating ones are those involving surface oxidation. The state-of-the art treatments, applied routinely to commercial CF, provide 2-4 fold increases in the fiber resin adhesion strength, both in shear and in tension. In composites, this adhesion normally is fully adequate for obtaining satisfactory axial tensile performance. However, it is too high for optimal impact resistance or fracture toughness, and too low for satisfactory transverse tensile strength. Consequently, in order to utilize effectively any further increases in the interfacial adhesion, the CF composites should be toughened first, e.g., by hybridization. Alternately, the transverse tensile strength might be increased by development of tougher and more compatible binder resins. Some promising approaches toward this objective have, in fact, been recently reported [56,57].

REFERENCES

[1] Anon. "Graphite composites from space to auto", Automot. Ind., 154, 9, p. 39, 1976.

[2] Clark, D., Wadsworth, N. J. and Watt, W., Carbon fibres, (Plastics & Poly. Conf. Suppl. No. 6), The Plastics Institute, London, pp. 44-51, 1974.

[3] Amon, F. H., US Pat. 2,439,442, April 13, 1948.

[4] Johnson, J. W., J. Appl. Poly. Sci., Applied Poly. Symposia, 1969, 9, pp. 229-243, Cunningham, A. L., US Pat. 3, 816, 598, June 11, 1974.

[5] Molleyre, F. and Bastick, M., 4th London Intern. Conf. on Carbon & Graphite, Sept. 23-27, 1974, Society of Chemical Industry, London, pp. 190-200, 1976.

[6] Kalnin, I. L., US Pat. 3,723,607, March 27, 1973.

[7] Goan, J. C., US Pat. 3,634,220, January 11, 1972; US Pat. 3,776,829, December 4, 1973.

[8] Wadsworth, N. J. and Watt, W., US Pat. 3,476,703, November 4, 1969.

[9] Barr, J. B., US Pat. 3,791,840, February 12, 1974.

[10] Goan, J. C., Joo, L. A. and Sharpe, G. E., Proc. Ann. Tech. Conf. of RP/Composites Inst., The Society of Plastics Ind., New York, Vol. 27, 21-E, 1972.

[11] Ray, J. D., Steingiser, S. and Cass, R. A., US Pat. 3,671,411, June 20, 1972.

[12] Chapman, D. R. and Paterson, W. C., US Pat. 3,759,805, September 18, 1973.

[13] Shyne, J. J. and Milewski, J. V., loc. cit. in [10], Vol. 24, 18-D, 1969.

[14] Pinchin, D. J. and Woodhams, R. T., J. Mater. Science, 9, pp. 300-306, 1974.

[15] Riess, G., Bourdeaux, M., Brie, M. and Jouquet, G., loc. cit. in [2], pp. 52-57.

[16] Fourdeaux, A., Perret, R. and Ruland, W., Carbon fibers, (Plastics & Poly. Conf. Suppl. No. 5), The Plastics Institute, London, pp. 59-67, 1971.

[17] Diefendorf, J. R. and Tokarsky, E., Poly. Eng. Sci., 1975, 15, (3), pp. 150-159; Lemaistre, C. W. and Diefendorf, R. J., SAMPE Quart., 4, (4), pp. 1-6, 1973.

[18] Barnet, F. R. and Norr, M. K., loc. cit. in [2], pp. 32-43.

[19] Johnson, D. J., Crawford, D. and Jones, B. F., J. Mater. Sci., 8, pp. 286-289, 1973.

[20] Wicks, B. J., ibid., 6, pp. 173-175, 1971.

[21] Laine, N. R., Vastola, F. J. and Walker, P. L., Jr., J. Phys. Chem., 67, pp. 2030-2034, 1963.

[22] Rand, B. and Robinson, R., Carbon, 15, pp. 257-263, 1977.

[23] Idem, pp. 311-315.

[24] Nichioka, Murayama, K. and Matsubara, I., (in Jap.), 1st Symposium on Composite Materials, Tokyo, September 19-20, 1975.

[25] Ehrburger, P., Herque, J. J. and Donnet, J. B., loc. cit. in [5], pp. 201-208.

[26] Moller, P. J. and Fort, T., Jr., Coll. & Poly. Sci., 253, pp. 98-103, 1975.

[27] Fowkes, F. M., J. Adhesion, 4, pp. 155-159, 1972.

[28] Kaelble, D. H., Physical Chemistry of Adhesion; Inter-Science, New York, pp. 139-170, 1971.

[29] Kaelble, D. H., Dynes, P. J. and Cirlin, E. H., J. Adhesion, 6, pp. 23-48, 1974.

[30]   Dietz, R. and Peover, M. E., J. Mater. Sci., 6, pp. 1441-1446, 1971.

[31]   McLean, A. F. and Kalnin, I. L., US Pat. 3,864,626, February 4, 1975.

[32]   Sharp, J. V., Burnay, S. G., Matthews, J. R. and Harper, E. A., loc. cit. in [2], pp. 25-31.

[33]   Thorne, D. J., J. Appl. Poly. Sci., 14, pp. 103-113, 1970.

[34]   Johnson, J. W., J. Appl. Poly. Sci., - Symposia, No. 9, pp. 229-243, 1969.

[35]   Schmitz, G. K. and Metcalfe, A. G., loc. cit. in [10], 20, 3-A, 1965.

[36]   Baer, E., Koenig, J. L., Lando, J. B. and Litt, M. H., loc. cit., in [10], Vol. 26, 20E, 1971.

[37]   Skudra, A. M. and Kirulis, B. A., Mekh. Polimerov, 10, pp. 246-251, 1974.

[38]   Berg, C. A., Tirsoh, J. and Israeli, M., in Composite Materials: Testing and Design, 2nd Conf., ASTM Spec. Tech. Publ. 497, pp. 206-218, 1972.

[39]   Chamis, C. C. and Sinclair, J. H., Exp. Mech., 17, pp. 339-346, 1977.

[40]   Broutman, L. J., loc. cit. in [10], Vol. 25, 13-B, 1970.

[41]   Andreevskaya, G. D. and Gorbatkina, Y. A., ibid., paper, 16-F.

[42]   Hawthorne, H. M. and Teghtsoonian, E., J. Adhesion, 6, pp. 85-105, 1974.

[43]   Favre, J. P. and Perrin, J., J. Mater. Sci., 7, pp. 1113-1118, 1972.

[44]   Andreevskaya, G. D., Gorbatkina, Y. D., Ivanova-Mumzhieva, V. G. and Epifanova, S. S., Mekh. Polimerov, 10, pp. 37-42, 1974.

[45]   Kelly, A., Strong Solids, 2nd ed., Clarendon Press, Oxford, G. B., p. 172 et ff, 1973.

[46]   Beaumont, W. R., J. Adhesion, 6, pp. 107-137, 1974.

[47]   Wadsworth, N. J. and Spilling, I., Brit. J. Appl. Phys. (J. Phys. D.), 2, Vol. 1, pp. 1049-1058, 1968.

[48]   Claus, W. C., loc. cit. in [10], Vol. 27, 9-C, 1972.

[49]   Hancox, N. L., J. Mater. Sci., 12, pp. 884-892, 1977.

[50]   Chamis, C. C., Composite Materials, Vol. 6, E. P. Plueddemann, ed., Academic Press, New York, p. 55 ff., 1974.

[51]   Greszczuk, L. B., Composite Reliability, ASTM Spec. Tech. Publ. 580, pp. 311-325, 1975.

384

[52] Hancox, N. L., J. Mater. Sci., 10, pp. 234-242, 1975.

[53] Hawthorne, H. M. and Teghtsoonian, E., J. Mater. Sci., 10, pp. 41-51, 1975.

[54] Bader, M. G., Bailey, I. E. and Bell, I., Brit. J. Appl. Phys., (J. Phys. D), 6, pp. 572-586, 1973.

[55] Marston, T. U., Atkins, A. G. and Felbeck, D. K., J. Mater. Sci., 9, pp. 447-455, 1974.

[56] Stevens, G. T. and Lupton, A. W., J. Mater. Sci., 11, pp. 568-570, 1976.

[57] McGarry, F. J., Rowe, E. H. and Riew, C. K., loc. cit. in [10], Vol. 32, 16-C, 1977.

SOME INTERESTING MECHANICAL BEHAVIORS OF FIBER COMPOSITE MATERIALS

T. T. Chiao

Lawrence Livermore Laboratory, University of California
Livermore, California 94550

ABSTRACT

In the course of our studies of the mechanical behaviors of aramid/epoxy and the high-strength glass/epoxy composites, we have observed two interesting but puzzling behaviors. First, when a sustained tensile load is removed any time before stress-rupture failure, the composites retain almost their full original strengths. Second, the contribution of a high-elongation, low-modulus matrix to the performance of a simple glass composite seems to be surprisingly little. These results have raised questions about the validity of the strength retention test as well as the usefulness of a low-modulus matrix in fiber composites.

INTRODUCTION

We at LLL have been studying the mechanical behavior of many composite material systems for nearly a decade. We have observed two puzzling but interesting behaviors of aramid/epoxy and glass/epoxy composites. First is the lack of correlation between the lifetime (also called stress-rupture, static fatigue, or sudden death) and the strength retention of a composite; a high retained strength does not guarantee long life of a composite. Second is that a matrix with long failure strain does not seem to improve the composite performance significantly, not even the failure strain of a composite in the transverse direction. In spite of the fact that composites are rapidly being used in large-volume applications, we do not believe that these two behaviors of composites are widely known. Here we will attempt to clarify the subject of strength retention vs stress-rupture and the role of high-elongation, low-modulus matrix for a composite.

STRENGTH RETENTION vs STRESS-RUPTURE

Strength retention is the ratio of remaining strength to the original strength of a material; it is a measure of short-time static material strength. This test is often associated with environments such as stress, temperature, humidity, and time, either individually or in various combinations. Stress-rupture, on the other hand, is a time-dependent property of a material under a sustained and constant load. Thus it is a measure of the lifetime of a material for a given stress and environment.

Simple fiber/epoxy composite strands were tested both for strength retention and for stress-rupture by Chiao and Moore 1,2 on the apparatus illustrated in Fig. 1. In the strength retention test, the composite specimen is held under a constant load for a given length of time. After the constant stress is removed from the composite specimen, the specimen is further loaded to failure on a separate universal test machine. In the stress-rupture test, the specimen is loaded to the constant stress and only the lifetime to failure under this sustained stress is recorded.

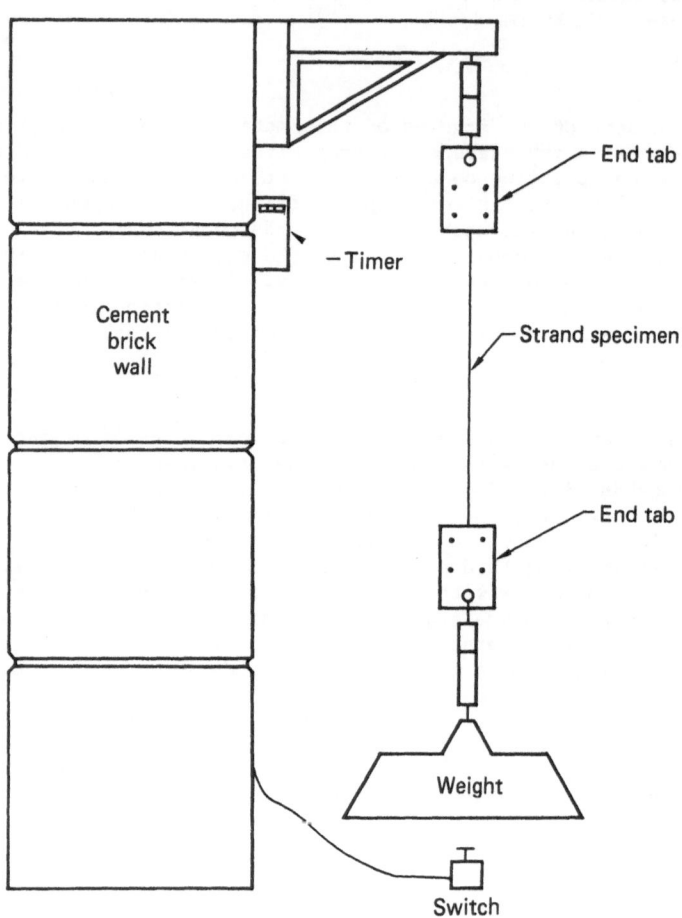

Fig. 1. Diagram of apparatus for applying sustained uniaxial tension to a simple fiber/epoxy composite strands.

Stress-rupture data [3,4] are represented by the shaded bands in Figs. 2 and 3. The width of the bands indicates the large scatter in time. Strength retention data [2,5] are superimposed over the stress-rupture data for S-glass (Fig. 2) and aramid (Fig. 3) composites. From these data, we note that:

● Regardless of the level of applied stress (30 to 80% of the composite strength) and of the time under load (days to years), the strength retention of these composites varied from a minimum of 93% to a maximum of 102%. Considering the material variability and some experimental error involved, this 93% retention really amounts to little strength degradation.

● In the strength retention test at high applied stress levels, the time under load of the composite specimens often penetrates the stress-rupture envelope. This indicates that these specimens have a very high probability of sudden death. However, when these specimens are unloaded and tested to failure, we do not see any degradation of their strength. This clearly indicates that the lifetime of a composite cannot be estimated by its high value of strength retention. Rather, a high strength retention value may give a false sense of security.

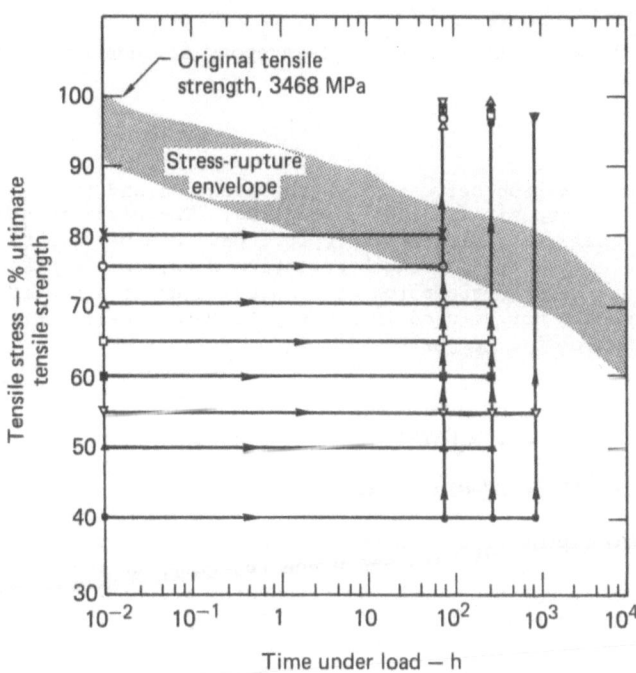

Fig. 2. Strength retention data of aramid/epoxy strands superimposed on the stress-rupture envelope.

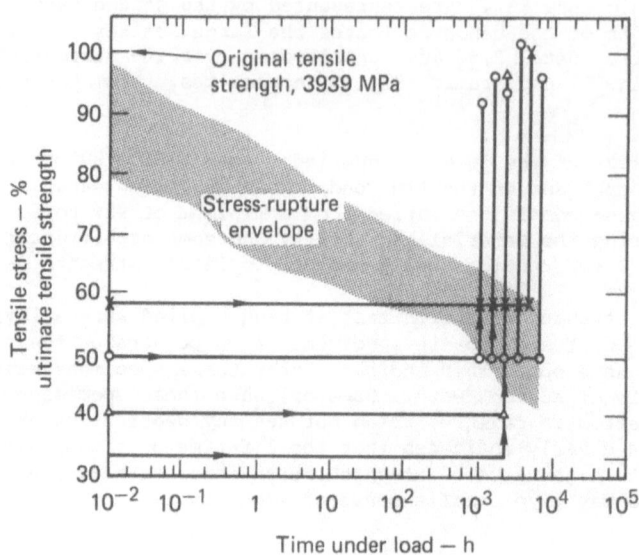

Fig. 3. Strength retention data of S-glass/epoxy strands superimposed on the stress-rupture envelope.

This lack of correlation between strength retention and the stress-rupture of a composite is indeed puzzling. The significance of this is that only extremely reliable stress-rupture data can be used to estimate the lifetime of a composite. Strength retention data, on the other hand, assess only the degree of degradation of a material under a specific environment and should not be used in any way to infer the lifetime or the long-term performance of a composite.

THE ROLE OF A HIGH-ELONGATION MATRIX IN A COMPOSITE

Transverse Strength of S2-Glass Composites

In an effort to improve the transverse properties of a composite for flywheel applications, Christensen and Rinde [6] compared four epoxy resins of widely different properties with the transverse properties of S2-glass composites made from the same four epoxies. The key results are summarized in Table 1 and Fig. 4. It is surprising to see that the composite transverse tensile failure strain is less than 1% when this low-modulus pure epoxy has a uniaxial tensile strain to failure of 69% (Fig. 4, curves D). Christensen [6] has attributed this to two main factors:

Table 1. Composition of the four epoxy matrix resins with varying degrees of low modulus and high elongation.

Resin	Components	Weight Ratio
A	XD 7818[a]/XD 7114[a]/Tonox 60-40[b]	100/30/30.7
B	Epon 871[c]/XD 7818/XD 7114/Tonox 60-40	50/30/20/16.4
C	Epon 871/XD 7818/XD 7114/Tonox 60-40	70/20/10/13.3
D	DER 732[a]/XD 7818/Tonox 60-40	70/30/15.0

[a]Dow Chemical Company. Reference to a company or product name does not imply approaval or recommendation of the product by the University of California or the U.S. Department of Energy to the exclusion of others that may be suitable.

[b]UniRoyal Chemical Company.

[c]Shell Chemical Company.

● Fracture-controlled composite failure; the mode of failure changed from "necking" and "flow" type failure for the pure resin to the brittle fracture type failure for the transverse composite.

● A strain concentration factor of over 10 in the composites. The strain concentration argument has been known for more than a decade but the fracture-mode of failure, as explained by fracture mechanics [6], is a new discovery.

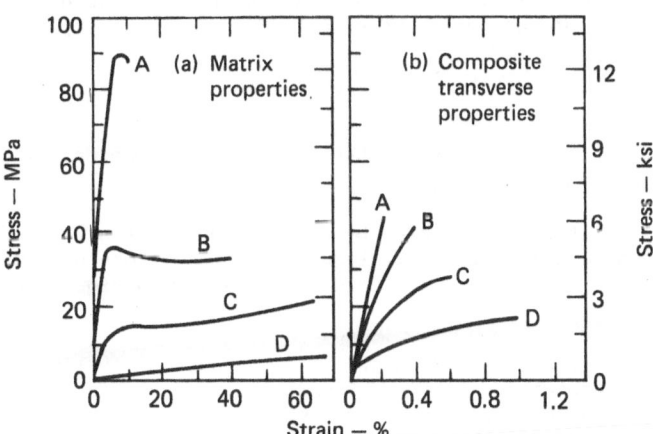

Fig. 4. Stress-strain curves of the four epoxy resins (a) and the corresponding S2-glass composites (b) tested in transverse tension.

In any event, for a unidirectional composite under transverse tension we can draw two conclusions. First, the low-modulus high-elongation resin matrix can increase the transverse failure strain of the composite several fold. However, this improved strain capability probably is not adequate for most composites; the level of improvement is far short of our general expectations. Second, other considerations such as the reduced service temperature of the matrix and the reduction of longitudinal properties of the composite may further discourage the use of such low-moduli matrices.

Longitudinal Strength of Composites

We also examined the effect of the matrix properties on the longitudinal properties of a unidirectional composite. Christensen and Rinde [6] showed conclusively for S2-glass composite (Fig. 5) that by decreasing the matrix modulus by a factor of 30, the longitudinal tensile strength of the composite strand decreased by 35%. Work by Chiao and coworkers on S-glass/epoxy strands and aramid/epoxy strands (61 to 65 vol% fiber) revealed that the effect of the nine different epoxies on fiber tensile strength was insignificant.

This apparent contradiction of results can be explained quite easily. In Ref. 7, Chiao et al., were looking for high-performance filament-winding epoxies. The epoxies examined were of slight variation of the type A as shown in Fig. 4a. For example, the matrix modulus varied only from 2.7 to 3.7 GPa, producing only a 10% effect on the fiber strength. This result compares well with that of Fig. 5: the two sets of data fit quite well. Thus we can conclude that matrix properties clearly affect the longitudinal strength of a unidirectional composite. Generally, the lower the epoxy modulus, the lower the strength when the composite is truly under uniaxial loading and the fiber content is controlled to within a narrow limit. Second, we conclude that within the same class of matrix material, when the properties of the matrix systems do not vary drastically (e.g., modulus and interlaminar shear properties), the effect of the matrix properties can be neglected.

SUMMARY

Based on the data accumulated at LLL over the past decade, we draw the following conclusions:

● There is little correlation between strength retention and the stress-rupture of a composite. For long-term applications of composites, valid stress-rupture data are much more valuable to a designer than strength retention.

● Strength retention test is of value in environmental degradation assessment of a composite. However, the designer must understand that although high strength retention is necessary for long composite life, it is not the only parameter on which the long-term performance of a composite should be based.

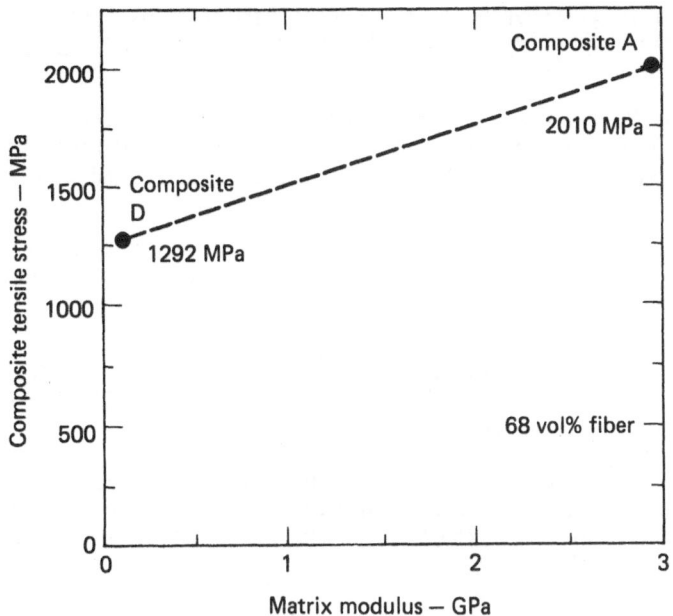

Fig. 5. Longitudinal tensile strength of S2-glass/epoxy composite strands.

● Low modulus, high-elongation epoxy resins in unidirectional composites are of very limited value; this is particularly true for the fiber-controlled failure of a composite.

ACKNOWLEDGMENTS

I thank my colleagues, R. M. Christensen and J. A. Rinde, as well as my wife, C. C. Chiao, for providing some of the data on which this paper is based. This work was performed under the auspices of the U.S. Department of Energy, at Lawrence Livermore Laboratory, under contract No. W-7405-Eng-48.

REFERENCES

[1]   Chiao, T. T., and Moore, R. L., "Stress-Rupture of S-glass/Epoxy Multifilament Strands," J. Composite Mat. 5, 1 (1971).

[2]   Chiao, T. T., and Moore, R. L., "Strength Retention of S-glass/Epoxy Composites," J. Composite Mat. 6, 156 (1972).

[3]   Chiao, T. T., Lepper, J. K., Hetherington, N. W., and Moore, R. L., "Stress-Rupture of Simple S-glass/Epoxy Composites," J. Composite Mat. 6, 358 (1972).

[4]   Chiao, T. T., Wells, J. E., Moore, R. L., and Hamstad, M. A., "Stress-Rupture Behavior of Strands of an Organic Fiber/Epoxy Matrix," in Composite Materials: Testing and Design, ASTM STP 546 (1974), p. 209.

[5]   Chiao, C. C., Sherry, R. J., and Chiao, T. T., "Strength Retention and Life of Fibre Composite Materials," Composites (April 1976), p. 107.

[6]   Christensen, R. M., and Rinde, J. A., Transverse Tensile Characteristics of Fiber Composites with Flexible Resins: Theory and Test Results, Lawrence Livermore Laboratory, Rept. UCRL-79983 (1977), submitted to Polymer Eng. Sci.

[7]   Chiao, T. T., Jessop, E. S., and Hamstad, M. A., "Performance of Filament-Wound Vessels From An Organic Fiber In Several Epoxy Matrices," in Proc. 7th Natl. SAMPE Tech. Conf. Albuquerque, NM, October 14-16, 1975.

NOTICE

THE FRACTURE CHARACTERISTICS AND PROPERTIES OF PAN GRAPHITE-ALUMINUM COMPOSITES

W. L. Lachman, President
Fiber Materials, Inc., Biddeford, Maine

ABSTRACT

The development of a liquid metal infiltration process and successful fiber-matrix interface barrier coatings has led to an exciting new class of graphite fiber reinforced metals. Composite wire preforms are continuously cast on a pilot production basis using the liquid infiltration technique. Aluminum composites reinforced with PAN graphite fibers have shown strengths in excess of 150,000 psi and modulus of up to 20 million psi. A review of the fracture, mechanical behavior and processing of graphite fiber reinforced aluminum is given in the paper below.

INTRODUCTION

The simultaneous advents of potentially low-cost, high-strength and stiffness multifiber graphite materials and new process technology to incorporate them in lightweight metals such as aluminum on a high volume production basis gives the potential for dramatic improvement in structural reinforcement without significant increase in cost. High performance polyacrylonitrile (PAN) precursor graphite fibers of moderate cost have been available for the past several years and have been widely used to reinforce organic resin matrices. Past attempts to incorporate these PAN fibers into aluminum have been unsuccessful due to a lack of chemical stability of the fibers causing severe degradation of strength. Recently, significant progress has been made in the development of new interface barriers for PAN graphite fiber reinforced aluminum.

A liquid metal infiltration process has been developed [1] which involves the use of very small concentrations of titanium and boron to promote the wetting of graphite fibers by commercial aluminum alloys. The graphite fibers are coated with a layer of Ti/B between 100 and 200 angstrom units thick to promote wetting and to protect them from attack by molten aluminum. This results in considerably less than 0.5 weight percent titanium and boron in the composite matrix. Similar concentrations of titanium and boron have been successfully used as grain refining additions to commercial aluminum alloys for the past 20 years and have beneficial effects by increasing alloy strength, ductility, and corrosion resistance. The titanium and boron are deposited on the fibers by the reduction of titanium tetrachloride and boron trichloride using zinc vapor, the zinc being continually extracted from the coating step of the process as zinc chloride gas. The coated fibers are infiltrated by drawing through a molten aluminum alloy bath producing a unidirectional graphite-aluminum composite wire. Significant advances have been made toward developing the Ti/B liquid infiltration process for future large-scale manufacturing (Figure 1). For example, the introduction of textile techniques for feeding fiber into the Ti/B coating chamber and the use of foundry style aluminum open melt techniques for infiltration both facilitate composite wire handling and show the way for the future continuous casting of shapes directly from the melt.

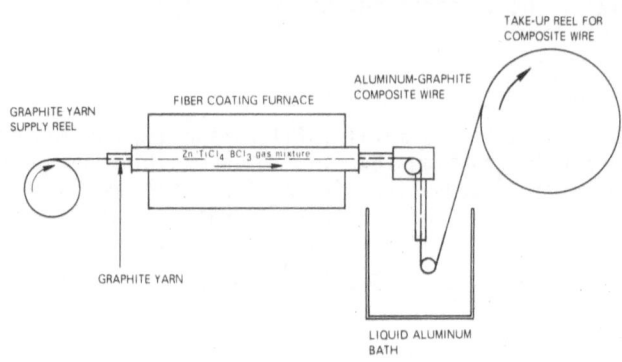

CONTINUOUS PROCESS FOR ALUMINUM-GRAPHITE COMPOSITE WIRE

Fig. (1)

Figure 2 shows a scanning electron micrograph of a polished end section of a graphite-aluminum composite wire containing approximately 11,000 filaments produced by the liquid infiltration unit. Good fiber distribution and excellent metal wetting characteristics are apparent. The graphite fiber in this composite is high modulus T50, a rayon precursor graphite fiber. Prior to 1976, rayon based graphite fibers were used almost exclusively in the development of metal matrix graphite fiber composites. Rule- of-mixture tensile strengths and elastic modulus have been attained utilizing a variety of aluminum alloys with these fibers. Typical fiber properties are shown in Table 1, and composite wire and fabricated bar properties are shown in Table 2.

Previous work using PAN based graphite fibers in metal matrices had shown PAN based graphite to be more reactive with liquid aluminum than rayon based graphite fibers. Therefore, poor translation of fiber strength and elastic modulus resulted in the composite, Table 3. During the past two years, a significant breakthrough has been made in research and development on PAN graphite fiber reinforced aluminum composites. Modified fiber barrier coatings have been developed and demonstrated which prevent attack and degradation of the fibers by aluminum alloys during liquid metal infiltration and subsequent fabrication of composite wire into shapes, Table 3. Both moderate and high modulus PAN fibers have been processed with the new coatings. Typical mechanical properties for the newly developed PAN based graphite fiber reinforced aluminum matrices are shown in Table 4. Note the high translation of tensile strength and elastic modulus for fabricated test bars. These bars were produced from PAN precursor

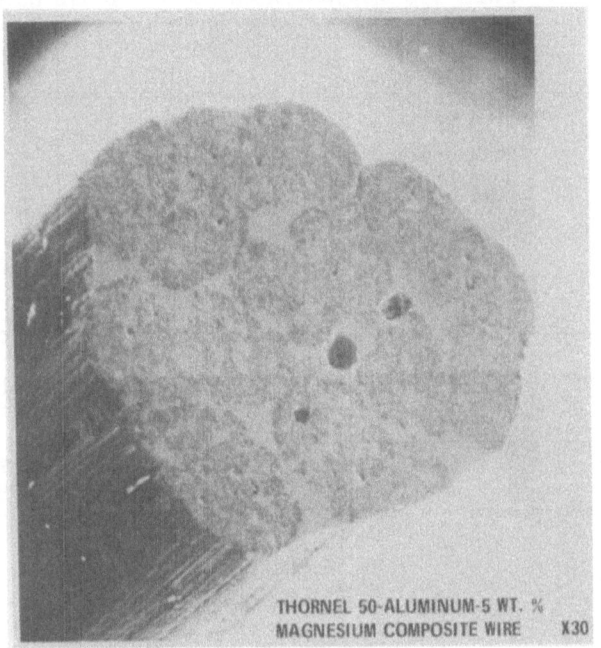

THORNEL 50-ALUMINUM-5 WT. %
MAGNESIUM COMPOSITE WIRE   X30

Fig. (2)

TABLE 1 - TYPICAL FIBER PROPERTIES

Fiber Designation	Strength (x 10^3 psi)		Elastic Modulus (x 10^6 psi)	
T50 (rayon)	315	(2172 MPa)	57	(393 GPa)
T50 (PAN)	300	(2068 MPa)	50	(345 GPa)
T300 (PAN)	360	(2482 MPa)	34	(234 GPa)
HMS, HM/PVA (PAN)	340 min.	(2344 MPa)	50-55	(345-379 GPa)

composite wire (approximately .035 in. diameter) using a hot pressing technique.

Graphite-Aluminum Fracture and Mechanical Behavior

The new fiber coating and bonding technology has resulted in the development

TABLE 2 - STATE-OF-THE-ART T50 GRAPHITE/ALUMINUM COMPOSITE MECHANICAL PROPERTIES

Property	35 v/o T50/A201 Wire	42 v/o T50/A201 Wire	30 v/o T50/A201 Test Bar
Tensile Strength	120 Ksi 828 MPa	150 Ksi 1034 MPa	90 Ksi 621 MPa
Tensile Modulus	27 Mpsi 186 GPa	30 Mpsi 207 GPa	20 Mpsi 138 GPa
Transverse Strength	5 Ksi 34.5 MPa	NOT TESTED	7.1 Ksi 49.0 MPa

TABLE 3 - MECHANICAL PROPERTY COMPARISON PAN BASED GRAPHITE-ALUMINUM COMPOSITES (PAST DATA VERSUS PRESENT STATE-OF-THE-ART)

Property	As Received T300 A201 Alloy (Wire)	Barrier Coated T300 A201 Alloy (Wire)
Tensile Strength	50.2 Ksi 346.4 MPa	180 Ksi 1240 MPa
Tensile Modulus	---------	20 Mpsi 138 GPa
Fiber Content	45.9 %	40.0 %

TABLE 4 - PAN BASED GRAPHITE FIBER REINFORCED ALUMINUM COMPOSITE PROPERTIES

Property	40 v/o T300 A201 (Wire)	30 v/o T300/A201 Test Bar	40 v/o T300/A201 Test Bar
Tensile Strength	180 Ksi 1242 MPa	125 Ksi 863 MPa	150- 160 Ksi 1035-1104 MPa
Tensile Modulus	20 Mpsi 138 GPa	------- -------	21 Mpsi 145 GPa
Flexure Transverse	-------- --------	16 Ksi 110 MPa	-------- --------
Tensile Transverse	-------- --------	------- -------	5 Ksi 34.5 MPa

of a ductile Thornel 300 (PAN based) graphite fiber reinforced aluminum composite. This achievement of ductility is extremely important since it implies that the toughness or forgiveness necessary to prevent sudden catastrophic failures in highly stressed hardware can be developed in graphite-metal composites. To illustrate this composite ductility effect observed in the newly developed PAN based graphite-aluminum composites, a comparison of the mechanical behavior of rayon based (T50) and polyacrylonitrile based (T300) graphite fiber reinforced aluminum composites is made below. Figure 3 illustrates some typical stress/strain data for rayon (T50) and PAN (T300) precursor graphite fiber reinforced aluminum matrices in the as-cast conditions [2]. See Table 1 and Table 5 for

STRESS STRAIN BEHAVIOR OF RAYON AND POLYACRYLONITRILE
GRAPHITE ALUMINUM AS-CAST COMPOSITE WIRE

Fig. (3)

fiber and matrix properties, respectively. The T50/A201 Al wire composite at 42 volume percent fiber shows a completely elastic stress/strain relationship having a modulus of elasticity of 30 Msi (207 GPa) up to the ultimate tensile strength of 160 Ksi (1103 MPa). The total strain-to-failure for this composite is 0.53%, which is representative of the strain range 0.45 to 0.55% usually observed for this system. The stress/strain behavior is typically characteristic of a high modulus high strength composite material failing in a brittle manner. In contrast, the T300 (PAN based)/356 Al composite at 35 volume percent fiber shows a pronounced increase in total strain-to-failure, (1.4%) and shows an initial high primary elastic modulus (typically 17-19 Msi, 117-131 GPa) with a transition at approximately 20 Ksi (138 MPa) stress level and 0.1% strain to a lower secondary modulus (typically 13-15 Msi, 90-103 GPa). The T300/356 Al composite exhibits a marked deviation from linear behavior prior to failure at the ultimate tensile strength of 166 Ksi (1145 MPa). Cyclic loading the T300/Al composites to their yield stress followed by unloading results in elimination of the primary modulus. The value of the resultant average elastic modulus obtained on cycling is typically 17 Msi (117 GPa). Similar ductile-like behavior is observed, but to a lesser degree for high modulus PAN based graphite fiber reinforced Al composites, Figure 4.

TABLE 5 - TYPICAL MATRIX PROPERTIES

Matrix Designation	Nominal Composition (Weight Percent)						Strength (x 10^3 psi)
	Cu	Ag	Si	Mg	Cr	Al	
A201	4.7	0.6	0.1	0.3	--	Bal	$20^{(1)}$(138 MPa) $50^{(2)}$(345 MPa)
356	0.2	--	7.0	0.3	--	Bal	$25^{(1)}$(172 MPa) $38^{(2)}$(262 MPa)
6061	0.25	--	0.6	1.0	0.20	Bal	$18^{(1)}$(124 MPa) $45^{(2)}$(310 MPa)

NOTE: (1) As cast; (2) Heat treated; (3) Elastic modulus of all alloys approximately 10 Msi (69 GPa)

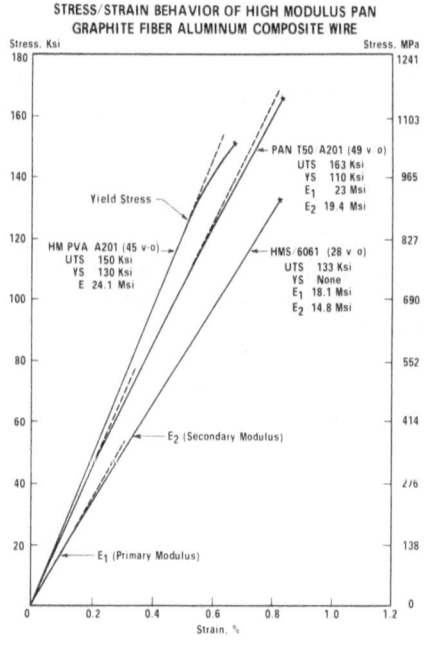

STRESS/STRAIN BEHAVIOR OF HIGH MODULUS PAN
GRAPHITE FIBER ALUMINUM COMPOSITE WIRE

Fig. (4)

   The linear stress/strain behavior of the T50 (rayon based) A201 Al composite
in the as-cast condition suggests that the fiber and matrix are essentially
strained to failure elastically. There appears to be no plastic contribution by
the matrix to the overall deformation behavior of the composite. This suggests
that, in the rayon based graphite-aluminum composite, the matrix is not bonded
to the fiber well enough for the matrix to contribute its plasticity to the composite when subjected to a tensile strain. The observed strength at which the
composite fails, 160 Ksi (1103 MPa) and strain (0.53%), is probably due entirely
to the fiber. That this is the case is supported by examining the tensile fracture of the as-cast T50/A201 composite, Figure 5. It is evident that extensive
fiber pullout has occurred, which is indicative of a weak fiber to matrix bond.

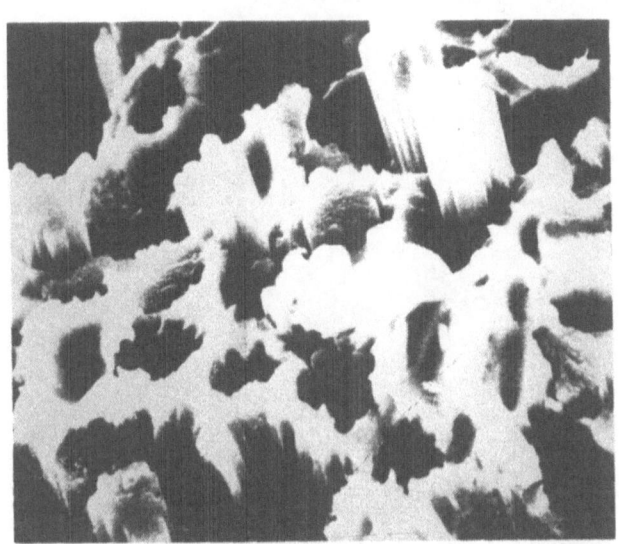

T50 (RAYON BASED)/A201 TENSILE FRACTURE
SHOWING FIBER PULLOUT

5μ

Fig. (5)

   The stress/strain behavior of the PAN precursor graphite fiber aluminum composite (T300/356 Al), however, indicates that both fiber and matrix are strained
elastically only up to a stress level of approximately 20 Ksi. At this point,
the elastic modulus of the composite changes from the primary E_1 (fiber and matrix) to the secondary E_2 stage (fiber elastic, matrix plastic), and this change
is attributed to the onset of microplastic deformation in the matrix. In addition, the T300/356 Al composite also shows a definite yield stress of 150 Ksi
(1035 MPa) at 1.1% strain, at which nonlinear behavior (macro-plasticity) is
evident. No plastic deformation occurs in T50 or T300 graphite fibers; thus,
the nonlinearity prior to failure in T300/Al is attributed to progressive fail-

ure of the fibers into discontinuous segments. This has been observed on tensile fracture edges and on chemically extracted fibers from tested tensile specimens. The total strain-to-failure of the composite is greater than the maximum
strain-to-failure (1.1%) of the T300 fiber because the matrix continues to deform
after fiber breakage but is still reinforced by the discontinuous broken fiber
segments. For this effect to occur, an improved fiber/matrix bond is a prerequisite and must be maintained during progressive fiber failure. Examination of tensile fracture surfaces (Figure 6) of as-cast composite wire specimens of T300/356
Al reveals little fiber pullout and pronounced matrix yielding (cellular structure)
around the fibers. These observations indicate that an improved bond has been achieved between the matrix and fiber which effectively transfers tensile load from
the matrix to the fiber by matrix shear.

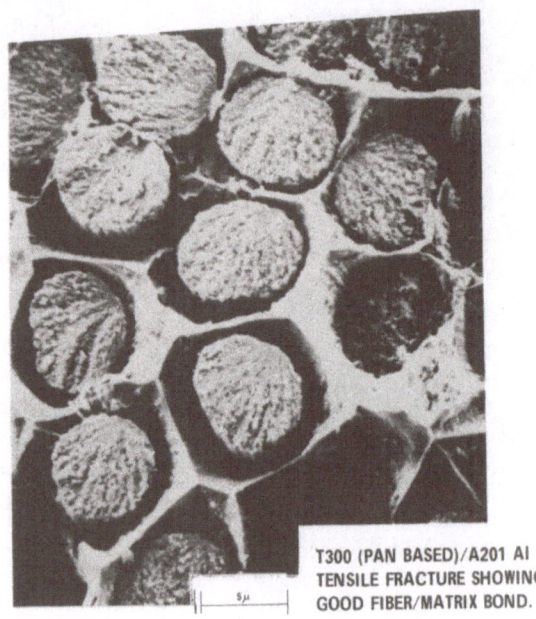

T300 (PAN BASED)/A201 Al
TENSILE FRACTURE SHOWING
GOOD FIBER/MATRIX BOND.

Fig. (6)

Additional evidence of composite plastic behavior is shown by examining progressive stress cycling data (Figure 7). Loading the composite to below 20 Ksi
(138 MPa) and unloading (cycle 1) shows only composite elastic behavior. Load-
ing/unloading above 20 Ksi (138 MPa) shows E_1/E_2 modulus transition and retain-
ment of permanent set (cycle 2 through 7) on unloading. The E_1/E_2 transition
point increases rapidly with small increments of strain indicating that a high
rate of strain hardening is occurring in the matrix (cycle 2 through 4). The
abrupt E_1/E_2 transition disappears at cycle 5 through 7. True composite plastic
behavior is evidenced when one considers the loading and unloading slopes of cy-

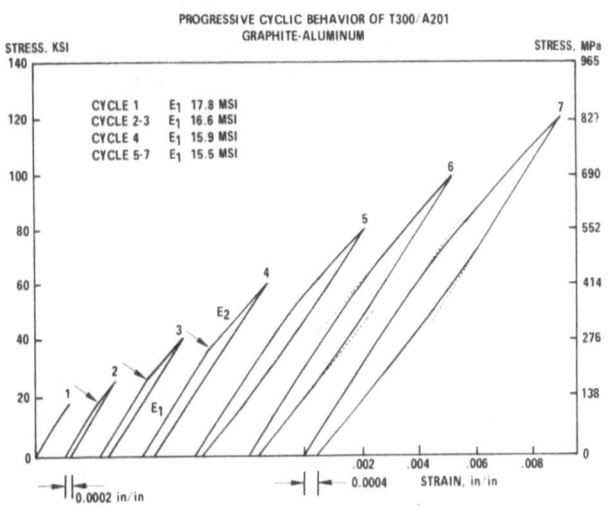

Fig. (7)

cles 5 through 7. On unloading, the initial slope is parallel to the initial
loading slope indicating that the first contraction on unloading is elastic. The
remainder of the slope on unloading to zero load is parallel to the secondary E_2

slope on loading. This indicates that the remaining contraction on unloading is
plastic. The elastic contractive force exerted by the fiber on the matrix results
in plastic compression of the matrix. In conventional and completely elastic materials, all the contraction would be elastic when the load is released. Cyclic
loading between 20 and 120 Ksi (138-827) shows a relatively small hysteresis loss
indicating potential high fatigue resistance of the composite.

SUMMARY

High strength aluminum and titanium alloys have been developed over the past
fifty years which are superior to steel on a strength to density basis and are
now in widespread use. The development of premium high strength and stiffness
graphite fibers over the past ten years surpasses the above development in the
strength and particularly the stiffness of structural materials. For example,
graphite fiber reinforced aluminum is two and three times better than commercial
high strength aluminum alloys on a strength and stiffness to density basis, respectively. An R&D data base has been obtained over the past four years with
graphite fiber reinforced aluminum and excellent mechanical properties have been
attained.

REFERENCES

[1] Kendall, E. G. and Pepper, R. T., Graphite reinforced metal-matrix compos-
     ites, U.S. Patent Application Serial No. 624,809, 1975.

[2] Gigerenzer, H., Zack, T. A. and Pepper, R. T., Observations on the mechani-
     cal behavior of rayon and polyacrylonitrile graphite-aluminum composites,
     Second International Conference on Mechanical Behavior of Materials, Boston,
     1976.

PREDICTION OF FRACTURE INITIATION IN COMPOSITE STRUCTURES

P. W. Mast, L. A. Beaubien, M. A. Clifford, D. R. Mulville, S. A. Sutton, R. W. Thomas, J. Tirosh and I. Wolock

Mechanics of Materials Branch, Naval Research Laboratory, Washington, D.C. 20375

ABSTRACT

A program is described for characterizing the fracture behavior of composite materials under complex loading. It consists of three parts: (1) the development of a suitable loading system for applying complex loading; (2) the development of failure criteria from the fracture studies conducted; (3) the demonstration of the validity of these failure criteria in laboratory studies on structural components.

INTRODUCTION

The need for experimental data on the fracture behavior of fiber-reinforced composites under complex loading has been recognized for some time. Recent publications have pointed out that the need still exists. In 1975, Rowlands [1] stated that "Most available strength information is based on uniaxial stress states, while practical applications invariably involve at least biaxial loading". More recently, Morris and Hahn [2] pointed out that "A review of the literature indicates that most of the work in fracture of composites has emphasized Mode I* loading. By comparison, mixed-mode fracture has received little attention. Most of the available mixed-mode data falls in the category of unidirectional composites, with the crack being parallel to the fiber". Guess and Gerstle [3] state much the same, "A survey of the literature for experimental data on composites revealed that most data are from uniaxial stress tests. However, the loading on composite structures will be, in general, more complex than uniaxial stress states". The reason, of course, for the scarcity of complex loading data is that it is much more difficult and more expensive to obtain. Numerous analytical criteria have been examined with regard to their ability to predict the strength behavior of composites under complex loading [1] in order to extend the limited data available. Unfortunately, these attempts have met with only limited success. For all practical purposes, we are dependent on experimental data to define the strength behavior of composites under complex loading.

---

*Mode I loading refers to cleavage or opening mode.

A program was initiated at the Naval Research Laboratory to characterize the fracture behavior of composite materials. As a result of experiences in the early stages of this study, several criteria were established. First, fracture mechanics concepts would be utilized, in that attention would be focused on defining the conditions under which fracture initiates from a controlled defect or stress concentrator. However, the study would not be limited to using existing fracture mechanics parameters such as strain energy release rate G, stress intensity factor K, or the J-integral, established for isotropic materials. Second, experimental studies should be conducted under general in-plane loading, to simulate service conditions as closely as practical. Third, with the very large number of variables involved in composite materials, tests should be relatively inexpensive to conduct. This means that the specimens should be of simple geometry to reduce machining costs and be small to reduce materials costs, and that the time to conduct a test should be relatively short. However the specimen dimensions should be large enough relative to fiber diameter or lamina thickness so that the material could be considered homogeneous and orthotropic for purposes of analysis. Fourth, analytic expressions should be developed to permit interpolation and extrapolation of data. Fifth, the data obtained should be usable by design engineers. Finally, laboratory tests should be conducted on structural prototypes to demonstrate that the predictions based on the laboratory coupons are indeed valid.

The program involved three parts: (1) the development of a suitable loading system for applying complex loading; (2) the development of failure criteria from fracture studies conducted under complex loading; (3) the demonstration of the validity of these failure criteria in laboratory studies on structural components.

GENERAL DESCRIPTION

The in-plane loader developed in this program (Figure 1) consists of three independent computer-controlled hydraulic actuators connected to a movable head.

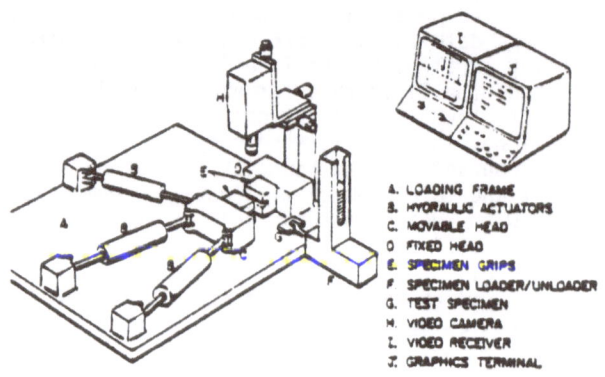

A. LOADING FRAME
B. HYDRAULIC ACTUATORS
C. MOVABLE HEAD
D. FIXED HEAD
E. SPECIMEN GRIPS
F. SPECIMEN LOADER/UNLOADER
G. TEST SPECIMEN
H. VIDEO CAMERA
I. VIDEO RECEIVER
J. GRAPHICS TERMINAL

Fig. (1) - Schematic of the in-plane loader

By programming the loading rate of each actuator, general in-plane loading consisting of combinations of tension, shear and in-plane bending can be applied to a test specimen (Figure 2). A single-edge-notch specimen was used 1 in. x 1.5 in. x 0.1 in. (25 mm x 3.8 mm x 2 mm) with the notch 0.6 in. (15 mm) long parallel to the 1-inch dimension. The specimen is held by a pair of hydraulically-controlled

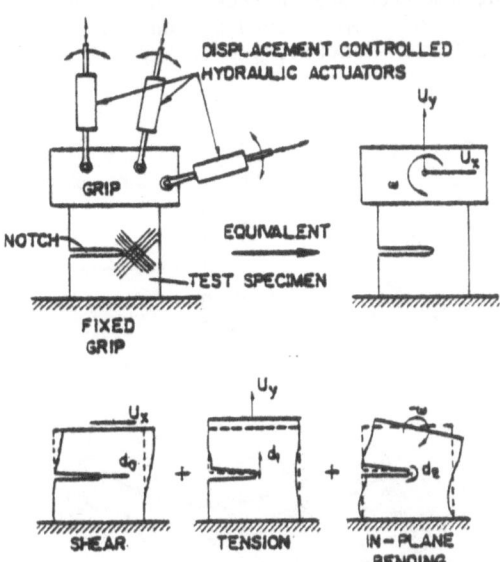

Fig. (2) - Loading conditions produced by the in-plane loader

grips in which the clamping pressure can be programmed. One grip is attached to a fixed head, and the other to the actuators by means of a movable head through which programmed translation and rotation motions are executed. The specimens are removed from a loading magazine and inserted in the grips by a mechanical feeding device. Initial data regarding specimen geometry, such as notch location, is obtained via a digitized video picture of the specimen in the grips and the data is then stored in a computer. A computer system is interfaced with these components and controls the tests, gathering the specimen geometry data, controlling the pressure in the grips, loading the specimen in a prescribed proportional loading path and gathering and storing data on force and displacement. Currently, the system requires the operator to manually operate the loader, digitize sample geometry using the video system, and input the test parameters to the computer controlled test system. Full automation could be accomplished with a limited amount of additional effort.

FRACTURE STUDIES

Fracture studies were conducted in the in-plane loader on a series of graphite/epoxy composites in which the angle of layup of the reinforcement was varied (included angles of 30°, 45°, 60° and 75°). Tests were repeated two and three times at selected points. The variability in failure loads for a given point was less than 5%. Tests results can be represented by plotting the critical displacements, i.e., those that produce fracture initiation, in spherical coordinates mapped onto a cartesian reference frame (Figure 3). This figure also shows the relation between the components of shear, tension and bending applied to the specimen, d_0, d_1, and d_2 respectively, and the spherical coordinates, θ_1, θ_2, and r used to represent the fracture surface. These results represent failure surfaces for a borad range of complex loading conditions. They can be used to predict fracture initiation in laboratory prototype structures.

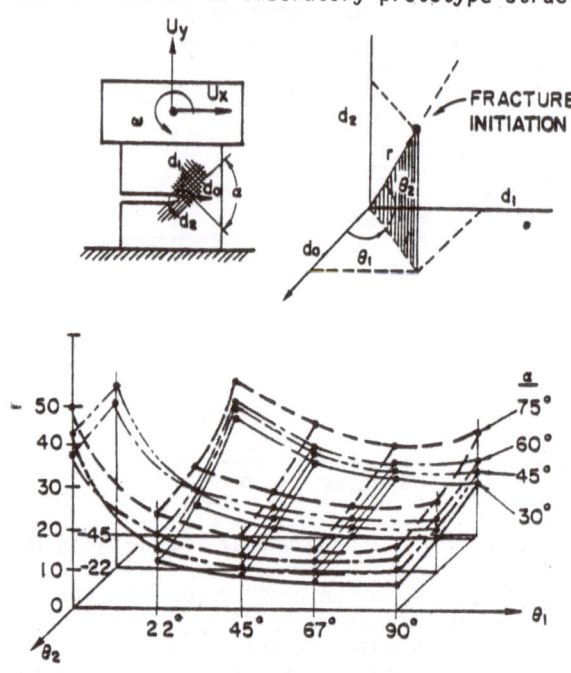

Fig. (3) - Failure surfaces for carbon/epoxy composite
tested under in-plane loading

PREDICTION OF FRACTURE INITIATION IN A STRUCTURAL SUBCOMPONENT

The next step in the process was the demonstration of the validity of the failure criteria developed in the in-plane loader, using a box beam as the prototype. The box beam was fabricated of aluminum to represent the type of construction used in aircraft structures (Figure 4). Aluminum skins 12 in. x 18 in. x 0.125 in. (305 mm x 460 mm x 3 mm) were bolted and bonded to 1.5 in (38 mm) deep aluminum channels. A series of loading holes were drilled in a fitting attached to the top and bottom of the box beam. By selection of the holes, various combinations of in-plane tension, shear and bending loads could be applied, using a uniaxial loading system. A 4.5 in. (114 mm) diameter, 0.1 in. (2 mm) thick disk of the test material was mounted over a cut-out in the center of one face of the box beam. This size disk was used rather than making the entire face of the test material in order to conserve material and reduce costs. A central notch was machined in the test specimen and the orientation of the specimen varied with respect to the loading axis.

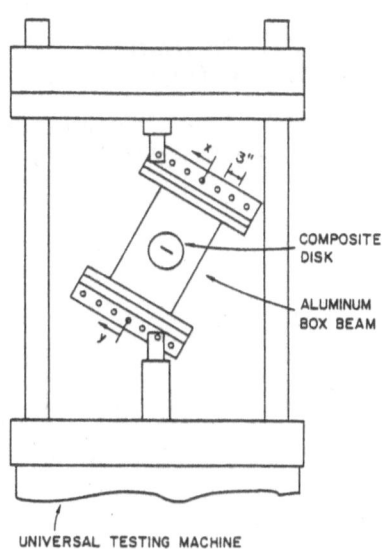

Fig. (4) - Box-beam loading arrangement

Prior to conducting fracture studies of the graphite/epoxy composite disk, an experimental stress analysis was performed on the test samples mounted in the box beam. The purpose of the analysis was to verify the numerical modeling of the stress and displacement field in the vicinity of the notch obtained with the finite element analysis program. A photoelastic coating was bonded to a graphite composite disk with a notch and combinations of in-plane loads were applied to the box beam containing the sample. A series of isochromatic patterns were observed for various combinations of in-plane loads. Satisfactory correlation was obtained between experimental and computed values of isochromatic patterns for various in-plane loading states. Confirmation of the adequacy of the numerical model is essential, since failure predictive capability is based on similarities in the deformation field near the notch tip between small laboratory test samples tested in the in-plane loader, and the box beam prototype. Deformation fields for both specimen geometries are obtained via the same structural analysis program.

Having verified the analysis method, experimental fracture studies were conducted on graphite composite disks mounted in the box beam and subjected to combinations of in-plane loads. Predictions of the load at crack initiation were made on the basis of similar deformation fields. Several tests conducted to date show good predictive capability for fracture initiation using this technique.

REFERENCES

[1] Rowlands, R. E., "Flow and failure of biaxially loaded composites: experimental-theoretical correlation", Inelastic Behavior of Composite Materials, AMD, Vol. 13, C. T. Herakovick, ed., ASME, p. 97, 1975.

[2] Morris, D. H. and Hahn, H. T., "Mixed-mode fracture of graphite/epoxy composites: fracture strength", J. Composite Materials 11, p. 124, 1977.

[3] Guess, T. R. and Gerstle, F. P., Jr., "Deformation and fracture of resin matrix composites in combined stress states", J. Composite Materials 11, p. 146, 1977.

LIST OF PARTICIPANTS

Ainbinder, S.
Institute of Polymer Mechanics
LSSR Academy of Sciences, USSR

Annin, B.
State University of Novosibirsk
USSR

Bayev, L.
Institute of Hydrodynamics
USSR Academy of Sciences, USSR

Bogdanovich, A.
Institute of Polymer Mechanics
LSSR Academy of Sciences, USSR

Bulavs, F.
Polytechnic Institute of Riga
USSR

Bunsell, A. R.
Ministere du Development Industriel
   et Scientifique
FRANCE

Chamis, C. C.
NASA - Lewis Research Center
USA

Chiao, T. T.
University of California
USA

Chou, S. C.
Department of the Army
USA

Crossman, F. W.
Lockheed Palo Alto Research Laboratory
USA

Dundurs, J.
Northwestern University
USA

Faitelson, L.
Institute of Polymer Mechanics
LSSR Academy of Sciences, USSR

Herrmann, K.
Gesamthochschule Paderborn
WEST GERMANY

Jirgen, L.
Institute of Polymer Mechanics
LSSR Academy of Sciences, USSR

Kalnin, I.
Celanese Research Company
USA

Kalninsh, M.
Polytechnic Institute of Riga
USSR

Knets, I.
Institute of Polymer Mechanics
LSSR Academy of Sciences, USSR

Kregers, A.
Institute of Polymer Mechanics
LSSR Academy of Sciences, USSR

Kuksenko, V.
Physico-Technical Institute
USSR Academy of Sciences, USSR

Latishenko, V.
Institute of Polymer Mechanics
LSSR Academy of Sciences, USSR

Lomakhin, V.
Moscow State University
USSR

Malmeister, A.
LSSR Academy of Sciences, USSR

Matiss, I.
Institute of Polymer Mechanics
LSSR Academy of Sciences, USSR

Maximov, R.
Institute of Polymer Mechanics
LSSR Academy of Sciences, USSR

Mileiko, S.
Institute of Physics of Solids
USSR Academy of Sciences, USSR

Molchanov, Y.
Institute of Polymer Mechanics
LSSR Academy of Sciences, USSR

Nemirovsky, Y.
Institute of Hydrodynamics
USSR Academy of Sciences, USSR

Nikitin, L.
Institute of Physics and Earth Science
USSR Academy of Sciences, USSR

Ovchinsky, A.
Institute of Metallurgy
USSR Academy of Sciences, USSR

Parfeyev, V.
Institute of Polymer Mechanics
LSSR Academy of Sciences, USSR

Perov, B.
All-Union Research Institute of
Aviation Materials, USSR

Portnov, G.
Institute of Polymer Mechanics
LSSR Academy of Sciences, USSR

Pozdnyakov, O.
A. F. Ioffe Physico-Technical Institute
USSR

Regel, V.
Physical and Technical Institute
USSR Academy of Sciences, USSR

Rikards, R.
Institute of Polymer Mechanics
LSSR Academy of Sciences, USSR

Rowlands, R.
University of Wisconsin
USA

Sarkisyan, N.
Institute of Mechanics and Mathe-
matics
Armenian Academy of Sciences
USSR

Sih, G. C.
Lehigh University
USA

Skudra, A.
Polytechnic Institute of Riga
USSR

Smith, C. W.
Virginia Polytechnic Institute
and State University
USA

Tamuzh, V.
Institute of Polymer Mechanics
LSSR Academy of Sciences, USSR

Tarnopolsky, Y.
Institute of Polymer Mechanics
LSSR Academy of Sciences, USSR

Teters, G.
Institute of Polymer Mechanics
LSSR Academy of Sciences, USSR

Tolks, A.
Institute of Polymer Mechanics
LSSR Academy of Sciences, USSR

Tutans, M.
Institute of Polymer Mechanics
LSSR Academy of Sciences, USSR

Urzhumtsev, Y.
Institute of Polymer Mechanics
LSSR Academy of Sciences, USSR

Vanin, G.
Institute of Mechanics
Ukranian SSR Academy of Sciences, USSR

Vasilyev, V.
Moscow Institute of Aviation Technology
USSR

Wu, E.
University of California
USA

Zinchenko, V.
Institute of Polymer Mechanics
LSSR Academy of Sciences, USSR